Focus on Organometallic Chemistry Research

FOCUS ON ORGANOMETALLIC CHEMISTRY RESEARCH

MARIN A. CATO
EDITOR

Nova Science Publishers, Inc.
New York

For permission to use material from this book please contact us:
Telephone 631-231-7269; Fax 631-231-8175
Web Site: http://www.novapublishers.com

LIBRARY OF CONGRESS CATALOGING-IN-PUBLICATION DATA

Focus on organometallic chemistry research / Marin A. Cato (editor).
p. cm.
Includes bibliographical references and index.
ISBN 1-59454-495-6
1. Organometallic chemistry--Research. I. Cato, Marin A.
QD411.F63 2005
547'.05--dc22 2005017850

Published by Nova Science Publishers, Inc. ✦New York

CONTENTS

PREFACE

Organometallic chemistry is based on the reactions and use of a class of compounds (R-M) that contain a covalent bond between carbon and metal. They are prepared either by direct reaction of the metal with an organic compound or by replacement of a metal from another organometallic substance. Research in organometallic chemistry is also conducted in the areas of cluster synthesis, main-group derivatives in unusual oxidation states, organometallic polymers, unstable organometallic compounds and intermediates in matrices, structure determination of organometallic compounds in the solid state [X-ray diffraction] and gaseous states [electron diffraction], and mechanisms of reactions of transient silylenes and related species. In addition to the traditional metals and semimetals, elements such as selenium, lithium and magnesium are considered to form organometallic compounds, e.g. organomagnesium compounds MeMgI, iodo(methyl)magnesium and diethylmagnesium which are Grignard reagents an organo-lithium compound BuLi butyllithium; Organometallic compounds often find practical use as catalysts, the processing of petroleum products and the production of organic polymers.

In chapter 1, conjugated molecules containing metal centers are an important class of molecular materials by virtue of the combined features of the chemical, electronic and optical properties of the metal groups to those of the organic chromophores. Within this framework, homopolymers and copolymers of fluorene derivatives have been extensively explored due to their thermal and chemical stability, their high emission quantum yields and synthetic accessability. Metal alkynyl complexes incorporating functionalized fluorene moieties form a unique family of new materials at the frontier of materials science. While there is great scope for chemical modifications of the metal groups and substituents on the fluorene ring, these organometallic complexes can exhibit rich photophysical and redox properties. The fundamental understanding on these properties of such metal-based alkynyls would lead to the production of novel functional materials and represent good model systems in the development of optoelectronic devices, molecular electronics and chemical sensing appliances. In this chapter, fluorene-derived metal alkynyl systems of the late transition metals are discussed, and particular attention is focused on the electronic absorption spectroscopy, electrochemical and photoluminescence behavior, and structural aspects of this class of metal-organic materials. A detailed account of the evolution of the first excited singlet and triplet state on the electronic structure of the fluorene spacer groups of some metal polyyne polymers will also be given.

AS reported in chapter 2, ethers can be easily cleaved with various electron-transfer reagents, i.e. alkali metals, alkali metal aromatic radical anions, alkalides, and low-valent transition metal compounds. Alkalides (M^-, M^+C) are salts containing metal anions and complexed metal cations (M: Na, K, Rb, and Cs; C: crown ether or cryptand). Some of them, e.g. K^-, K^+(18-crown-6) **1** or K^-, $K^+(15\text{-crown-}5)_2$ **2**, are recently used for the reductive C-O bond cleavage in linear ethers, oxiranes and other cyclic ethers, as oxetane and tetrahydrofuran. Moreover, crown ethers, which are necessary to prepare concentrated solutions of alkalides are also destroyed by metal anions during the self-decomposition process. Organometallic compounds being intermediate products of these reactions are generally unstable at ambient temperature. They react rapidly with crown ether complexing the metal cation and/or with ether substrate; some of them eliminate unsaturated hydrocarbon, i.e. ethylene or cyclohexene. The first type of the reaction leads to the crown ring cleavage. Potassium tetraethylene glycoxide vinyl ether is its product if 15-crown-5 is used as a ligand. In the system containing oxiranes the organometallic intermediate can also eliminate potassium alkoxide or potassium phenoxide. Particularly interesting is the reaction of **2** with (benzyloxymethyl)oxirane. The selective scission of the linear ether bond occurs in the latter without the opening of its oxirane ring. Benzylpotassium formed as the intermediate cleaves immediately the same kind of the ether bond in the next (benzyloxymethyl)oxirane molecule. Bibenzyl and potassium glycidoxide are the final reaction produts. The alkoxide obtained can be applied to the synthesis of glycidyl ethers, glycidyl esters, crown-like oligomers with alkoxide groups as well as star-shaped and hyperbranched polymers. An exception to the rule of instability of the organometallic compounds makes triphenylmethylpotassium formed by the linear ether bond cleavage in (triphenylmethoxymethyl)oxirane. It does not take part in any further processes. The appropriate reaction mechanisms of the formation and decay of organometallic intermediates in the systems containing ethers and potassium alkalides are discussed in this chapter. A qualitative method of identification of the short-lived species is proposed.

As reported in chapter 3, tellurium ligands of different types viz. RR'Te, RTe^-, Te^{2-}, halotellurium ligands, polytelluride ligands and tellurium containing macrocycles have attracted inorganic chemists during last three decades. Many organometallic compounds (mononuclear as well as polynuclear) containing such ligands have been designed. They include metal carbonyls, half sandwich compounds and metal alkyl based derivatives. Tellurium ligands have been found very good building blocks for designing metal carbonyl clusters. Unique structural features observed in solid state as well as in solutions for many such compounds are attractive. The secondary Te····Cl interactions are capable of generating supramolecular assemblies. In this paper the chemistry of organometallic compounds containing Te ligands is reviewed. In general metal tellurium bonds are not very stable as tellurium is noblest metalloid. Some metallic centers like mercury(II) promote the cleavage of Te- C bonds when organotellurium ligands coordinate with them. The greater emphasis in this review is laid on compounds for which single crystal structure determination by X-ray diffraction or detailed NMR studies have been reported. The trends in the reactivity due to the presence of Te ligands are highlighted. The possibilities of employing Te ligands for the further growth of organometallic chemistry are envisaged.

In chapter 4 cyclometalation of new pincer ligands constituted of centered toluene unit and two phosphine sulfides afforded platinum and palladium pincer complexes with η^3-S,C,S type coordination. Reaction of other pincer ligands, 1,3-benzenedicarbothioamides, with $PtCl_2(PhCN)_2$ also gave new η^3-S,C,S pincer platinum complexes. The molecular structures of the complexes were determined by X-ray crystallography. The pincer complexes exhibit strong emission in a glassy frozen state as well as in the solid state. Light-emitting diodes using the complexes as an active layer displayed red or reddish orange electroluminescence.

As reported in chapter 5, up to date, β–diketonate complexes have been studied thoroughly and have been successfully used in various fields due to their high stability originated from chelate and resonance effects. β-Ketoiminate complexes have recently drawn interest because of their higher thermal stability than the β–diketonate analogues and their versatility, by changing the imine substituents, for tailoring their reactivity and volatility. In addition, polydentate β-ketoiminate derivatives have been adopted for the saturation or stabilization of coordination spheres to control the thermal and chemical stability. Easy control of charge on the ligand is another merit. Various designed complexes containing β-ketoiminate ligands have been successfully exploited as precursors for metal or metal oxide depositions and catalysts. In this monograph, preparation of ligands and complexes, characteristics, and applications of metal complexes as MOCVD (metalloorganic chemical vapor deposition) precursors and catalysts will be reviewed.

In chapter 6 the spectacular development of modern coordination chemistry relies upon programmed ligands, which are organic molecules designed to generate complexes with pre-established properties. This paper reviews our work on a family of tetradentate unsymmetrical Schiff-base ligands, which have been employed to synthesize homo- and heteropolynuclear complexes with various nuclearities and topologies of the metallic centers. Such ligands are obtained by reacting the 3-formyl-salycilic acid with organic molecules bearing primary and tertiary amino groups [*e. g.* 2-(aminomethyl)-pyridine; 2-(2-aminoethyl)-pyridine; N,N-dimethyl-ethylenediamine]. The resulted complexes have relevance in molecular magnetism, as they have been used as models to study the factors governing the nature (ferro- or antiferro), and magnitude of the exchange interactions.

Chapter 7 describes aminophosphonic and aminophosphonous acids and esters, which bear furan moiety substituted in the '2' or '2' and '5' positions. It demonstrates the importance of aminophosphonic acids and esters basing on the most important reviews treating with them – especially reviews by Pudovik and Konovalova or by Drabowicz and Mikołajczyk. I made the attempt to present the furan-containing drugs in pharmacy in order to demonstrate the biological importance of furan compounds. This chapter describes all achievements concerning the chemistry of aminophosphonates bearing furan moiety in the field of synthesis, chemical conversions as well as the applications in the chemistry of polymers and crop protection. The literature is covered since 1952 (Ellis Fields' work) till today and works of such distinguished specialists in organophosphorus chemistry as Boduszek, Kraicheva or Finnocchiaro are described. The stereochemical aspect of their synthesis is also discussed. The importance and methods for synthesis of aminophosphonous acids is also described and such papers as Baylis' et al. one, Grobelny's one or Xiao et al. one are quoted. The synthesis of title furan-derived aminophosphonous acids and formation of their ester with adenosine and cholesterol is discussed, the discussion covers also the stereochemical aspect of these syntheses. This topic has never been described and no review has been written about furyl aminophosphonic derivatives.

In: Focus on Organometallic Chemistry Research
Editor: Marin A. Cato, pp. 1-52
ISBN 1-59454-495-6

Chapter 1

METAL ALKYNYL COMPLEXES OF FLUORENE DERIVATIVES

Wai-Yeung Wong* and Suk-Yue Poon
Department of Chemistry and Centre for Advanced Luminescence Materials,
Hong Kong Baptist University, Waterloo Road, Kowloon Tong, Hong Kong, P.R. China.

ABSTRACT

Conjugated molecules containing metal centers are an important class of molecular materials by virtue of the combined features of the chemical, electronic and optical properties of the metal groups to those of the organic chromophores. Within this framework, homopolymers and copolymers of fluorene derivatives have been extensively explored due to their thermal and chemical stability, their high emission quantum yields and synthetic accessability. Metal alkynyl complexes incorporating functionalized fluorene moieties form a unique family of new materials at the frontier of materials science. While there is great scope for chemical modifications of the metal groups and substituents on the fluorene ring, these organometallic complexes can exhibit rich photophysical and redox properties. The fundamental understanding on these properties of such metal-based alkynyls would lead to the production of novel functional materials and represent good model systems in the development of optoelectronic devices, molecular electronics and chemical sensing appliances. In this chapter, fluorene-derived metal alkynyl systems of the late transition metals are discussed, and particular attention is focused on the electronic absorption spectroscopy, electrochemical and photoluminescence behavior, and structural aspects of this class of metal-organic materials. A detailed account of the evolution of the first excited singlet and triplet state on the electronic structure of the fluorene spacer groups of some metal polyyne polymers will also be given.

---

* E-mail: rwywong@hkbu.edu.hk

ABBREVIATIONS

The following abbreviations are used throughout the text.

CV	cyclic voltammetry
DFT	density functional theory
DP	degree of polymerization
EFISH	electric-field-induced second harmonic generation
FABMS	fast atom bombardment mass spectrometry
Fc	ferrocenyl group $(\eta^5\text{-}C_5H_5)(\eta^5\text{-}C_5H_4)Fe$
Fc/Fc^+	ferrocene/ferrocenium couple
GPC	gel permeation chromatography
HOMO	highest occupied molecular orbitals
ICT	intramolecular charge-transfer
IR	infrared spectroscopy
ISC	intersystem crossing
LED	light-emitting diodes
LUMO	lowest unoccupied molecular orbitals
MLCT	metal to ligand charge-transfer
M_w	weight-average molecular weight
M_n	number-average molecular weight
NLO	non-linear optics
NMR	nuclear magnetic resonance
PL	photoluminescence
T_{decomp}(onset)	onset decomposition temperature
TLC	thin-layer chromatography
UV/Vis	ultraviolet-visible

1. INTRODUCTION

The past few decades have witnessed a rapid expansion of the chemistry of organic based materials [1]. Although the vast majority of these conjugated systems are based on purely organic backbones, there has been a push around the world to incorporate transition metal complexes into electronically extended frameworks [2-4]. A key concern in this research area is to determine the nature and extent of the electronic coupling between the metal and conjugated backbone units, therefore, much attention has been focused on building and studying model complexes in which metal groups are tethered to conjugated fragments via different linkages, and examining the nature of the coupling in these systems. This chapter is intended to provide a comprehensive account of alkynyl complexes of transition metals incorporating one type of organic backbone, namely, fluorene, all of which can exhibit rich photophysical as well as redox properties. Other metallated alkynylfluorene systems are also surveyed herein.

The chemistry of metal alkynyls has continued to attract enormous research attention and these complexes hold a fascination for synthetic chemists, structural chemists and materials scientists alike [5,6]. The linear geometry of the alkynyl unit and its unsaturated character have made the organometallic alkynyls attractive and versatile building blocks for molecular wires and organometallic oligomeric and polymeric materials which can possess unique properties, such as luminescence, liquid crystallinity, optical non-linearity and electrical conductivity. Despite the large body of work on metal acetylides spaced by carbocyclic and heterocyclic units (e.g. oligothienyl, oligopyridyl, thienyl-pyridine, phenylene, etc.), relatively little was known for fluorene-based materials of this kind. The use of fluorenyl chromophores and other related derivatives such as carbazoles offers exciting perspectives for the design of new molecular and polymeric materials for various optoelectronic applications [7]. The ease of modification and knowledge of the structure-property relationship of polyfluorene homopolymers and copolymers continue to make the fluorene-functionalized compounds very attractive candidates in the development of new functional materials. It was demonstrated that substituted poly(2,7-fluorene) derivatives (Figure 1) are attractive as active components of organic light-emitting diodes (LEDs) due to their thermal and chemical stability and their high emission quantum yields [7-12]. The fluorene structural moiety provides a rigidly planar biphenyl unit within the molecular backbone. Substituent derivatization at the C-9 position of the monomeric fluorenes provides the prospect of controlling polymer properties such as solubility, emission wavelengths as well as processability and mediating potential interchain interactions in films [13-16]. Moreover, substituted fluorenes are widely used as electron acceptors in charge transfer complexes and electron transport materials [17,18]. More recently, a number of elegant synthetic means have been employed to generate fluorene-containing materials with enhanced device performance [7]. However, to our knowledge, related work on metal-based functional materials of fluorenes is rarely studied. In this context, we have directed our research specifically to the molecular design, synthesis and materials properties of new metal alkynyl complexes of fluorene derivatives. The fluorenyl linkers **L_1–L_9** widely employed in our investigations are shown in Figure 2.

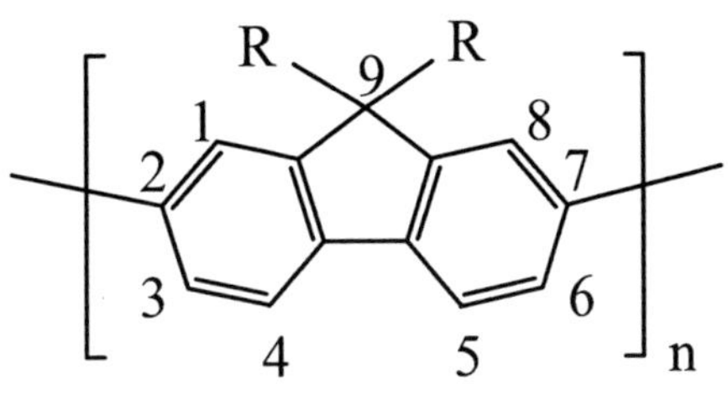

Figure 1. Chemical structure of poly(2,7-fluorene) derivatives

2. Metallated Complexes of Ethynylfluorene Derivatives

2.1. General Background

In general, there are three general ways of incorporating transition metals into conjugated organic backbones (Figure 3). The metals can exist as components inserted into the conjugation main chain (Type I), electronically isolated pendant groups (Type II), or electronically coupled pendant groups (Type III). Due to the interruption of the π-conjugation in Type I materials, the influence of the metal group in this category is typically more significant than in Type III materials, where the metal may only indirectly affect the conjugated backbone via a combination of steric and inductive effects. In the following account, we will focus on some recent advances in the synthesis and chemistry of a wide range of metallated complexes of acetylenic fluorenes, grouped according to the nature of the metal moieties.

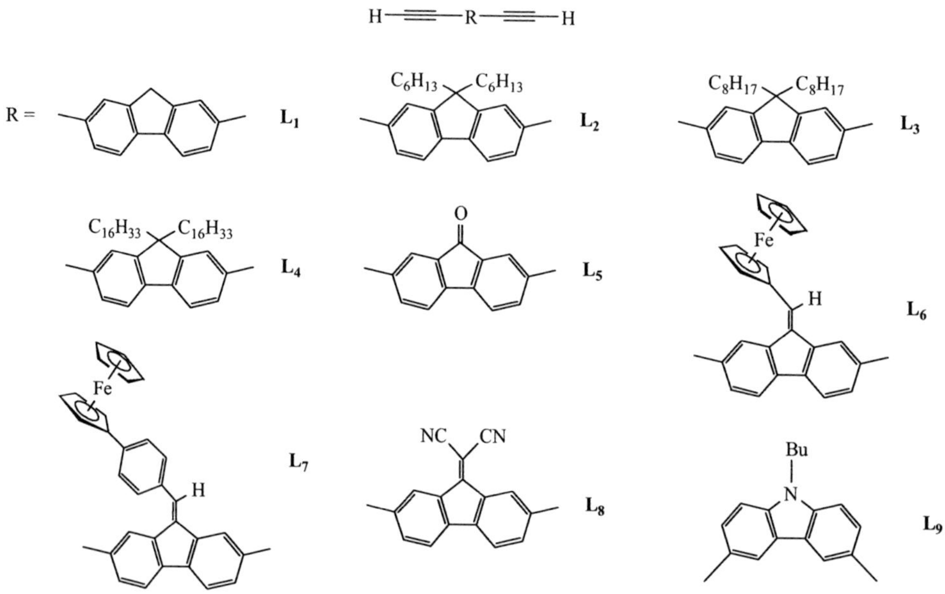

Figure 2. Structures of substituted 2,7-diethynylfluorene derivatives.

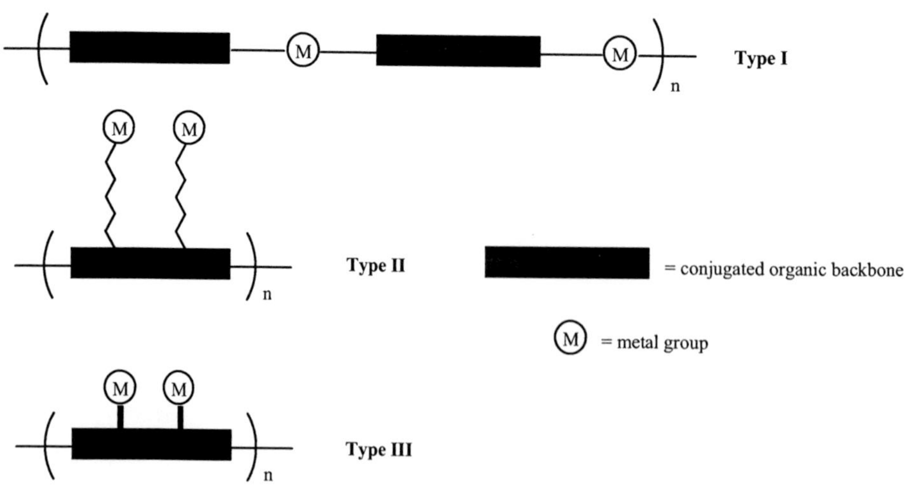

Figure 3. Three common types of metal-containing conjugated molecules.

2.2. Ferrocenyl Acetylide Complexes

There is an upsurge of interest in the research community in the design and synthesis of carbon-rich bi- or multimetallic assemblies containing π-conjugated chains [19-21]. It has been demonstrated that molecular wires comprising mixed-valence bimetallic fragments or remote redox-active organometallic building blocks linked by all-carbon chains could be used in molecular electronics, optoelectronic devices and chemical sensing appliances [22-25]. Two common models are used to evaluate the capability of electronic communication between two metal centers "M_1" at the termini made feasible by an organic spacer "M_1–spacer–M_1", and the effect of another organometallic fragment "M_2" in a conjugated organic chain "M_1–spacer–M_2–spacer–M_1" [26-35]. Several systems where redox active sites are strongly coupled electronically through the polyynyl fragments are known [26-35]. Although the first report of the mixed-valence ion of diferrocenylacetylene was reported as early as 1974 [36], interest in metal-capped linear polycarbon chains as models for materially useful compounds is still very intense and much research efforts have been devoted to the molecular design of ferrocene-derived homo- and heterometallic molecular scaffolds connected by oligoene, oligoyne, thiophene, furan, phenylene, phenyleneyne, thienyl ethynyl and furanyl vinyl units [37-49]. The introduction of a metallocene unit into rigid-rod one-dimensional oligomers and polymers may give rise to a range of properties (e.g. redox, optical, electrical and catalytic properties) that differ from those of conventional organic polymers.

New bis(ferrocenylethynyl) complexes with fluorenyl-based spacers **1–3** were prepared by the Sonogashira coupling reactions of ethynylferrocene with the corresponding dibromofluorene derivatives in a 2:1 stoichiometry (Scheme 1) [50]. The product yields of **1–3** are very high (>90%) and formation of the homocoupling compound, 1,4-diferrocenylbutadiyne, can be minimized under strictly anaerobic conditions. Compounds **1–3** are soluble in common organic solvents and characterized by satisfactory microanalysis, FABMS, IR and NMR spectroscopies. The crystal structures of **1** and **2** as determined by X-ray diffraction method are depicted in Figures 4 and 5, respectively.

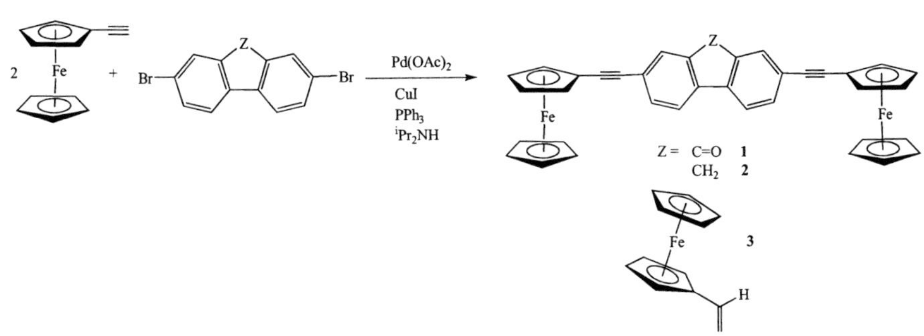

Scheme 1.

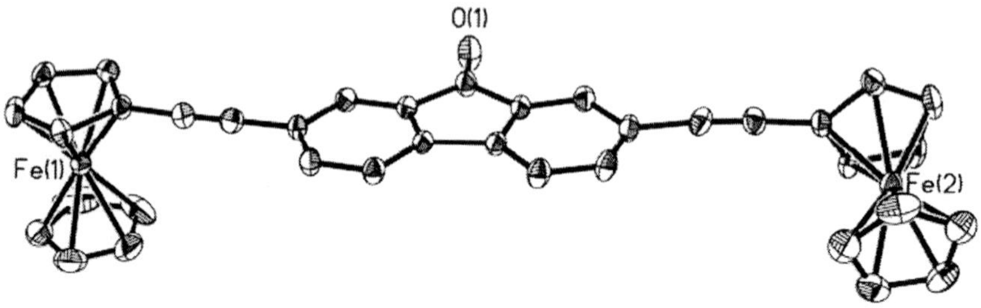

Figure 4. X-ray structure of **1**.

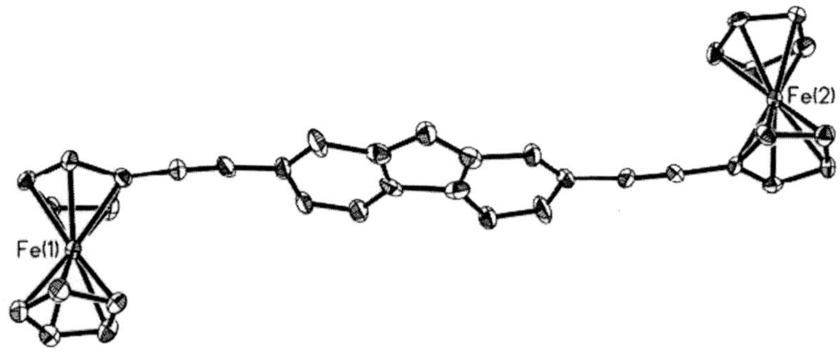

Figure 5. X-ray structure of **2**.

The two ferrocenyl end groups only weakly communicate through the fluorenyl ring system as revealed by cyclic voltammetry (CV). Both complexes **1** and **2** were found to undergo a single-step two-electron oxidation involving the concomitant oxidation of the two ferrocenyl subunits whereas complex **3** showed a one-step three-electron oxidation when these compounds were subjected to the electrochemical measurements with a scan rate of 100 mV s^{-1}. The half-wave potential of the terminal ferrocenyl units shows a gradual increase in the order **3** < **2** < **1** upon changing the substituent at the 9 position of the fluorenyl ring from the electron-donating ferrocenyl group in **3** to an electron-withdrawing oxo group in **1**. The potential is anodically shifted by ca. 110 mV from **3** to **1** and the low value for the oxidation potential in **3** shows that such a triferrocenyl species is electron-rich. Loss of electron density from the ferrocenyl donor unit to the net electron-accepting fluorenone moiety through the alkynyl linking unit makes the oxidation wave appear at a more positive potential in **1**. For **1**, an additional irreversible wave was also observed at –0.95 V, which was attributable to the reduction of the fluorenone moiety.

New ferrocenyl heterometallic systems featuring such fluorene units were also investigated. In a multistep fashion, fluorenone-containing ferrocenylacetylene complex **4** was prepared which can provide a direct access to novel mixed-metal complexes **5–8** via the CuI-catalyzed dehydrohalogenation reactions with appropriate metal chloride compounds (Scheme 2) [51]. Complexes **5–8** dissolve well in common organic solvents and their formulas were successfully established by the intense molecular ion peak in each of the FAB mass spectra. The solid-state structures of **6–8** were confirmed by X-ray crystallography (Figures 6–8). For **6**, there is a centrosymmetry at the platinum center and the crystal structure shows a diferrocenyl end-capped species in which two deprotonated forms of **4** are bonded to a $Pt(PBu_3)_2$ unit in a *trans* geometry to afford a rigid-rod, nano-sized molecule with an iron–iron through-space separation of about 34 Å. The corresponding mean Fe–Pt distance is ca. 17 Å. The platinum center is square-planar with four ligand groups in the coordination sphere. The alkynyl bond lengths span the narrow range 1.169(3)–1.189(3) Å, which are comparable to those found in the bithienyl-linked analogue reported recently [52]. The nearly linear bond angles around the alkyne units conform to the rigid and straight-chain nature of the molecule. The structures of **7** and **8** belong to new heterometallic compounds in which the isoelectronic fragments $Au(PPh_3)$ and MeHg are attached to the terminal alkyne unit of **4**, respectively. The geometry about the Au(I) and Hg(II) centers is essentially linear and the mean P–Au–C angle of 178.1(3)° and Au–C–C angle of 177.7(8)° in **7** as well as the C–Hg–C angle of 179.0(4)° and Hg–C–C angle of 178.5(9)° in **8** are indicative of sp hybridization in the metal atom and acetylenic carbon.

Introduction of the platinum(II) segment in **5** and **6** was found to red-shift the lowest energy bands and to increase the absorption intensity, suggesting an enhancement in the extent of π-delocalization through the platinum conjugated system that is fully consistent with the IR data. A bathochromic shift of about 20 nm is observed from **4** to **5** and **6**. However, apart from an increase in the molar extinction coefficients, there is no significant variation in the lowest energy peak from **4** to **7** and **8**. Thus, delocalization of π electrons continues through the metal center to a different extent and the conjugation is greater for the Pt(II) complexes than the Au(I) and Hg(II) species. As far as the electrochemical properties of these complexes are concerned, each of them is electroactive and the cyclic voltammogram is dominated by one quasi-reversible ferrocenyl oxidation wave. An anodic shift of the ferrocene-ferrocenium (Fc/Fc^+) couple relative to the ferrocene standard is caused by the unsaturation of the ethynyl groups which renders the removal of electrons more difficult. When the conjugation length is increased upon capping the end group with an organometallic segment from **4** to **5–8**, oxidation is favored by the delocalization of charge along the main chain through a $d\pi \rightarrow p\pi$ interaction which leads to a lowering of the oxidation potentials. The most significant cathodic shift of ca. 0.06 V is noted in the most extended heterotrimetallic complex **6**, however, the two ferrocenyl end groups do not exhibit an observable electronic communication through the fluorenyl ring system in such case. Complex **6** only undergoes a single-step two-electron oxidation involving the concomitant oxidation of the two ferrocenyl subunits at a scan rate of 100 mV s^{-1}. Non-interacting diferrocenyl systems are known for *trans*-$[FcC{\equiv}CPt(PR_3)_2C{\equiv}CFc]$ [53,54], *trans*-$[FcC{\equiv}C(C_4H_2S)_mC{\equiv}CPt(PBu_3)_2C{\equiv}C(C_4H_2S)_mC{\equiv}CFc]$ (m = 1–3) [52] and *trans*-$[Pd(PBu_3)_2(C{\equiv}CC_6H_4C{\equiv}CFc)_2]$ [26], where Fc = $(\eta^5\text{-}C_5H_5)(\eta^5\text{-}C_5H_4)Fe$. The $E_{1/2}$ value is slightly more anodic in **8** than in **7** because of the higher formal oxidation state of mercury(II) than that of gold(I), which would hinder the back-donation to π^* orbitals. Another irreversible redox event due to the $Pt^{2+} \rightarrow Pt^{3+}$ oxidation was also detected at higher positive potentials in **5** and **6**.

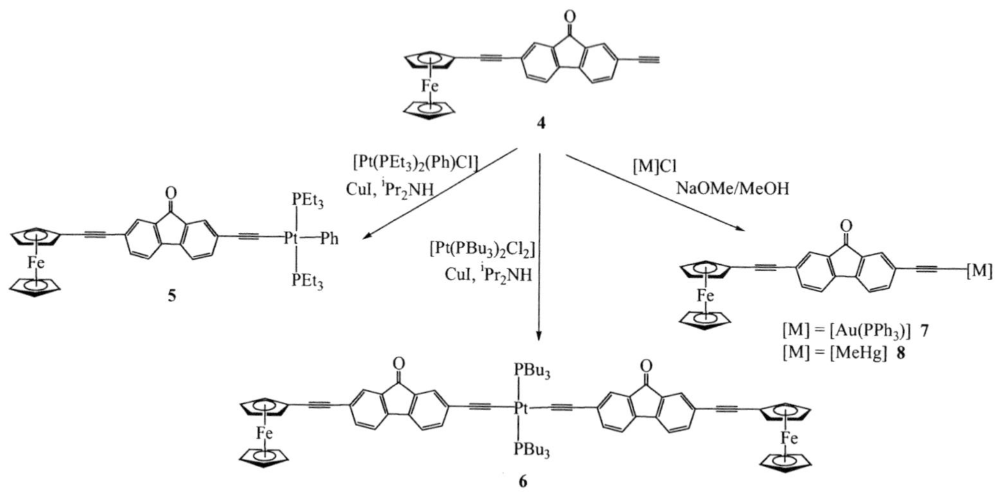

Scheme 2.

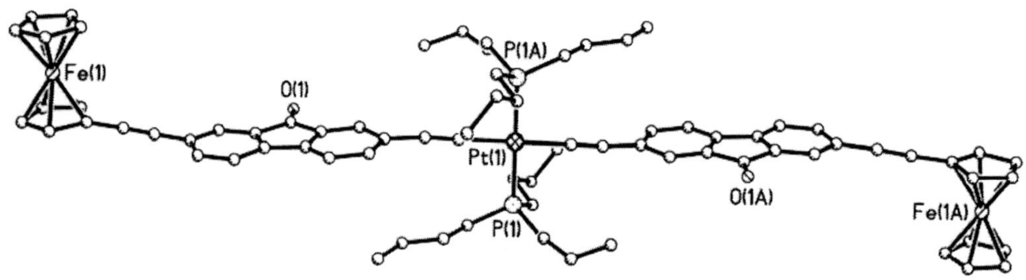

Figure 6. X-ray structure of **6**.

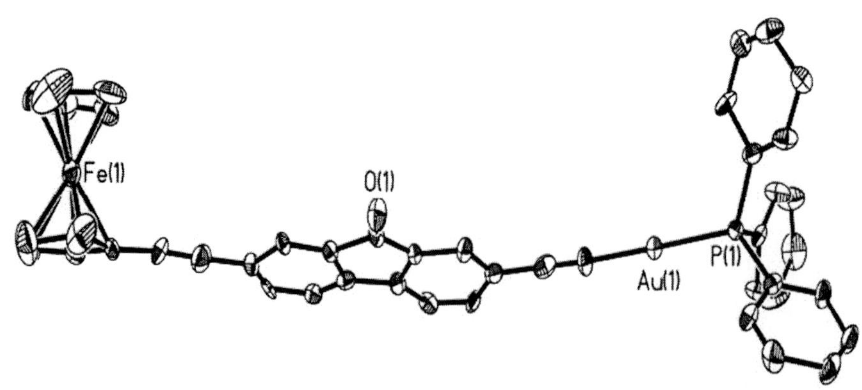

Figure 7. X-ray structure of **7**.

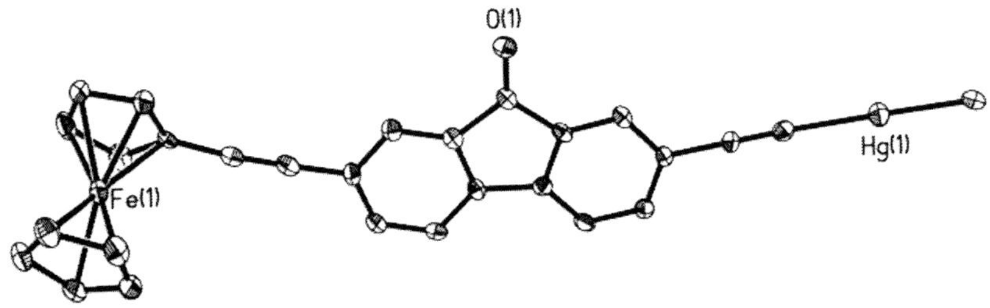

Figure 8. X-ray structure of **8**.

In the quest for new alkynylferrocene-oriented materials, we also developed new molecular architectures based on the more extended congener, 4-ethynylphenylferrocene, and such a study should represent an interesting extension to the chemistry of related systems. Here, 2-ethynyl-7-(4-ferrocenylphenylethynyl)fluoren-9-one **9** was prepared via three steps from 4-ethynylphenylferrocene and can be used for the build-up of new multimetallic systems. Scheme 3 outlines the synthetic routes to the various bi- and trimetallic complexes **10–13** [55]. The classical dehydrohalogenation reactions between **9** and *trans*-$[Pt(PEt_3)_2(Ph)Cl]$ or *trans*-$[Pt(PBu_3)_2Cl_2]$ led to new platinum(II) alkynyl complexes **10** and **11**. New σ-acetylide complexes of Group 8 metals **12** and **13** were made by reacting *cis*-$[M(dppm)_2Cl_2]$ (M = Ru, Os) with $NaPF_6$ to give first the vinylidene intermediates which can be deprotonated by DBU to furnish the desired compounds. The redox properties of **9–11** are very similar to those derived from ethynylferrocene. The structural properties of **11** and **13** were studied (Figures 9 and 10). For **11**, the X-ray structure shows a diferrocenyl end-capped species in which two deprotonated forms of **9** are σ-bonded to the central $Pt(PBu_3)_2$ unit in a *trans* configuration to afford a one-dimensional molecular rod of defined length and the through-space separation between Fe(1) and Fe(2) is estimated to be 42 Å in the solid state. Hence, extension of the dimensions of these metal-capped molecules from molecular scale to nanoscale becomes viable based on this synthetic strategy.

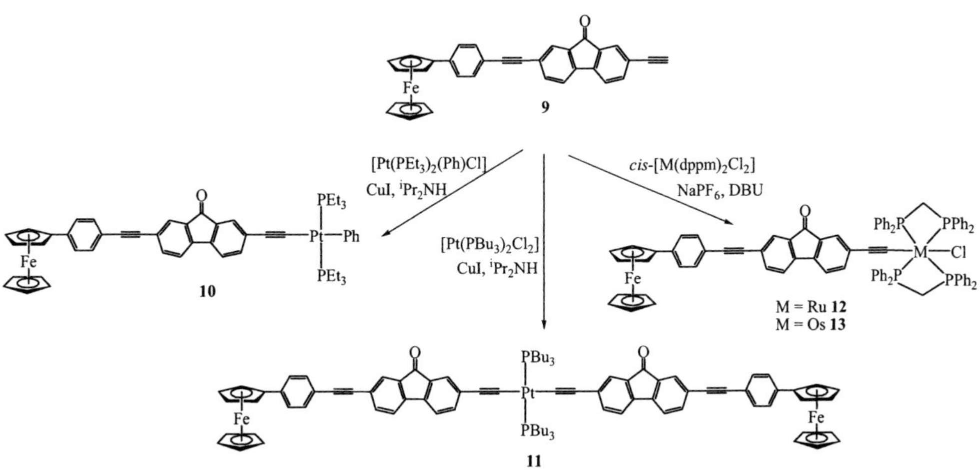

Scheme 3.

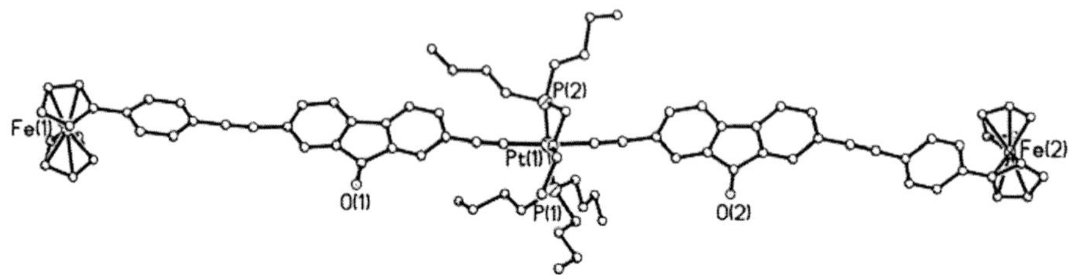

Figure 9. X-ray structure of **11**.

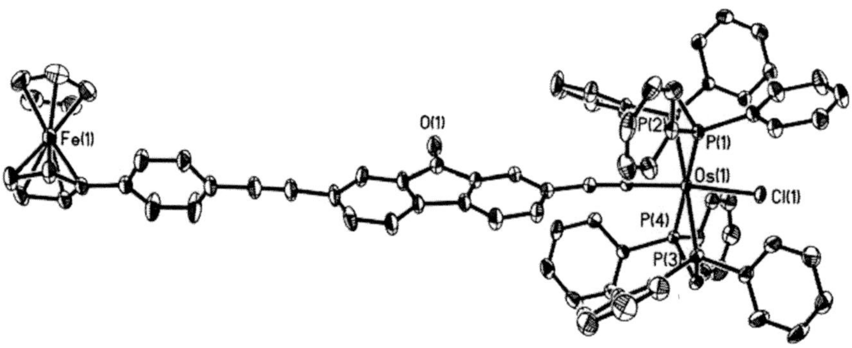

Figure 10. X-ray structure of **13**.

Molecular orbital calculations at the B3LYP level of density functional theory (DFT) for complexes **11** and **13** based on their experimental geometries obtained from the X-ray data have been carried out [56-58], and the results are compared to those of **9.** We found that the three HOMOs for **9** correspond to those derived from the mixing of the highest occupied orbitals of the conjugated organic structural unit and the two Fe(d_δ) orbitals. The δ-type is defined by viewing the orbitals along the Ct-Fe-Ct axes (Ct: the center of each cyclopentadienyl ligand). When the Ct-Fe-Ct axis is defined as the z axis, the two d_δ orbitals correspond to the d_{xy} and d_{x2-y2} orbitals. The LUMOs for **9** correspond to the π* orbital in the structural unit having a carbonyl group. Carbonyl groups normally have lower π* orbitals because of the electronegative oxygen atom. An orbital correlation diagram shown in Figure 11 is used to illustrate the change in the orbital patterns in **11** and **13** upon incorporation of different metal groups to **9**. The frontier orbitals of **11** consist of the orbital combinations of the two ferrocenyl-containing alkynyl ligands. The two HOMOs are contributed from the highest occupied π orbitals of the two ligands whereas the two LUMOs arise from the lowest unoccupied π* orbitals of the organic portion in the two ligands. The two HOMOs of **13** are derived from the mixing of the Os metal d orbitals (i.e., d_{xz} and d_{yz} when the Os–Cl bond direction is defined as the z axis) and π orbitals of the Os-bonded C≡C unit. The LUMOs correspond to the lowest π* orbitals of the carbonyl-containing conjugate structural moiety. Because of π-donating properties of the chloride, there are π*-antibonding characters between Os and Cl in the two HOMOs. Therefore, the absorption band corresponding to the HOMO–LUMO transition probably has the MLCT character and the calculated HOMO–LUMO gap (2.69 eV) for **13** is considerably smaller than that for **9** (3.03 eV). The three HOMOs in **9** correspond to the 3rd to 5th HOMOs in **13.** Likewise, the smaller HOMO–LUMO gap for **11** (3.01 eV) as compared to **9** is in line with their observed UV–vis data which indicates the possibility of π-conjugation through Pt.

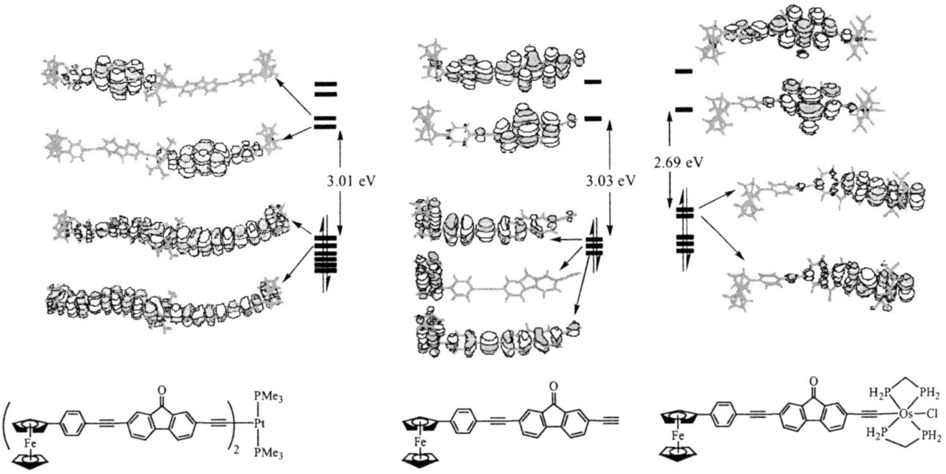

Figure 11. Contour plots of the HOMOs and the LUMOs for **9**, **11** and **13**. In the calculations, the PBu_3 ligands in **11** were modeled by PMe_3 groups whereas the phenyl rings in **13** were modeled by H atoms.

In a further step, the coupling reaction of 2,7-dibromo-9-(dicyanomethylene)flurorene with ethynylferrocene was studied. Although the palladium-catalyzed Sonogashira coupling of aryl halides with terminal acetylenes has been shown to be a versatile method for introducing an ethynyl group into organic molecules [59], no direct coupling product was obtained in this case but instead an unsymmetrical mono-cyano substituted fluorene compound **14** formed in which one of the cyano groups is being expelled by the incoming ferrocenylacetylide moiety [60]. This reaction also works very well for organic terminal acetylenes. Attempts to make the symmetrical coupling product **15** require the utilization of the corresponding fluoren-9-one derivative in the first step in such a way that coupling reaction must take place before the condensation reaction with malononitrile (Scheme 4). The crystal structure of **14** depicted in Figure 12 represents the first structurally characterized example of a fluorene derivative possessing an σ-acetylide group at its 9-methylene position. The electronic spectra of **14** and **15** essentially show the structureless $\pi \rightarrow \pi^*$ transition bands of the corresponding fluorene π skeleton as well as metal to ligand charge-transfer (MLCT) transitions at the low-energy regime, which were supported by DFT calculations. We note a bathochromic shift of ca. 130 nm in λ_{max} in going from **1** to **15**. Comparing the data for **1** and **15**, the presence of the two π*-accepting cyano groups lowers the optical gap value significantly by as much as 0.60 eV since these groups can stabilize the LUMO to a greater extent relative to that of the carbonyl group in **1.** One remarkable feature in this study is the observation of solvatochromic phenomenon for the lowest energy MLCT band in **14** (a red shift of 30 nm from hexane to CH_2Cl_2), consistent with a dipolar excited state due to intramolecular charge-transfer (ICT) transition from the electron-donating ferrocenyl group to the electron-withdrawing cyano group through the conjugated system. It is due to such MLCT band in **14** which accounts for the deep blue colour of its solution in CH_2Cl_2. In contrast, the higher-energy absorptions of **14** are relatively insensitive to variation of solvent polarities (Table 1). Analogous electron donor-π-acceptor chromophores **16a–b** have also been reported by Perepichka and co-workers which are obtained from the reaction between formylated ferrocenylacetylene and substituted fluorenes [61]. Compounds **16a–b** show ICT bands from the donor onto the acceptor moiety at 1.88 and 2.01 eV, respectively, and moderate second-order non-linear optical (NLO) properties as determined by the EFISH technique at $\lambda = 1.54$ μm are observed for **16b** [$\mu\beta(0) \approx 170 \pm 30 \times 10^{-48}$ esu]. They have demonstrated that the polynitrofluorene moiety is much more favorable as an acceptor terminal group in NLO chromophores than the dicyanomethylene moiety in these circumstances.

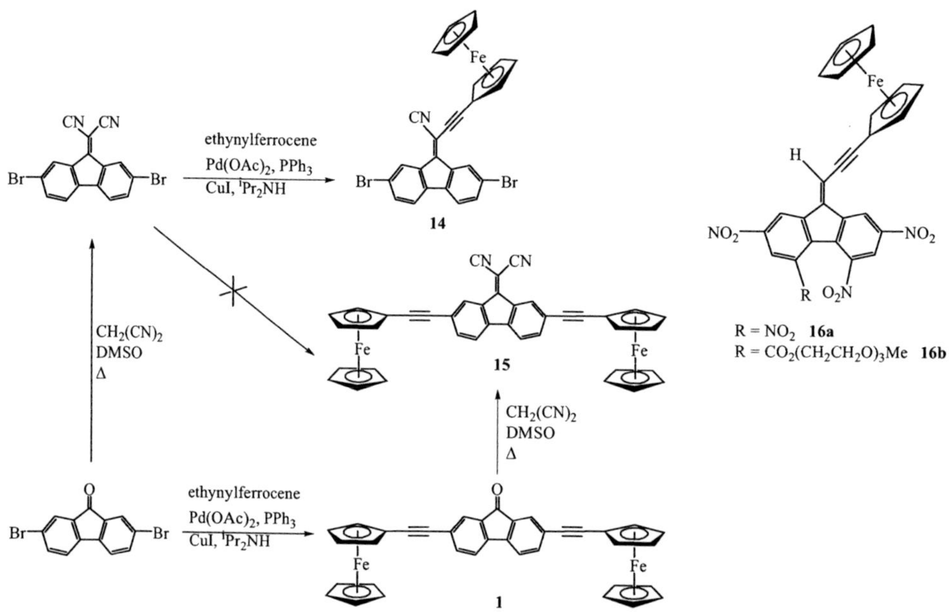

Scheme 4.

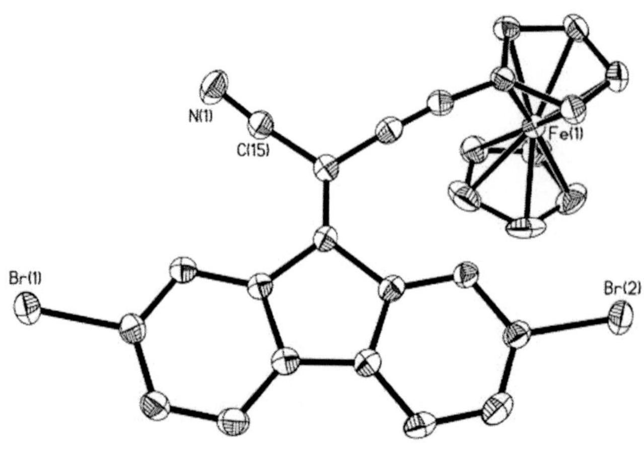

Figure 12. X-ray structure of **14**.

Table 1. Electronic Absorption and Redox Data for 1, 14 and 15

compound	λ_{max}/nm ($\varepsilon \times 10^{-4}$/$M^{-1}cm^{-1}$)[a]	optical gap/eV[b]	color	E_{ox} (ΔE_p)[c]	E_{red}[d]
1	288 (8.7), 349 (4.7), 493 (0.8)	2.13	red	0.14 (159)	
14	270 (5.0), 284 (4.0), 387sh (1.6), 404 (1.7), 592 (0.4)[e]	1.74	dark blue	0.27 (119)	
15	320 (5.4), 461sh (0.3), 623 (0.1)	1.62	green-blue	0.14 (99)	−1.13

[a] All absorption spectra were recorded in CH_2Cl_2 at 293 K. [b] Estimated from the absorption edge. [c] All the potential values are referenced to the external ferrocene standard. $E_{ox} = (E_{pc} + E_{pa})/2$ for reversible oxidation. ΔE_p in mV. Scan rate = 100 mV s^{-1}. [d] Irreversible wave. [e] Data in ethanol (nm): 267, 365, 400, 583; in hexane: 257, 266, 274, 383, 400, 562.

2.3. Ruthenium and Osmium σ-Acetylides

New terminal acetylide complexes of ruthenium and osmium **17** and **18** were made through the formation of vinylidene intermediates followed by deprotonation [26,62,63]. The reaction of *cis*-[M(dppm)$_2$Cl$_2$] (M = Ru, Os) with one equivalent of **L$_5$** and two equivalents of $NaPF_6$ generated vinylidene complexes which were not isolated but instead, were deprotonated in situ by one equivalent of DBU (Scheme 5) [62]. In a similar manner, homometallic diyne complexes **19** and **20** and a mixed metal diyne complex of ruthenium and osmium **21** have been synthesized via the $NaPF_6$/DBU method involving *cis*-[Ru(dppm)$_2$Cl$_2$] and **18** in a 1:1 molar ratio. All of these compounds were isolated as bright to dark green solids.

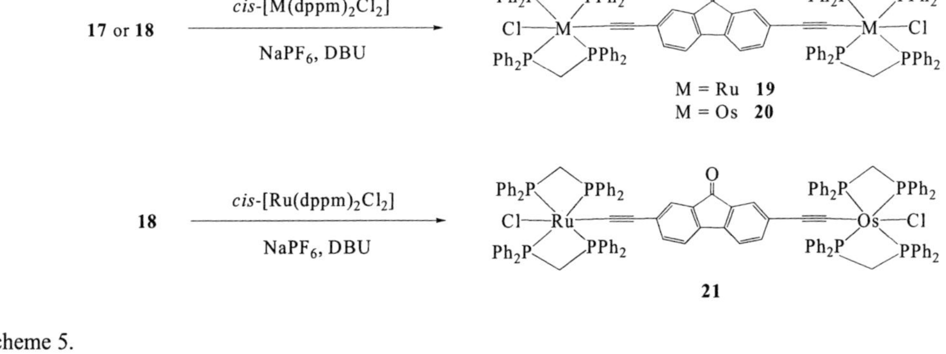

Scheme 5.

Complexes **17–21** are intensely colored compounds and their absorption spectra are dominated by a strong featureless absorption band around 400 nm and a low-energy band beyond 400 nm. These bands are ascribed to the ligand-based $\pi \rightarrow \pi^*$ and the MLCT transitions, respectively. We observe that both transitions appear at longer wavelengths for the osmium complexes than those in the ruthenium counterparts.

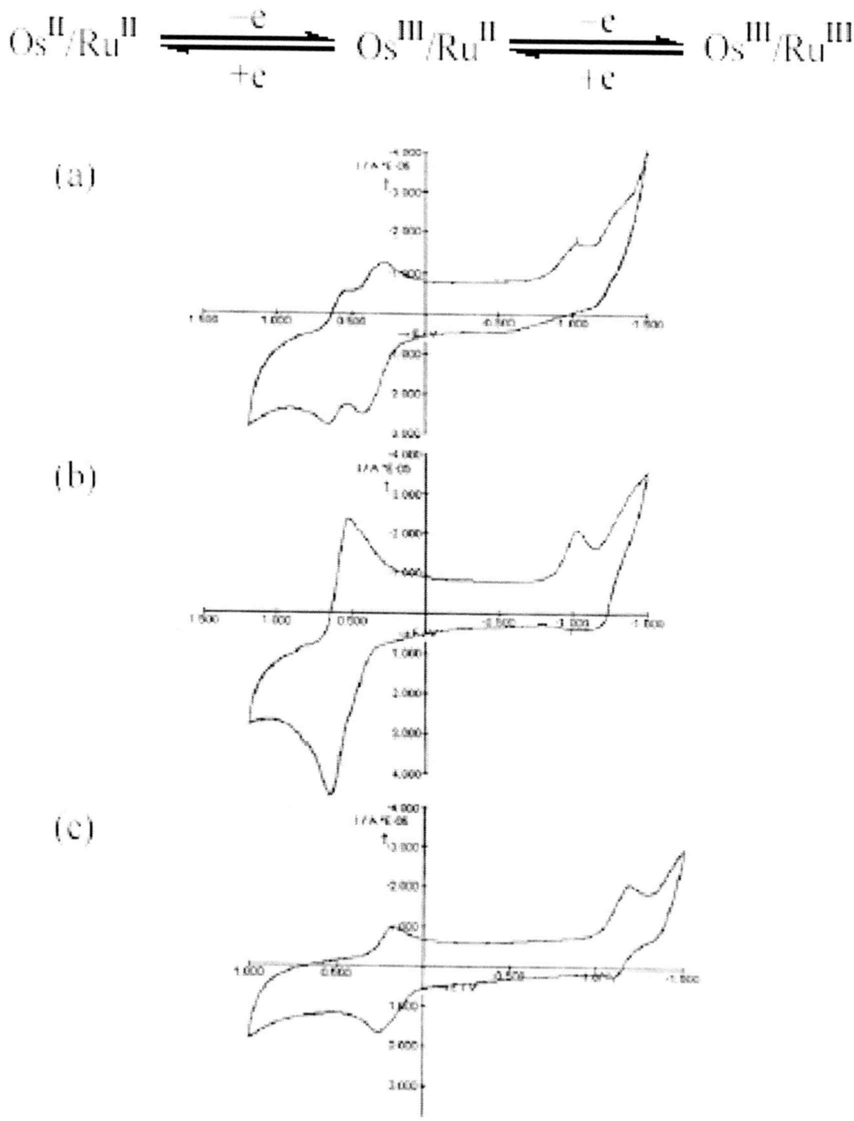

Figure 13. Cyclic voltammograms of complexes (a) **21**, (b) **17** and (c) **18**.

In the cyclic voltammogram, complex **21** displays two well-separated ($\Delta E_{1/2}$ = 0.24 V) quasi-reversible oxidation processes centered at –0.31 and –0.07 V, attributed to the two one-electron oxidations at the osmium and ruthenium centers, respectively (Figure 13a). The terminal reference compounds **17** and **18** are quasi-reversibly oxidized at 0.0 and –0.20 V, respectively (Figures 13b and 13c). From the CV of **21**, it is shown that the ruthenium center releases electron density to the osmium unit via the central diethynylfluorenonyl bridge and thus the electrode potential of the Os^{III}/Os^{II} couple is lowered significantly (from –0.20 to –0.31 V). Electrochemical investigations of the homometallic dimeric complexes **19** and **20** have shown a similar trend and they show two successive, closely-spaced quasi-reversible oxidation waves which could be ascribed to the formation of the mono- and dications respectively, corresponding to the M^{III}/M^{II} and M^{III}/M^{III} (M = Ru, Os) couples. In each case, a ligand-based irreversible reduction wave was also observed. Similar to the complexes with the $-C{\equiv}C(p\text{-}C_6H_4)C{\equiv}C-$ linker [64], the redox processes can be interpreted as:

2.4. Cobalt(0) Carbonyl Alkyne Complexes

Oligomers and polymers containing the acetylide groups are under intensive study because of the possibility of charge transfer along the conjugated backbone and their potential application as building blocks in the design of conjugated materials [65,66]. It is well-known that the alkyne unit is a versatile functional group which undergoes many useful transformations [67]. Acetylenes are important precursors for the molecular carbon rod and cyclo[*n*] species [68]. The use of alkyne complexation to facilitate ring formation is exemplified by the successful isolation of a hexacobalt complex of cyclo[18]carbon from $[Co_2(CO)_6(Me_3SiC_2H)]$ [69] and an octacobalt complex of [8,8]paracyclophaneoctayne [70]. The role of cobalt-alkyne complexes in organic synthesis is also well-established [71-74].

Electrochemical studies on many transition metal carbonyl clusters have shown that they are electron reservoirs with tunable redox behavior dependent on the coordination sphere of the metal core [75,76]. Such cluster unit, when incorporated to other redox centers such as ferrocene, can also participate in electronic interactions leading to classical mixed-valence molecules [77]. The $Co_2(CO)_6$ unit readily coordinates with the C≡C bond in an η^2-fashion [78-82]. Redox studies of these compounds suggested that the coordination of cobalt carbonyl groups maintains the rigidity of the system and still allows for potential electronic delocalization down the chain [82]. A series of organometallic dimers of the form $[\{Co_2(CO)_6\}_2(\text{diyne})]$ **22–27** have been prepared from the reactions of two molar equivalents of octacarbonyldicobalt(0), $[Co_2(CO)_8]$, with the appropriate diynes in *n*-hexane at 0 °C (Scheme 6) [83]. They all gave satisfactory spectroscopic data and the single-crystal structures of **23** and **25–27** were established. Representative examples are shown in Figures 14 and 15. Structurally, each of these displays two 'Co_2C_2' cores adopting the usual pseudo-tetrahedral geometry with the alkyne bond lying essentially perpendicular to the Co–Co vector and there is an asymmetry of the Co–C(alkyne) distances with each Co_2C_2 unit.

The $Co_2(CO)_6$ entity bound to an alkyne was reported by Robinson et al. to undergo an irreversible reduction at potentials in the range –0.65 to –1.10 V *vs.* Ag/AgCl in CH_2Cl_2 at room temperature [84,85]. We observed this expected reduction at potentials from –1.56 to –1.65 V *vs.* Fc/Fc^+ couple in CH_2Cl_2 at room temperature for complexes **22–24**, which is believed to be cobalt cluster based. At ambient temperature, each of them undergoes an irreversible reduction process followed by an oxidation event in the potential range 0.62–0.88 V due to $[Co(CO)_4]^-$ resulting from the decomposition of the electrogenerated monoanion (Table 2) [53,84,85]. At first sight, it seems that there is no electrochemically detectable electronic communication between the Co_2 cores. In fact, it was shown that the fast chemical decomposition following the redox process in all these compounds prevents proper electrochemical analysis.

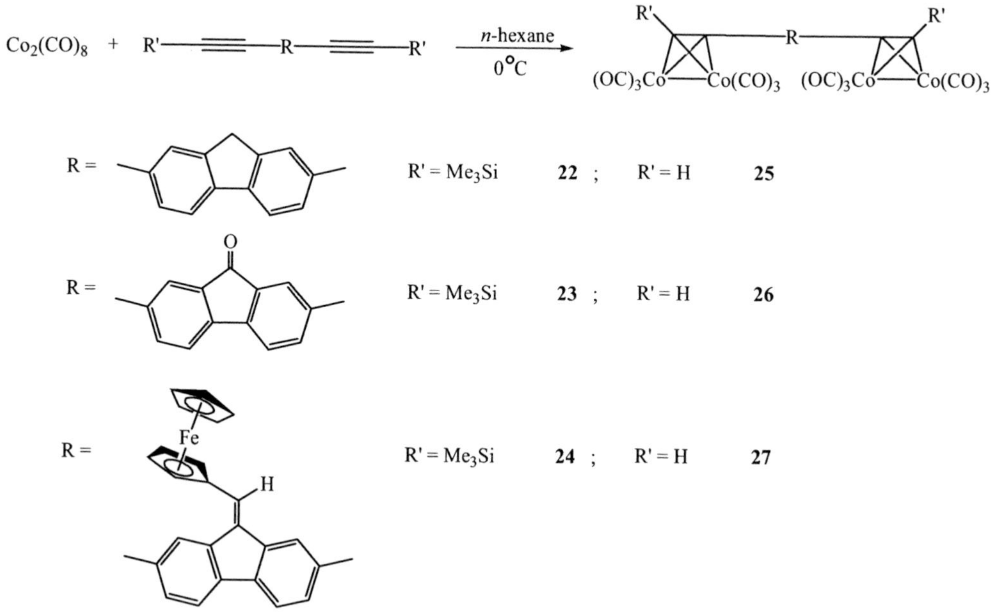

Scheme 6.

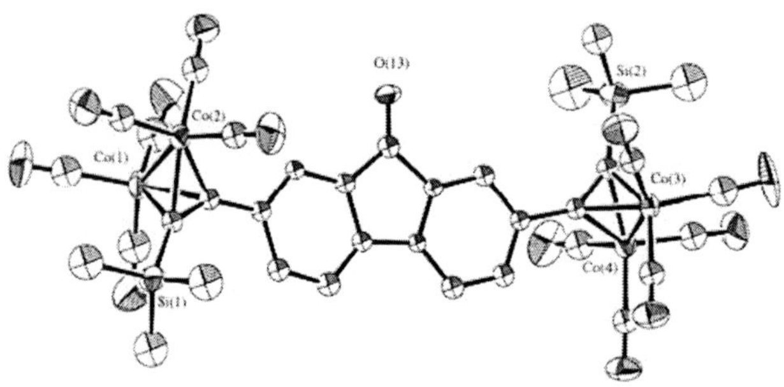

Figure 14. Molecular structure of **23**.

At –78 °C, complications were completely quenched so that chemical reversibility was achieved. The reduction waves for **22–24** gradually split into two apparently chemically reversible one-electron transfer peaks, with a splitting of 470 mV for **22**, 230 mV for **23** and 510 mV for **24**. Such an observation suggests that, despite the two redox centers being identical, they are coupled from an electronic perspective. These values may be compared with the smaller splitting of 220 mV found in $[\{Co_2(CO)_6\}_2(PhC{\equiv}C\text{-}C{\equiv}CPh)]$ [53] and this indicates that the presence of aromatic entities within the diyne ligands tends to increase the electronic communications between the two Co_2 moieties. Moreover, it is interesting to note that attachment of an electron-donating ferrocenyl group contributes to the largest splitting as in **24** whereas the smallest peak separation in **23** is achieved with an electron-withdrawing fluorenone moiety. It is therefore envisaged that the coupling between the redox centers can be modulated by adjustment of the donor-acceptor characteristics of the central linker.

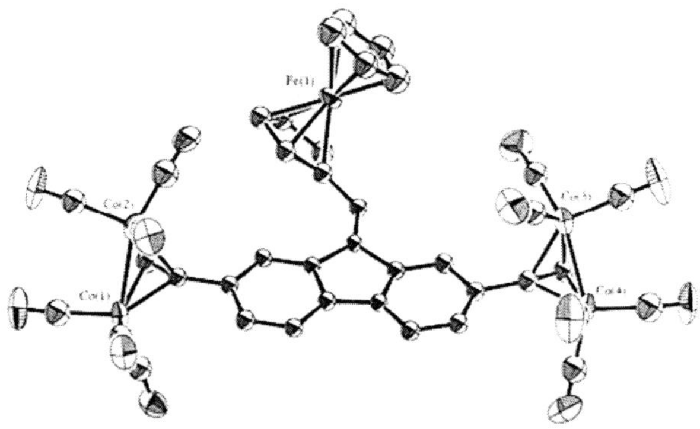

Figure 15. Molecular structure of **27**.

The ferrocenyl residue in **24** is characterized by a reversible one-electron oxidation wave at 0.23 V *vs.* Fc/Fc^+ couple in CH_2Cl_2. When compared to the free diyne complex, this wave is anodically shifted by ca. 90 mV. The shift is ascribed to conjugation of the ferrocenyl group to the electron-withdrawing $Co_2(CO)_6$ unit via the fluorene bridge, which makes the removal of the electron from the ferrocene core more difficult.

Table 2. Electrochemical Data for Complexes 22–24 in CH_2Cl_2.[a,b]

complex	room temperature	−78 °C
22	−1.65,[c] 0.77[c]	−2.04 (95), −1.57 (80), 0.66 (200)
23	−1.57,[c] 0.88[c]	−1.75 (85), −1.52 (80), 0.75 (200)
24	−1.56,[c] 0.23 (87), 0.78[c]	−2.08 (100), −1.57 (100), 0.23 (87), 0.69 (150)

[a] The electrochemical measurements were made at a glassy carbon working electrode containing 0.1 M $[NBu_4]PF_6$ as base electrolyte, using a scan rate of 100 mV s^{-1}. All potentials were quoted in volts *vs.* Fc/Fc^+ couple which was used as an internal standard. [b] Difference between the anodic and cathodic peak potentials for a reversible wave is shown in parentheses. [c] Irreversible wave.

2.5. Platinum(II) σ-Acetylides

Since the pioneer work by Friend et al. in Cambridge, great efforts have been devoted to the design and synthesis of light-emitting organic polymers in the scientific community because of their potential applications in optoelectronic devices [65] such as LEDs [86], lasers [87] and photocells [88,89]. One of the major issues in organic LEDs is the ratio of 3:1 for the generation of non-emissive triplet to emissive singlet excitons based on the spin statistics [90]. Yet the percentage of emissive singlet excitons determines the overall efficiency of the LED. Higher efficiencies can be realized by harvesting the energy of the non-emissive triplet excitons [91-94]. This requires detailed knowledge of the energy levels of the triplet state. To study the triplet excitons directly, conjugated polymers of the type *trans*-[$-ML_2C{\equiv}CRC{\equiv}C-$]$_n$ (Figure 16) have attracted the attention of a number of research groups [5,6]. They can be used as excellent model systems to obtain a clear picture of the spatial extent of the singlet and triplet manifolds and provide important information on the photophysical processes that occur in organic conjugated polymers [95-100]. The prototypical polymer for much of this work is *trans*-[$-Pt(PBu_3)_2C{\equiv}C(p\text{-}C_6H_4)C{\equiv}C-$]$_n$, and more recently, a wide variety of derivatives have been prepared where alkynyl units are connected to a variety of aromatic ring systems such as the phenyl, pyridyl, oligothienyl and anthryl units. The electronic properties of these organometallic polymers are primarily governed by the nature of the metal, the auxiliary ligands L or the spacer group R. What is particularly appealing to us is that the strong spin-orbit coupling associated with these heavy metals enables significant mixing of singlet and triplet states, which renders the spin-forbidden emission from the triplet excited state (i.e. phosphorescence or triplet emission) partially allowed. In contrast to hydrocarbon conjugated polymers, the triplet excited state is thus accessible to experimentation by various optical methods and significant advances have been realized in this area. Another attractive feature of this class of materials is that there is scope for chemical modification of the organic ligands. For the commercial exploitation of these materials and direct application-oriented synthesis, a thorough understanding of the structure-property relationship is necessary such that suitable modification of the chemical structure may fine-tune their optoelectronic properties.

Recently, we explored the chemistry and photophysics of new organometallic polyynes containing fluorene auxiliaries [62,101-103]. The synthesis of polymeric derivatives of fluorene has attracted increasing research attention as an interesting approach to the development of efficient blue-light-emitting polymers. Here we have investigated the dependence of the first excited singlet and triplet electronic state on the electronic structure of the fluorene spacer groups of some platinum(II) polyynes. By varying the substituents at C-9 position of fluorene ring, the optical absorption and emission properties of these metal-containing polymers can be chemically tailored.

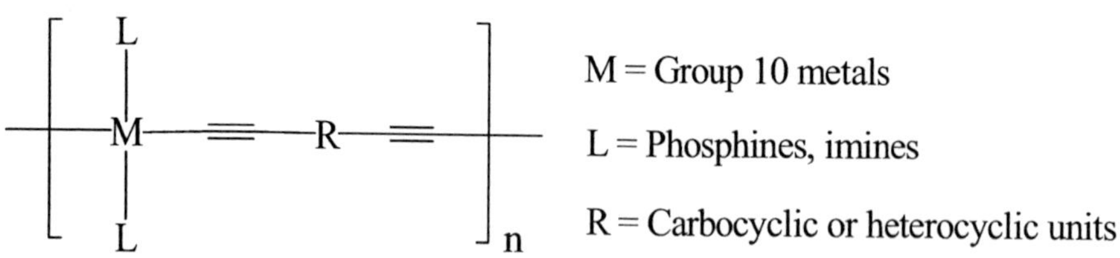

Figure 16. Basic structure of metal polyyne polymers.

The syntheses of selected diethynylfluorene ligand precursors in our studies are outlined in Schemes 7 and 8. The diethynylfluorene precursors can be used to form monomeric (**28–33**) and polymeric complexes (**34–39**) of platinum by the dehydrohalogenation procedures using the CuI/^{i}Pr$_2$NH catalytic system. Scheme 9 outlines some of these reactions. The feed mole ratio of the platinum chloride precursors and the diethynyl ligands were 2:1 and 1:1 for the monomer and polymer syntheses, respectively. The metal diynes were isolated in fairly good yields by preparative TLC on silica. Purification of the polymers was accomplished by silica column chromatography using CH_2Cl_2 as eluent and they were obtained in high purity and yields (75–80%). All these platinum complexes are air-stable and readily dissolve in aprotic solvents such as THF, CH_2Cl_2, $CHCl_3$ or toluene. The polymers are easily processable from organic solvents and exhibit good film-forming properties. The molecular weights of polymers were determined in THF by GPC using polystyrene as a standard (Table 3). It should be noted that GPC does not give absolute values of molecular weight but provides a measure of hydrodynamic volume. Rodlike polymers in solution possess different hydrodynamic properties than flexible polymers. So, calibration of the GPC with polystyrene standards is likely to inflate the values of the molecular weights of the polyynes to some extent. However, the lack of discernible NMR end group resonances provides support for the view that there is high degree of polymerization in these polyynes.

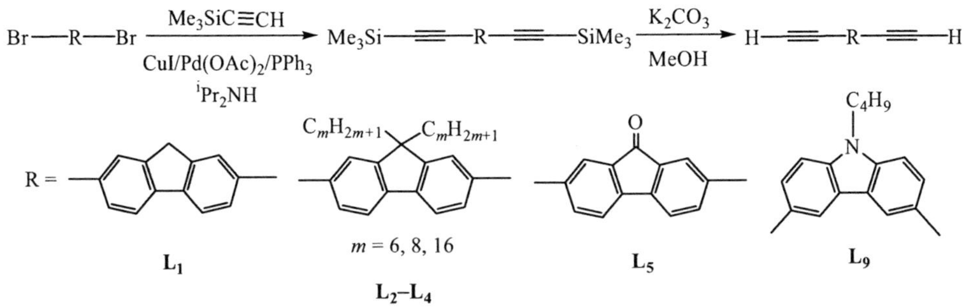

Scheme 7.

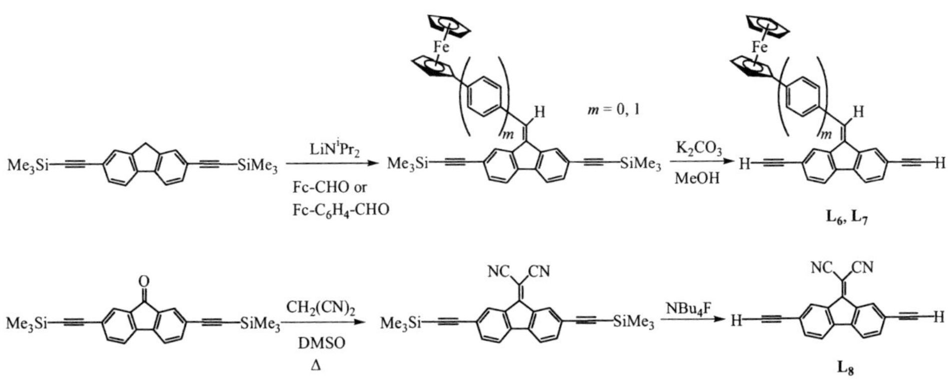

Scheme 8.

The IR spectra of these new platinum(II) complexes display a single ν(C≡C) absorption, which together with the singlet signal observed in each ^{31}P $\{^1H\}$ NMR spectrum reveal a *trans* geometry of the $Pt(PR_3)_2$ (R = Et or Bu) unit. The ν(C≡C) stretching frequencies for the polymers are lower than those for the free ligand precursors, in line with a higher degree of conjugation in the former. NMR analyses clearly indicate that a well-defined structure has been obtained in each compound. Notably, two distinct ^{13}C NMR signals for the individual sp carbons in these monomers and polymers were observed, in accord with their formulations and they are shifted downfield with respect to the free ligands. The aromatic region of their ^{13}C NMR spectra also gives more precise information about the regiochemical structure of the main chain skeleton and reveals a high degree of structural regularity in the polymers. The three-dimensional structures of **28** and **31–33** have been confirmed by single-crystal X-ray analyses (Figures 17–20). Generally, model diynes can be considered as building blocks for the high molecular weight polymers and valuable information concerning their molecular and electronic properties can be obtained through the studies of these model compounds. The coordination geometry at each Pt center is square-planar with the two PEt_3 groups *trans* to each other. The C≡C bond length is characteristic of metal-acetylide σ bonding and the bond angle close to linearity for the fragment Pt–C≡C conforms to the rigid-rod nature of these monomers.

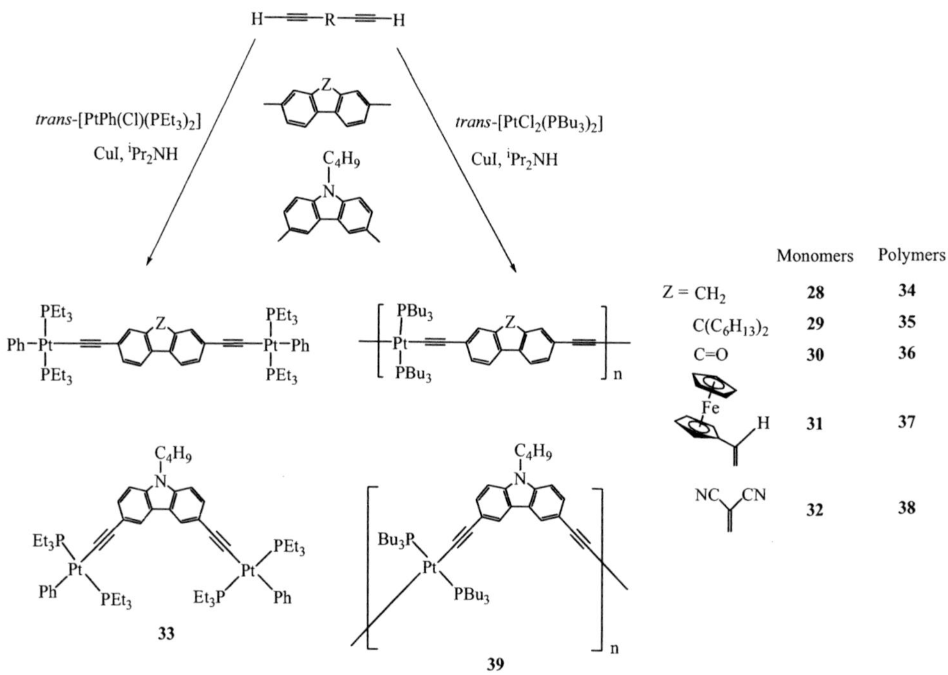

Scheme 9.

Table 3. Structural and Thermal Properties of Platinum(II) Polyynes

polymer	M_w	M_n	M_w/M_n	T_{decomp}(onset) (°C)
34	38,960	21,410	1.82	295 ± 5
35	28,900	16,900	1.71	349 ± 5
36	76,460	38,820	1.97	294 ± 5
37	34,540	17,970	1.92	250 ± 5
38	116,910	101,600	1.15	352 ± 5
39	19,100	8,920	2.14	315 ± 5

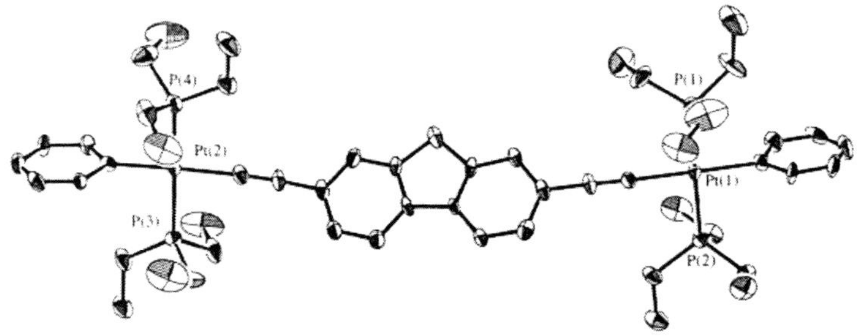

Figure 17. Molecular structure of **28**.

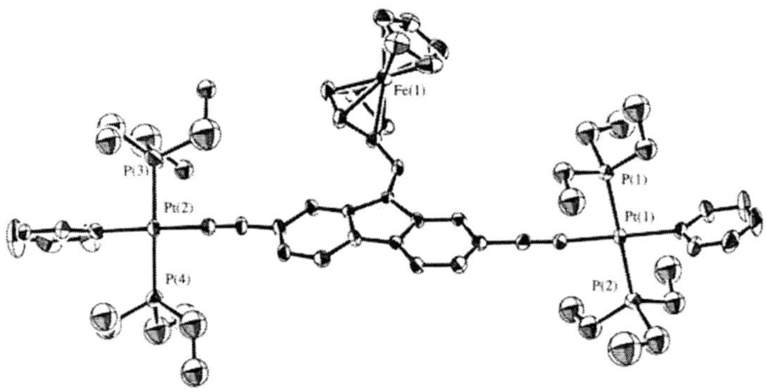

Figure 18. Molecular structure of **31**.

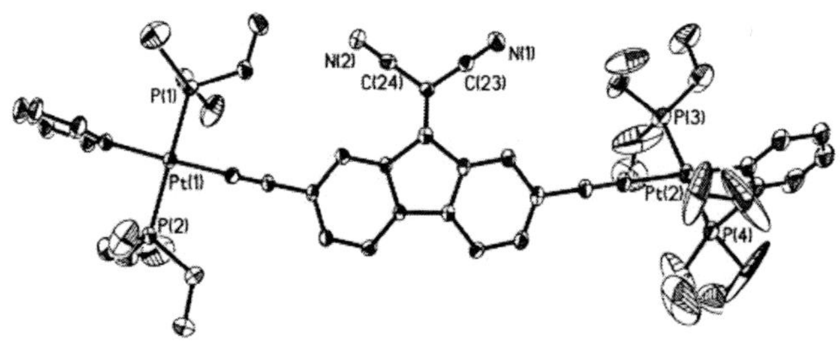

Figure 19. Molecular structure of **32**.

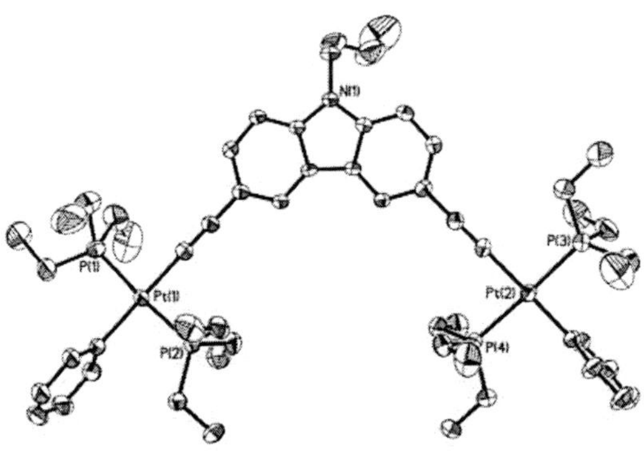

Figure 20. Molecular structure of **33**.

As measured from CV in CH_2Cl_2, only two irreversible oxidation waves were observed for the monomers **29** and **33**, attributable to the stepwise two-electron oxidation at the platinum center [$Pt^{2+} \rightarrow Pt^{3+} + e^- \rightarrow Pt^{4+} + 2e^-$] (Table 4). The polymers **35** and **39** only display a single irreversible oxidation wave at about +0.40 and +0.27 V, respectively, corresponding to the one-electron oxidation of the metal residue. The first oxidation potential was shown to be slightly more anodic for the monomers than for the polymers. Each of the compounds **31** and **37** shows a reversible ferrocene-centered oxidation and an irreversible oxidation wave at greater potentials due to the one-electron oxidation of the platinum residue. The half-wave potential of the ferrocene moiety appears slightly more anodic in **31** and **37** than in the free alkyne because of the loss of electron density from the ferrocenyl donor unit to the net electron-accepting platinum center through the acetylide bridge. Polyynes **32** and **38** both show a reversible reduction wave, attributable to the reduction of the 9-dicyanomethylenefluorene unit. The half-wave potential associated with this reduction event is more cathodic in platinum compounds than in the free ligand, indicative of electronic communication between the $Pt(PBu_3)_2$ group and the net electron-accepting 9-dicyanomethylenefluorene moiety through the alkyne bridge. Compounds **32** and **38** also display an irreversible oxidation wave at 0.68 V due to the one-electron oxidation of the platinum group.

Table 4. Electrochemical Data for Platinum(II) Monomers and Polymers.

monomer	E_{ox}	E_{red}	polymer	E_{ox}	E_{red}
29	+0.46, +0.70		**35**	+0.40	
31	+0.67, +1.16		**37**	+0.71, +1.34	
32	+0.67	−1.27	**38**	+0.68	−1.26
33	+0.30, +0.78		**39**	+0.27	

Analysis of the thermogravimetric traces indicates that decomposition starts at 349 and 315 °C for **35** and **39**, respectively. The onset of decomposition is very similar for **35** and **38**, but **34, 36** and **37** have lower decomposition temperatures. The decomposition step was assigned to the removal of two PBu_3 groups from the polymers. For instance, we note a 42% weight loss between 350–545 °C for **35** whereas 46% of the weight is lost between 315–515 °C for **39**. Another endotherm observed at 213 and 200 °C may be attributed to the melting of **35** and **39**, respectively. The polymer **35** also shows an exothermic phase transition at 137 °C.

Optical absorption data of **29**, **33**, **35** and **39** taken at room temperature in CH_2Cl_2 solutions and as thin films are gathered in Table 5. Each of the metal diynes and polyynes essentially shows absorption peaks in the near UV region due to π–π* transitions of the bridging ligand. We associate the lowest energy absorption peak with the $S_0 \rightarrow S_1$ transition from the HOMO to the LUMO, which are delocalized π- and π*-orbitals. We find that the position of the lowest energy absorption band is red-shifted when the platinum group is introduced. This reveals that π-conjugation of the ligands extends through the metal center. In the same way, the transition energies of the polymers are lowered with respect to those of the corresponding monomers, suggesting that the first optical absorption transition in the polymers arises from an electronic excitation that is delocalized over more than one monomer unit. This agrees with the results of other materials of this class having pyridine, phenylene or thiophene spacer [95-98] but is in contrast to **38** where the lowest singlet excited state is only confined to a single repeat unit [102]. In energy terms, the HOMO–LUMO energy gap (E_g) as measured from the onset wavelength in solid film state is 2.92 eV for **35** and 3.10 eV for **39**, and that of **37** with a ferrocenyl pendant is 0.8 eV lower than the parent fluorene polymer **34** (Table 5). According to the identity of the Z unit, the optical gap of these polyynes follows the experimental order: $N(C_4H_9) > C(C_6H_{13})_2 \geq CH_2 > C{=}O \approx C{=}C(H)\{(C_5H_4)Fe(C_5H_5)\} > C{=}C(CN)_2$. Due to the presence of the strong electron-withdrawing cyano substituents in **38**, the lowest energy absorption peak exhibits a considerable red-shift of ca. 154–266 nm compared to other fluorene derivatives exhibiting varied electronic properties. It is particularly encouraging to find that there is a dramatic decrease in the optical gap upon substitution of the cyano functionality to the fluorene unit, featuring the importance of donor-acceptor interaction in this class of materials. The bandgap of **38** is estimated to be 1.58 eV and, to our knowledge, this band gap value is currently the lowest reported for metal polyyne polymers. Replacing the fluorene spacer by a 9-dicyanomethylenefluorene unit lowers the optical gap substantially by 1.3 eV. So, electron-withdrawing substituents at the periphery exert a more pronounced influence on narrowing the band gaps of these polymers than electron-donating groups. With reference to the parent fluorene derivative **34**, we note that the 9-dicyanomethylene group enhances conjugation while both $C(C_6H_{13})_2$ and $N(C_4H_9)$ units reduce conjugation in the present system. Hence, the energy of the S_1 singlet state depends significantly on the nature of the Z moiety of the ring and is highest for **35** and **39**.

At 290 K, polymers **35** and **39** emit in the blue region ranging from 411 to 428 nm in CH_2Cl_2 solutions and from 418 to 432 nm in solid films (Table 6). Analogous to the absorption spectra, these emission features are shifted to longer wavelengths from monomers to polymers. Upon cooling, two main emission bands appear in each case. For **35**, we attribute the feature peaking at 433 nm to emission from singlet excited state (fluorescence $S_1 \rightarrow S_0$) at 11 K, due to the small Stokes shift (ca. 0.20 eV) between the bands in the absorption and the emission spectra. The fluorescence quantum yields measured in CH_2Cl_2 at 290 K for the monomers lie in the range of 0.30–0.48 and the lower values associated with their polymeric counterparts are of the same order of magnitude as other reported Pt σ-acetylide polymers [104]. At 11 K, the intense feature at the low-energy regime of the emission spectrum of **35** with the maximum at around 554 nm is due to emission from a triplet excited state (phosphorescence $T_1 \rightarrow S_0$) and such a triplet emission shows a vibronic structure at low temperature. The large energy shift observed between absorption and emission features (~1.0 eV) ascertains this emitting state as a T_1 excited state (Figure 21). This assignment can be further rationalized in terms of the observed temperature dependence of the emission data. The photoluminescence (PL) spectra of **35** as a function of temperature are shown in Figure 22. The triplet emission is found to be strongly temperature dependent in contrast to the singlet emission. From 150 to 11 K, the singlet emission peak intensity increases only by a factor of 1.3 for **35**. However, the intensity of the lower-lying emission increases by a factor of 63.5 and such an increase in emission intensity indicates a long-lived excited state that is quenched by thermally activated diffusion to dissociation sites at room temperature. Similarly, the emission pattern for **39** consists of a $S_1 \rightarrow S_0$ transition at 426 nm and a $T_1 \rightarrow S_0$ peak at the lower energy side (462 nm) with a large Stokes shift of ca. 0.9 eV for the latter. In this case, the intensity of the singlet and triplet PL bands increases by a factor of 1.4 and 68.0, respectively, as the temperature is lowered from 150 to 11 K. These triplet emissive peaks were observed at 554 and 459 nm for **29** and **33**, respectively. The lowest triplet excited state remains strongly localized, as can be inferred from the small energy difference between triplet emissions in the monomers and in the polymers. To evaluate the relative efficiency of intersystem crossing triggered by the heavy metals, the peak height ratio from triplet emission to singlet emission at 11 K, $\Delta E(T_1 \rightarrow S_0, S_1 \rightarrow S_0)$, was taken as a good indicator (Table 5). Clearly, the intersystem crossing rate is higher for the polymers than that for the monomers in each system. In particular for **39**, we observe an unusually high efficiency of triplet emission. These data provide an important insight of the effect of the fluorene substituents in improving the efficiency of intersystem crossing in these polyynes. While triplet emission is also detected at around 549 nm for **34** at 11 K, there is no evidence of detecting any $T_1 \rightarrow S_0$ transitions over the measured temperature range for **36–38**. One salient feature here is that a reduction in the HOMO–LUMO energy gap leads to a lowering of the intersystem crossing efficiency and hence the principal emission arises from a $S_1 \rightarrow S_0$ transition. This implies that reduced conjugation in these 9-anchored biphenyl structures effectively facilitates intersystem crossing and phosphorescence can easily be measured. The use of conjugation-interrupting segments such as $N(C_4H_9)$ and $C(C_6H_{13})_2$ groups in the polymer main chain can control the effective conjugation length and offers a good way to enhance the intersystem crossing rate, especially at low temperatures.

Table 5. Optical Data of Various Platinum(II) Fluorene-based Materials

compound Z =	color	largest λ_{max} in CH_2Cl_2/nm	E_g/eV	ref.
CH_2	off-white	394	2.90	62
$C(C_6H_{13})_2$	off-white	399	2.92	101
C=O	red	506	2.10	62
C=C(H)Fc	red	437	2.10	103
$C=C(CN)_2$	blue	660	1.58	102
$N(C_4H_9)$	off-white	344	3.10	101

Table 6. Absorption, Emission and Redox Data for 29, 33, 35 and 39

	λ_{max} (nm)[a]		λ_{em} (nm)			
	CH_2Cl_2	film	CH_2Cl_2 (290 K)[b]	film (290 K)	film (11 K)	$\Delta E(T_1 \rightarrow S_o, S_1 \rightarrow S_o)$[c]
29	309 (0.2), 358 (1.0), 376 (1.4)	311 359 378	411 (0.30)	396*, 418, 552, 598*	396*, 418*, 554, 604*	6.6
35	307 (0.4), 399 (2.9)	309 404	421, 555* (0.01)	432, 552*	416*, 433*, 554, 589*, 606*	18.0
33	260 (4.1), 304 (2.1), 333 (3.7)	263 304 343	410*, 426 (0.48)	408*, 426	407*, 425*, 459, 474*	11.0
39	258 (9.2), 301 (4.0), 344 (7.0)	261 304 347	412*, 428 (0.01)	404*, 427	403*, 426*, 462, 497*, 511*	136.0

[a] Extinction coefficients (10^4 M^{-1} cm^{-1}) are shown in parentheses. [b] Fluorescence quantum yields shown in parentheses are measured in CH_2Cl_2 relative to anthracene. [c] Ratio of the intensities of triplet emission to singlet emission at 11 K. * Emission peaks appear as shoulders or weak bands.

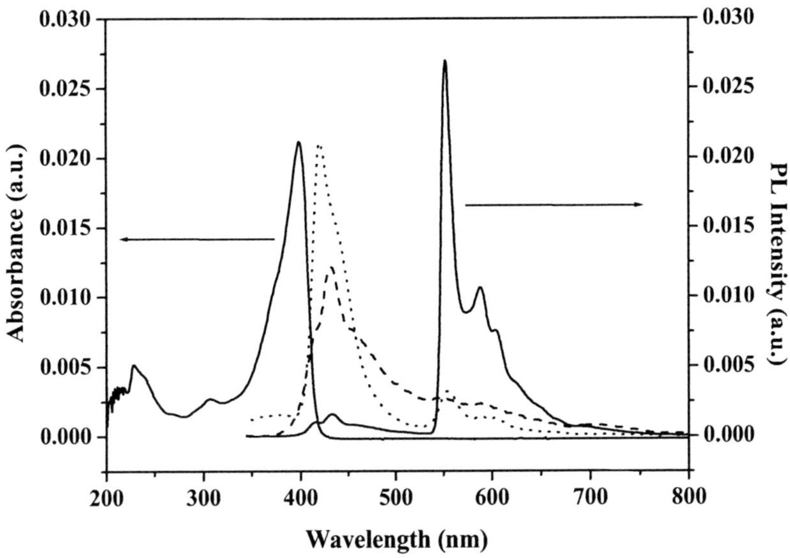

Figure 21. The room-temperature optical absorption spectrum and the PL spectrum (11 K) of **35** in solid (solid line). The dotted and the dashed lines correspond to the PL spectra in CH_2Cl_2 solution and solid film at 290 K, respectively.

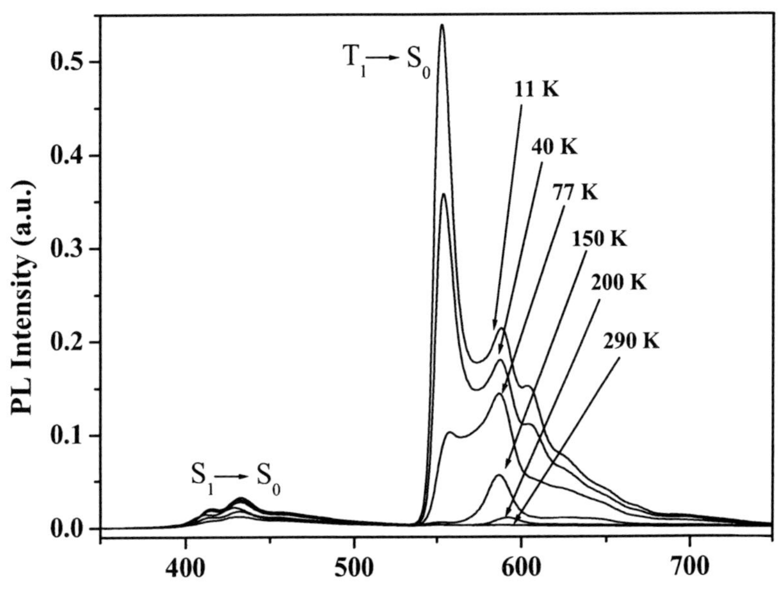

Figure 22. Temperature dependence of the PL of **35**.

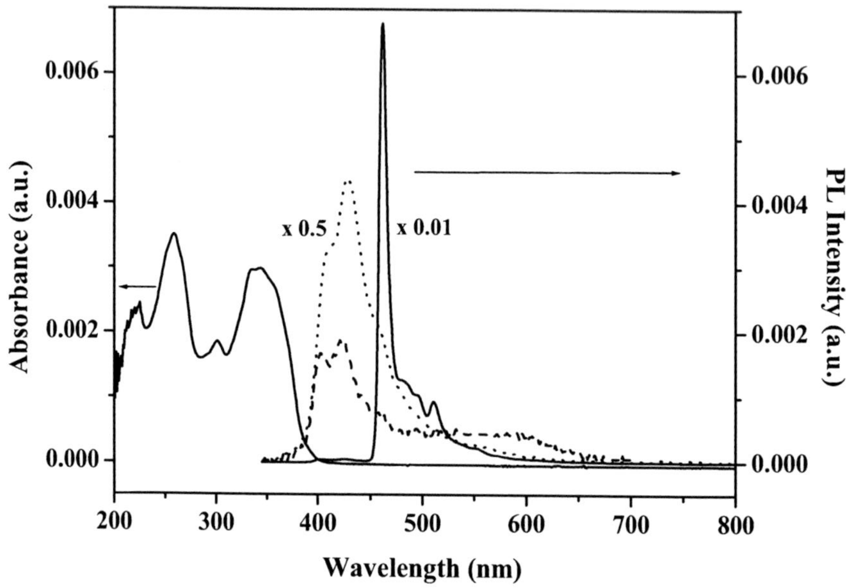

Figure 23. The room-temperature optical absorption spectrum and the PL spectrum (11 K) of **39** in solid (solid line). The dotted and the dashed lines correspond to the PL spectra in CH_2Cl_2 solution and solid film at 290 K, respectively.

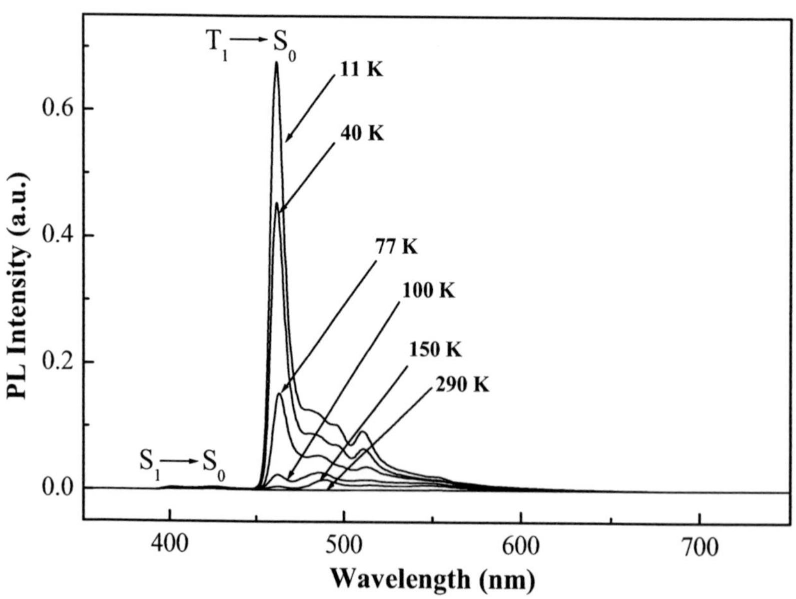

Figure 24. Temperature dependence of the PL of **39**.

The Cambridge group recently also undertook a systematic study on the synthesis and optoelectronic properties of several platinum(II) diynes and polyynes **40**–**42** and their organic counterparts **43** and **44** containing 9,9-dioctylfluorene unit in the backbone [105]. The crystal structure of the model complex **41** was determined in order to underpin the spectroscopic characterizations of the polymer precursors. Moreover, addresses have been made to compare the photophysics of **34** and **36** with their organic co-polyynes in their studies. The emission of **36** at 641 nm is unstructured at room temperature as for the organic analogue and shows only a slight red-shift together with an increased intensity at a lower temperature. Comparison of **44** with **42** shows that the inclusion of a heavy platinum atom into the main chain of an organic polyyne increases spin-orbit coupling so that triplet state emission can be observed. The T_1 state is found to be 0.7 ± 0.1 eV below the S_1 state for **40**–**42**. Since the non-radiative decay rate from the T_1 triplet state is equal or larger than the radiative decay rate, the PL quantum efficiencies of the platinum polyynes are reduced from those for the organic polymers. Optical spectroscopic measurements reveal that the attachment of octyl side-chains on the fluorenyl spacer reduces interchain interaction in the polyynes while a fluorenonyl spacer creates a donor-acceptor interaction along the rigid backbone of the organometallic polyynes and organic co-polyynes. Conjugated organic copolymer **45** possessing a pendant ferrocene moieties was also known and its spectroscopic, redox and thermal behavior have been investigated [106,107].

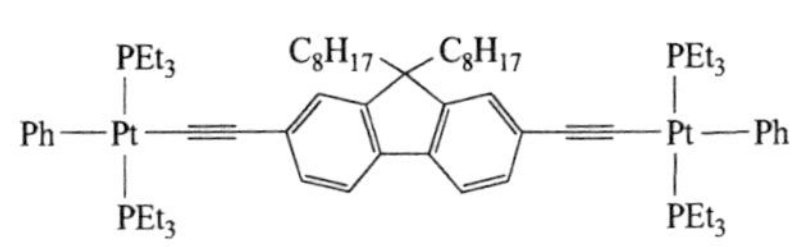

40

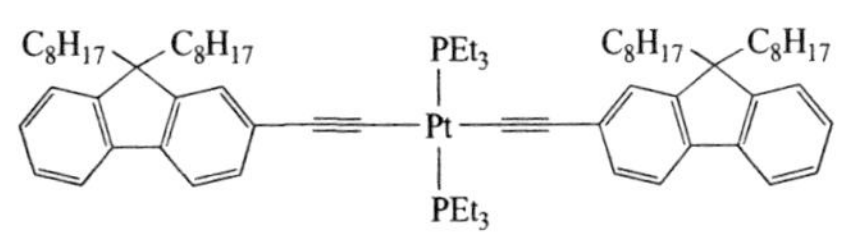

41

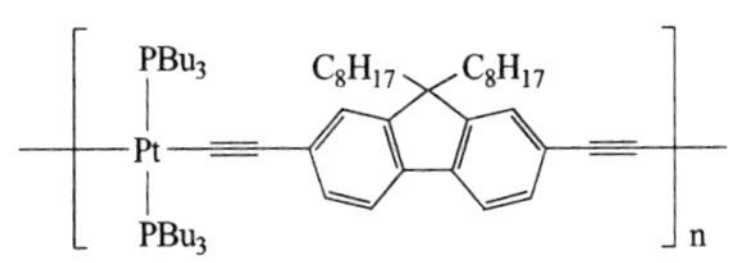

42

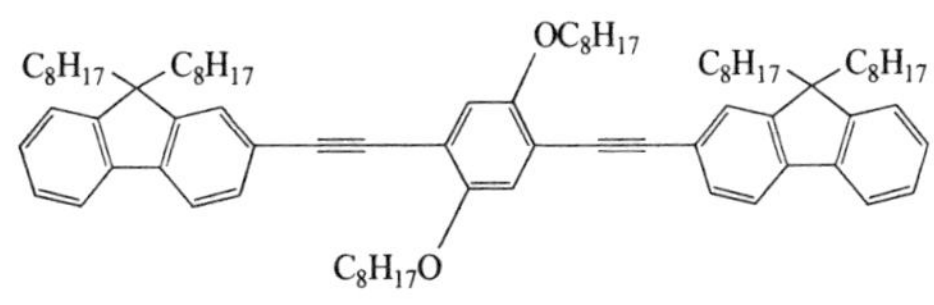

43

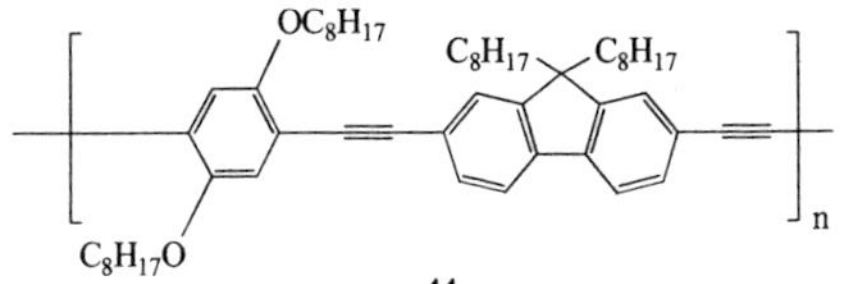

44

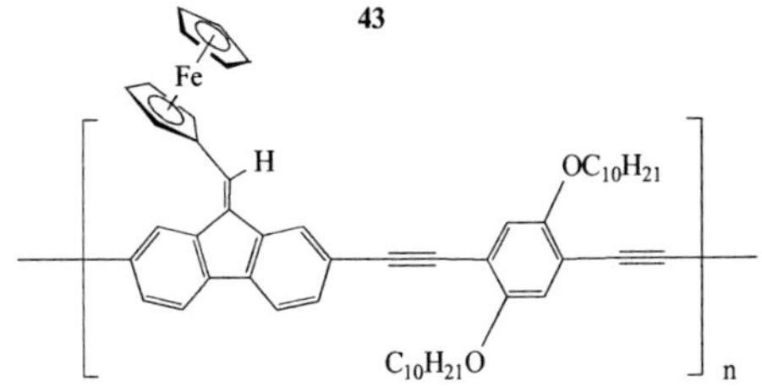

45

Fabrication of single-layer photocells in the sandwich-type structure ITO/polymer/Al (ITO = indium-tin-oxide) was carried out for the polymer samples **35**, **38** and **39** in order to study their photoconductive properties. The device structure is depicted in Figure 25 with the ITO electrode as the anode. The experiment was performed by using a lock-in-amplifier to measure the voltage drop across a resistor resulting from the photocurrent through the polymer film so that the signal due to the dark current can be eliminated. These polymers show moderate photoconductivity and Figure 26 depicts the logarithmic plots of the photocurrent of photocells ITO/**35**/Al and ITO/**39**/Al against the applied electric field across the polymer. The photocurrent of the devices is field-dependent and it increases with increasing bias voltage. The curve is steeper at higher voltages, typical of other known organometallic photoconductors [108]. Both polymeric materials show a photocurrent quantum yield of approximately 0.01% at 400 and 370 nm for **35** and **39**, respectively, which is a common value for single layer devices. There is no significant difference in the quantum efficiency with variation of the central fluorene ring and a similar quantum yield was also determined for **36**.

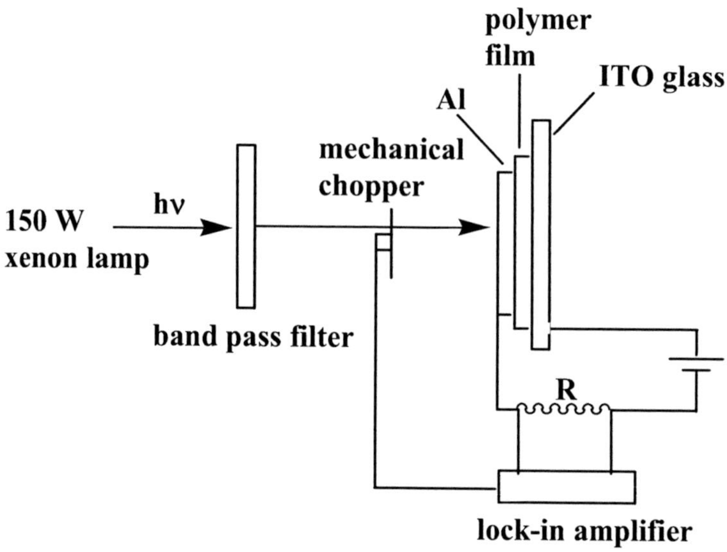

Figure 25. Device structure for photoconductivity measurements.

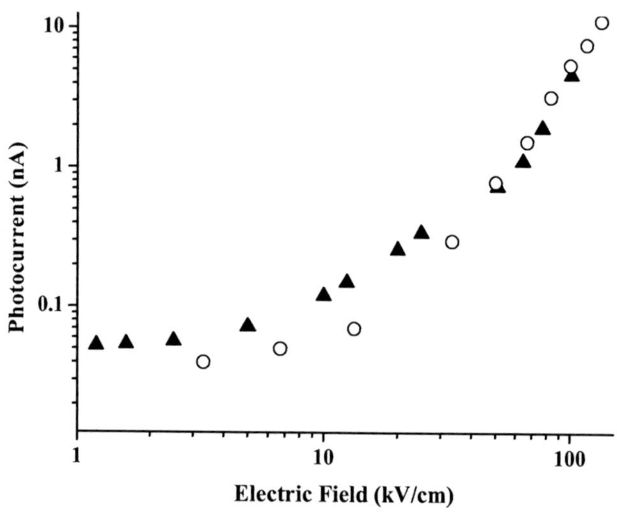

Figure 26. Logarithmic plots of photocurrent *vs.* external electric field for **35** (▲) and **39** (○).

2.6. Gold(I) and Mercury(II) σ-Acetylides

2.6.1. Binuclear Gold(I) and Mercury(II) Alkynyl Complexes

The chemistry of alkynylgold(I) complexes has also received much current attention in the study of rigid-rod metal acetylide materials. The preference of d^{10} gold(I) to form linear two-coordinate species, along with the linearity of the C≡C bond in alkynyl ligands, make the –(Au–C≡C)– moiety an attractive and promising candidate for the design of metal-containing materials with extensive electronic conjugation [109-120]. It has been shown that some of these gold(I) complexes show liquid-crystalline properties [121-123] or NLO behavior [124,125]. In addition, many alkynylgold(I) derivatives exhibit diverse and interesting luminescent properties [116-120]. In particular, alkynylgold(I) compounds with tertiary phosphine ligands display rich photochemistry and belong to a class of luminophores that may find applications in electronic devices [126-136]. In view of the fact that the $[R'Hg]^+$ unit (R' = alkyl or aryl) is isoelectronic and isolobal with the $[(Ph_3P)Au]^+$ fragment [137-140], a comparative study of the alkynylmercury(II) complexes to that of the gold(I) counterparts represents a challenging area of research. Despite the vast body of work on the chemistry of alkynylgold(I) species, much less is known for the mercury(II) congeners and related studies on the isoelectronic mercury(II) systems are still scarce. The first synthesis, characterization, redox and luminescence behavior of some novel bis(alkynyl) gold(I) and mercury(II) complexes **46–57** with electron-rich and electron-deficient fluorenes were accomplished.

Scheme 10 outlines the reactions leading to the new fluorene-linked digold(I) and dimercury(II) complexes **46–57** [141] following the classical dehydrohalogenation reaction between dialkynes and chlorogold(I) phosphine or alkylmercury(II) chloride in the presence of a base [142-147]. The product yields of **46–57** are very high (80–95%). However, all attempts to prepare the bis(mercurial) analogue of **52** failed. The direct reaction of MeHgCl and $\mathbf{L_9}$ in basic MeOH only led to the isolation of compound **56**, probably due to the attack of MeO^- on the $C{=}C(CN)_2$ group. Using the $CuI/^iPr_2NH$ system, no sign of the formation of bis(mercurial) species could be detected and the reaction only occurred with partial decomposition to metallic mercury. The reaction between **56** and malononitrile in DMSO also gave an inseparable complex mixture. Some of these complexes were structurally characterized. Perspective views of **49** and **56** are depicted in Figures 27 and 28. We are aware that there are only few structurally characterized examples of digold(I) diacetylide complexes in the literature [118-120,147], the first one being $[(Me_3P)AuC{\equiv}C(C_6H_2Me_2)C{\equiv}CAu(PMe_3)]$ reported by Puddephatt and co-workers [118]. In contrast to many Au(I) compounds, where there frequently exists short Au···Au intermolecular contacts, no such weak interactions are observed in **49** in the solid state, as the sterically bulky phenyl groups of the phosphine ligands prevent their close approach. The nearest intermolecular contact in **49** is attributed to the hydrogen bonding interaction between O(1) and H(49A) on C(49) of the phenyl ring [O(1)···H(49A) = 2.399 Å], and an intermolecular Au(1)···H(17A) interaction [3.004 Å] is also observed. The crystal structure of **56** is the first example of dimercury(II) diacetylide complex to be characterized crystallographically. Reported structures of mononuclear Hg(II) acetylides are also limited [148-152]. Reminiscent of the P–Au–C≡C fragment, the isoelectronic MeHg–C≡C entity displays similar structural motifs and geometry in **56**.

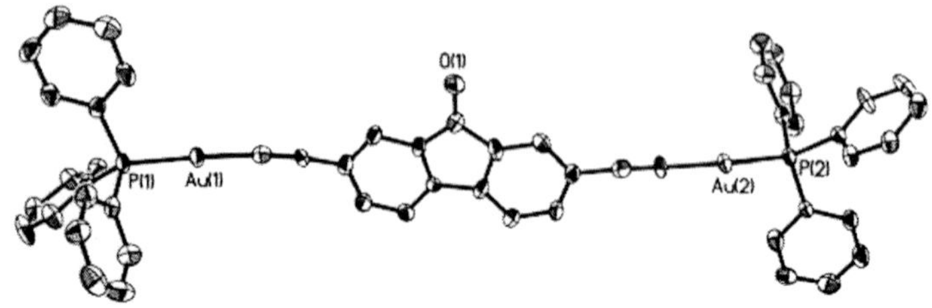

Figure 27. X-ray structure of **49**.

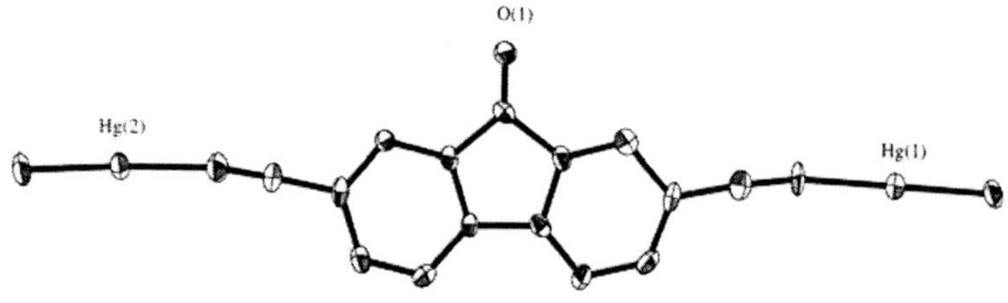

Figure 28. X-ray structure of **56**.

These complexes have been found to exhibit rich photophysical properties. The electronic absorption and emission data for compounds **46–57** in CH_2Cl_2 are presented in Table 7, together with data for their corresponding bis(acetylide) fluorenyl ligands. The digold(I) and dimercury(II) complexes display similar structured ligand-localized $\pi \rightarrow \pi^*$ transitions in the near UV region (ca. 231–389 nm) and, in most cases, at the low-energy regime of the visible spectrum (ca. 422–579 nm). The dependence of the absorption energies on the nature of the fluorene spacer suggests that the transitions are most likely dominated by $\pi \rightarrow \pi^*$ transitions of the organic bridging units. However, the involvement of a $\sigma \rightarrow \pi^*$ (aryl) transition should not be overlooked in some cases. It has been suggested that the lowest-lying absorption is caused by the $\sigma \rightarrow \pi^*$ (anthryl) transition in [Au(PPh_3)(C≡CAn)] and [Au($PAnPh_2$)(C≡CPh)] [117] and the $\sigma \rightarrow \pi^*$ (naphthyl) transition in other related alkynylgold(I) naphthyl-substituted phosphine systems [116]. We note that the absorption bands show bathochromic shifts when the d^{10} metal fragments are introduced at the terminal groups, and the red-shift is more pronounced with the gold(I) species. These observations are consistent with the increased π-delocalization through the metal groups due to metal to ligand back donation to π*(C≡CR). Due to the absorption feature at ca. 579 nm, the solution of **52** has an intense blue-green color in CH_2Cl_2. Because of the strong electron-withdrawing C≡N groups in **52**, the lowest energy absorption peak exhibits a considerable red-shift of ca. 114–220 nm as compared to other diaurated fluorene derivatives **46–51** displaying different electronic properties.

$$2\,[M]Cl + H{-}C{\equiv}C{-}R{-}C{\equiv}C{-}H \xrightarrow[\text{MeOH}]{\text{NaOMe}} [M]{-}C{\equiv}C{-}R{-}C{\equiv}C{-}[M]$$

R =

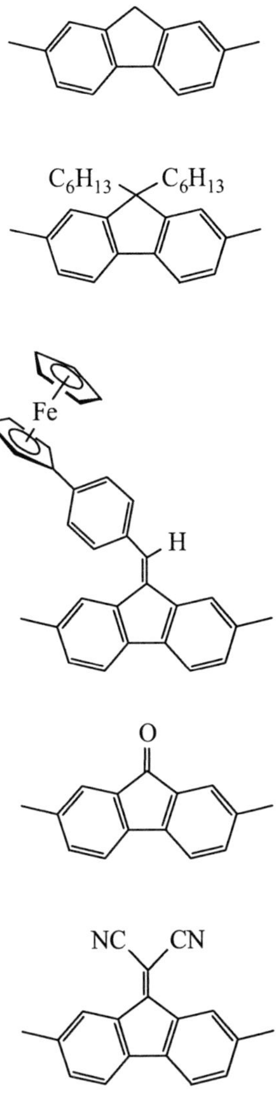

; $[M] = [AuPPh_3]^+$ **46**
$[M] = [MeHg]^+$ **53**

; $[M] = [AuPPh_3]^+$ **47**
$[M] = [MeHg]^+$ **54**

; $[M] = [AuPPh_3]^+$ **48**
$[M] = [MeHg]^+$ **55**

; $[M] = [AuPPh_3]^+$ **49**
$[M] = [AuPMe_3]^+$ **50**
$[M] = [AuPAnPh_2]^+$ (An = anthracenyl) **51**
$[M] = [MeHg]^+$ **56**
$[M] = [PhHg]^+$ **57**

; $[M] = [AuPPh_3]^+$ **52**

Scheme 10.

In such study, the energy gap series **46** > **47** > **48** > **49** ≈ **50** ≈ **51** > **52** and **53** > **54** > **55** > **56** ≈ **57** are obtained. Replacing the unsubstituted fluorene spacer in **46** by a 9-dicyanomethylenefluorene unit lowers the HOMO–LUMO gap substantially by 1.51 eV. For each particular spacer unit, there is a sequential lowering in the energy gap from the ligand to the Hg(II) complex and the Au(I) counterpart, indicative of the metal to ligand back π bonding associated with the metal centers and the stronger interaction in the gold center with +1 formal oxidation state gives a smaller optical gap. Molecular orbital calculations of **49** and **56** reveal that the molecular orbitals in the frontier region are mainly related to the π and π* orbitals of the bridging conjugated units. Figure 29 depicts the contour plots of the HOMO and LUMO for **49**, showing the common bonding feature in the frontier molecular orbitals of these three complexes. We can see that the LUMO has the π* character involving the carbonyl unit. By virtue of the accessible low-lying π* orbitals of CO, the fluorenone-containing organometallic complexes are likely to exhibit smaller HOMO–LUMO gaps. Based on the bonding feature of the frontier orbitals, we expect that metal fragments at both ends of complexes **49** and **56** affect more on the HOMO and less on the LUMO. Indeed, the LUMOs of **49** and **56** are similar in energy (–0.07904 and –0.08055 au, respectively). Because of the greater electronegativity and higher formal oxidation state, Hg(II) in **56** pulls down the HOMO's orbital energy more significantly when compared to Au(I) in **49**, leading to a larger HOMO–LUMO gap.

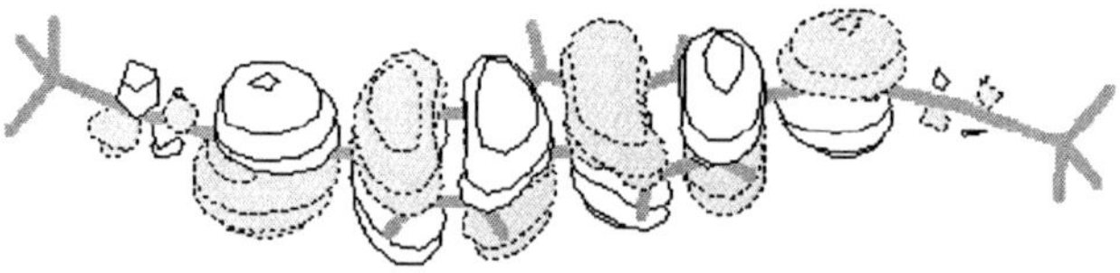

HOMO

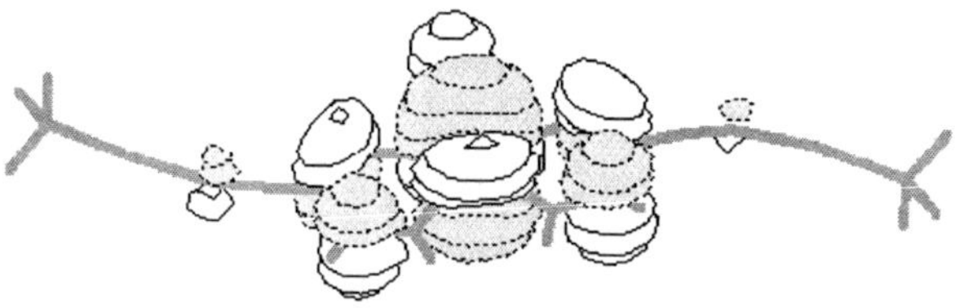

LUMO

Figure 29. Spatial plots of the HOMO and the LUMO for **49.** In the calculations, PPh_3 ligands were modeled by PH_3 groups.

Systematic investigations were carried out to correlate the emission properties with the nature of metal groups and their auxiliary ligands and the structure of spacer unit. The results have shown that electronic absorption and emission spectral features of these compounds extend over a wide range. With reference to spectroscopic studies on $[(Me_3P)AuC{\equiv}CRC{\equiv}CAu(PMe_3)]$ (R = C_6H_4, $C_6H_2Me_2$) and other similar alkynylgold(I) complexes, we tentatively assign the emissive bands to triplet π–π* or triplet σ–π* excited states, or a combination of both, associated with the digold or dimercury diacetylide groups [116-120]. Clearly, these luminescence bands are red-shifted as compared to those for the free alkynes. Within each metal class, the emission energy was found to change upon variation of both the fluorene moieties R and the auxiliary groups on Au(I) or Hg(II). Figure 30 shows the solution emission spectra of $\mathbf{L_5}$, **49** and **57** at 290 K. Their similar spectral pattern is suggestive of the ligand-dominating emissive state and the spectra display featureless luminescence peaks centered at ca. 531 ($\mathbf{L_5}$), 575 (**49**) and 543 nm (**57**). By comparing with $\mathbf{L_5}$, the bands are red-shifted by 12 nm for **57** and 44 nm for **49**. The corresponding shifts in the solid state are even larger at 42 and 73 nm, respectively We attribute the bathochromic shift to an increased delocalization of π electrons into the aromatic segment by virtue of the metal to ligand back π bonding interaction in the presence of filled metal d orbitals, leading to a stabilization of the first π* orbital of the acetylide groups. The more marked effect for Au(I) is consistent with the lower oxidation state of +1 in **49**, which would enhance the back donation to π*. The same trend is observed for other complexes and the red shift falls in the range 34–73 nm (43–52 nm) for Au(I) and 32–42 nm (12–24 nm) for Hg(II) in the solid state (in CH_2Cl_2). The solid-state emission spectra of **46** and **49** with different fluorene moieties show very similar emission patterns with energy bands at 583 (for **46**) and 589 nm (for **49**). The corresponding organomercurial complexes **53** and **56** emit at ca. 567 and 597 nm, respectively, with a wavelength shift of 30 nm. The difference can be accounted for by the higher degree of π-conjugation of the fluorenone unit which increases the overlap with π*(C≡CR) to stabilize the low-lying π* orbital. At 290 K, the emission energies of the complexes roughly follow the same order as the absorption maxima. The emission of **50** is red-shifted relative to **49** since PMe_3 ligand is electron-donating in nature, which renders the Au(I) center more electron-rich and hence favors the π-mixing to facilitate stabilization of the first π* energy level of the C≡C π system. The emission energies were found to follow the order **49** > **51** > **50**, which is in line with the σ-donating ability of the phosphines, i.e. PMe_3 > $PAnPh_2$ > PPh_3. Similarly, an order of emission energies of **57** > **56** is expected based on the different σ-donating capacity of the coordinated hydrocarbyl groups, i.e. Me > Ph (Figure 31).

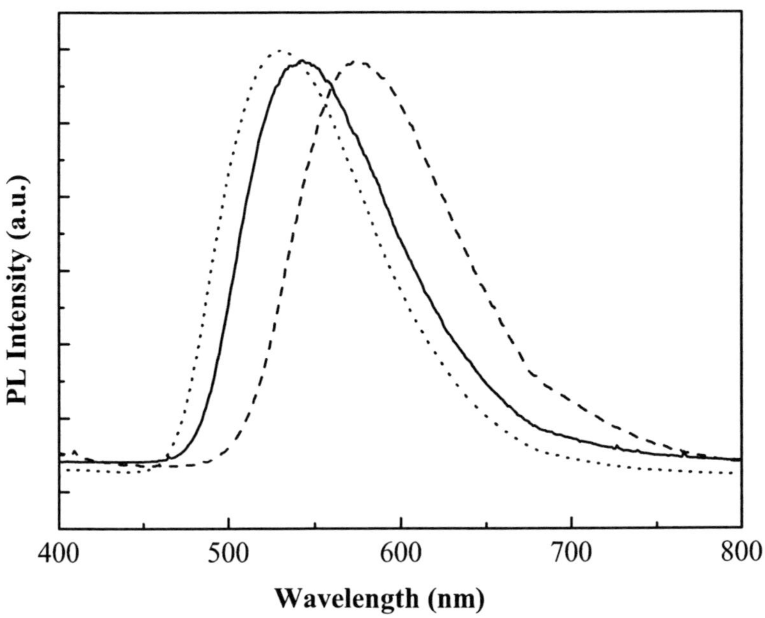

Figure 30. Emission spectra for L_5 (dotted line), **49** (dashed line) and **57** (solid line) in CH_2Cl_2 solutions at 290 K.

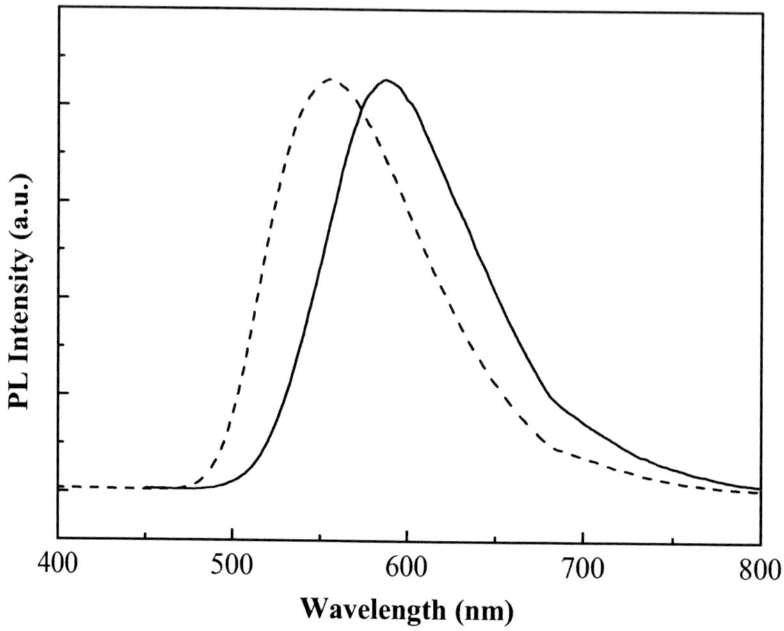

Figure 31. Emission spectra of **56** (solid line) and **57** (dashed line) at 290 K.

The redox data for electroactive complexes in CH_2Cl_2 are summarized in Table 8. Each of the cyclic voltammograms of **48** and **55** displays one quasi-reversible ferrocenyl oxidation couple with the half-wave potentials in the order **55** > **48** > $\mathbf{L_7}$. This is in accord with the unsaturation of the organometallic end-capped ethynyl bridges, which would make the ferrocenyl center less electron-rich. The $E_{1/2}$ value is substantially more anodic in **55** than in **48** because of the higher formal oxidation state of mercury(II) than that of gold(I). For **49**, **50**, **56** and **57**, a quasi-reversible reduction wave is observed in each case (–1.61 to –1.70 V), which is assigned as the fluorenone-centered reduction. The observation of a less cathodic potential in $\mathbf{L_5}$ is a manifestation of a more π-delocalized system in the Au(I) and Hg(II) complexes, and the reduction potentials are more negative for Au(I) species than those for the Hg(II) counterparts. Variation of the PR_3 ligands in the Au(I) compounds or changing the alkyl by an aryl group in the Hg(II) complexes incurs a slight shift in the reduction potentials, in agreement with the absorption and emission data. The good σ-donating PMe_3 groups enhance the σ(P–Au) back donation to π* in **50**, which leads to a slight positive shift of 0.02 V as compared to the relatively π-accepting PPh_3 ligands. Likewise, the purely σ-donating Me group in **56** makes the reduction potential of **56** appear at a more negative value. Complex **51**, however, only underwent an irreversible ligand-based reduction event at –1.57 V. Two reversible reduction waves, attributable to the central 9-dicyanomethylenefluorene unit, are noted for **52**, and they occur at potentials more cathodic than those of the diethynyl ligand $\mathbf{L_9}$.

Table 7. Absorption and Emission Data for Compounds 46–57 and their Free Alkynes

complex	absorption[a] λ_{max}/nm (ε/10^4 dm^3 mol^{-1} cm^{-1})	energy gap[b]/eV	emission[c] medium (*T*/K)	λ_{max}/nm
46	242 (4.4), 325 (4.4), 360 (2.6)	3.35	solid (290) solid (11)	583 551, 592
47	275 (1.2), 307sh (2.3), 319 (4.4), 343 (9.7), 359 (9.0)	3.28	solid (290) solid (11)	567 549, 597
48	236 (4.8), 276 (2.4), 339 (3.5), 355 (4.7), 422sh (0.5)	2.82	solid (290) solid (11)	vw vw
49	239 (4.3), 291 (4.3), 337 (3.8), 355 (2.3), 458 (0.3)	2.34	CH_2Cl_2 (290) solid (290) solid (11)	575 589 590
50	295 (5.8), 336 (4.0), 351 (4.1), 461 (0.4)	2.35	CH_2Cl_2 (290) solid (290) solid (11)	583 628 633
51	268 (1.2), 301 (1.3), 338 (0.7), 354 (1.0), 389 (0.3), 409 (0.2), 465 (0.1)	2.36	CH_2Cl_2 (290) solid (290) solid (11)	577 596 604
52	236 (11.9), 268 (2.7), 276 (2.8), 323 (17.9), 351sh (9.9), 579 (0.2)	1.84	solid (290) solid (11)	736 737
53	231 (1.5), 258sh (0.5), 311 (3.0), 344sh (2.3)	3.52	solid (290) solid (11)	567 564
54	296sh (2.0), 311 (4.0), 330 (6.4), 346 (8.2)	3.45	solid (290) solid (11)	444, 457, 548sh 444, 457, 555
55	238 (3.0), 286 (3.3), 326 (3.3), 460sh (0.3)	2.82	solid (290) solid (11)	vw vw
56	292 (3.4), 315 (1.3), 329 (2.0), 336 (1.4), 343 (2.1), 439 (0.2)	2.50	CH_2Cl_2 (290) solid (290) solid (11)	555 597 599
57	231 (2.2), 286 (3.9), 315 (1.9), 329 (2.6), 337 (2.1), 343 (2.4), 437 (0.4)	2.48	CH_2Cl_2 (290) solid (290) solid (11)	543 587 590
L_1	314 (1.2), 319 (1.1), 324 (1.1), 327 (1.2)	3.54	solid (290)	429sh, 505
L_2	291 (3.3), 302 (4.7), 316 (4.0), 323sh (3.8), 330 (7.1)	3.67	solid (290)[d]	492
L_5	283 (4.4), 308 (1.7), 321 (2.3), 334 (1.6), 421 (0.2)	2.59	CH_2Cl_2 (290) solid (290)	531 555
L_7	236 (4.3), 276 (4.5), 316 (3.6), 328 (4.2), 357 (1.8)	2.88	solid (290)	vw
L_8	296 (2.5), 325 (1.8), 347 (1.5), 364 (1.2), 505 (0.1)	2.10	solid (290)	562

[a] In CH_2Cl_2 at 290 K. [b] Estimated from the onset of absorption. [c] Excitation wavelength is 325 nm. [d] Sample as a thin film was used. sh = Shoulder. vw = Very weak.

Table 8. Redox Data for 48–52 and 55–57 and their Ligand Precursors

compound	E_{ox}/V (ΔE/mV)	E_{red}/V	compound	$E_{1/2}$/V (ΔE/mV)	E_{red}/V
48	0.03 (160)		56		−1.64
49		−1.70	57		−1.61
50		−1.68	L_5		−1.55
51		−1.57*	L_7	0.02 (115)	
52		−1.12, −1.75	L_8		−1.02, −1.65
55	0.10 (108)				

* Irreversible wave

2.6.2. Mercury(II) Polyynes

In the past two decades, while much is known for the photoluminescent behavior of oligomeric and polymeric acetylides of platinum(II) and gold(I) complexes where spin-forbidden triplet emissions can be identified clearly [95-101,153-155], we have relatively little understanding on the d^{10} mercury(II) system. To our knowledge, there is no literature report on the photophysics and structural properties of soluble mercury-based polyynes. Although linear polymeric copper and mercury acetylides of the form [−MC≡C(*p*-C_6H_4)C≡C−]$_n$ were reported as early as 1960, these materials were often found to be insoluble and intractable which hampered their purification and characterization. Group 12 heavy mercury atom, with its propensity to enhance spin-orbit coupling, could be an ideal candidate for harvesting the energy of triplet excitons. We have prepared the first examples of soluble well-defined high molecular weight mercury(II) polyyne polymers with 9,9-dialkylfluorenes from which we have been able to directly trigger $^3(\pi\pi^*)$ emissions localized on the organic system through efficient intersystem crossing (ISC) by ligation to the Hg(II) moiety.

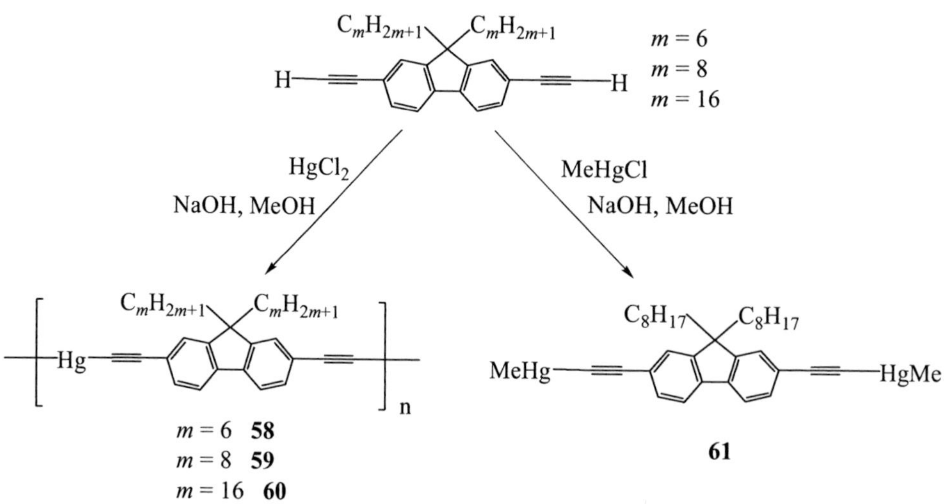

Scheme 11.

The synthesis of Hg(II) polyyne polymers **58–60** is shown in Scheme 11. Mercuration of 9,9-dialkyl-2,7-diethynylfluorenes $\mathbf{L_2}$–$\mathbf{L_4}$ with $HgCl_2$ using methanolic NaOH at room temperature produced a pale yellow suspension and the solids were collected by filtration to afford off-white air-stable products in high yields and purity [156]. Their excellent film-forming properties suggest high molecular weights of the materials and their good solubilities in CH_2Cl_2 and $CHCl_3$ render them amenable to spectroscopic studies and readily solution-processable for optical characterizations. Estimates of the molecular weights using GPC in THF indicate a high degree of polymerization (DP = 24–47) in these polyynes (Table 9) and the absence of end group NMR resonances supports their macromolecular nature. Similar synthetic route using 2,7-diethynylfluorene (m = 0 in Scheme 11) only afforded an intractable solid and thus it is critical to use long alkyl chains in achieving high solubility of the polymers. Complex **61**, as a model complex of **59**, was also synthesized by treatment of $\mathbf{L_3}$ with two equivalents of MeHgCl, in which one coordination site is protected by a Me group, under similar basic medium. The symmetrical nature of the molecules was confirmed by NMR studies. Two distinct ^{13}C NMR signals for the sp carbons of the acetylenic units are observed for **58–60** and they are shifted downfield with respect to the free alkynes, consistent with the formation of Hg–C(sp) bond. Only six peaks appear in the aromatic region of their ^{13}C NMR spectra, related to the 12 aromatic carbons in the structures and these data indicate high structural regularity in the polymers. Monomer **61** displays a single ^{199}Hg NMR peak at $\delta = -453$ (c.f. $\delta = -847$ for MeHgCl), reflecting polarization of the Hg–C≡ bonds [157]. The X-ray crystal structure of **61** also helped to establish polymer structures in the solid state and to correlate the photophysical properties with the structural data. The structure of **61** consists of dinuclear molecules in which the mercury centers adopt a two-coordinate linear geometry in a rigid-rod manner. The lattice structure is highlighted by the presence of weak intermolecular noncovalent Hg···Hg interactions (3.738 and 4.183 Å), which links the molecular units together to form a loose polymeric structure in a 3D network (Figure 32). The Hg···Hg interaction distances in **61** are comparable to the values of 3.71–4.25 Å for $[Hg(C{\equiv}CR)_2]$ (R = Ph, $SiMe_3$) and are toward the upper limit of those accepted as representing metallophilic interactions [158-161].

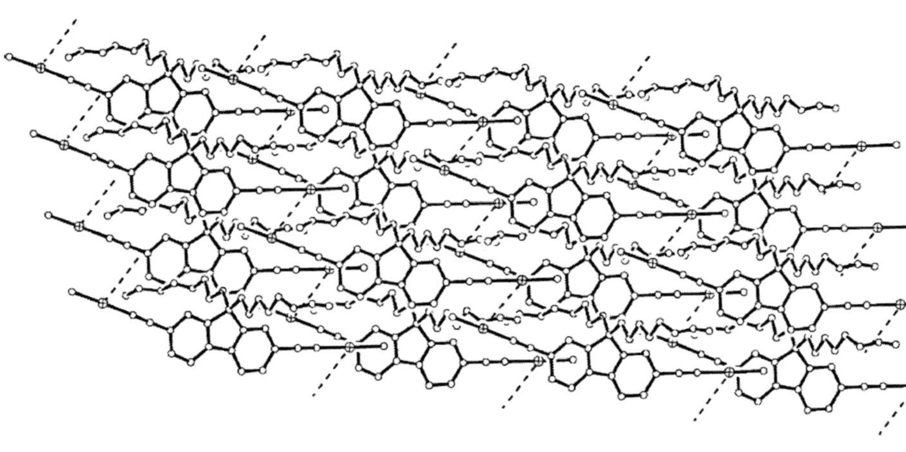

Figure 32. Crystal packing diagram for 61. The weak intermolecular mercuriophilic contacts in a 3D network are represented by the dashed lines.

Polymers **58–60** display moderate thermal stability with decomposition temperatures > 200 °C (Table 9) and exhibit an exotherm coincident with mass loss due to decomposition.

Polymer **60** shows the lowest onset decomposition temperature, while **58** and **59** with shorter alkyl chains exhibit increasing decomposition temperatures with decreasing *m* value. Each shows two stepwise losses in the thermogravimetric curve, corresponding to the elimination of two alkyl groups from their molecules.

Table 9. Structural and Thermal Properties of 58–60

polymer	M_w	M_n	M_w/M_n	DP	T_{decomp} (onset) (°C)
58	28,720	27,100	1.06	47	282 ± 5
59	18,320	15,090	1.22	24	220 ± 5
60	38,650	36,250	1.07	42	200 ± 5

Table 10. Photophysical Data for 58–61

	λ_{max}/nm	λ_{max}/nm	λ_{em}/nm		
	CH_2Cl_2	film	CH_2Cl_2 (290 K)[b]	film (290 K)	film (11 K)
58	313, 341, 355	316, 345, 365	381, 406* (0.003)[c]	475br	427, 570, 610*
59	313, 340, 355	316, 345, 364	382, 406* (0.004)[c]	477br	426, 583, 630*
60	313, 338, 355	316, 345, 364,	385*, 409 (0.007)[c]	478br	429, 590, 634*
61	311 (4.7), 330 (4.9), 346 (5.0)[a]	316, 339, 358	374 (0.02)[d]	417, 435, 477br	416, 438, 544*, 591*

[a] logε. [b] Fluorescence quantum yields in parentheses are measured relative to quinine sulfate in 0.1 N H_2SO_4 (Φ = 0.54). [c] λ_{ex} = 328 nm. [d] λ_{ex} = 340 nm. * Emission peaks appear as shoulders or weak bands. br = broad.

The photophysical data of **58–61** are shown in Table 10. They all display similar structured absorption bands in the near UV region. These bands are mainly associated with the organic $^1(\pi\pi^*)$ transitions, possibly with some admixture of metal orbitals, and the 0–0 absorption peak is assigned as the $S_0 \rightarrow S_1$ transition. We note a red-shift of absorption and emission bands in **58–60** after the inclusion of heavy metal atom which indicates an increase in π-conjugation. The C-9 substituents on the fluorene ring exert minimal effect on the absorption energies and comparable optical gaps (E_{gap}) are noted for **58–60** (Figure 33a). The transition energies of **59** are lowered with respect to those of the monomer **61**, suggesting a well-extended singlet excited state in the polymers. Red shifts of about 10 nm are observed for the lowest energy absorption bands of **58–60** in the solid state. Examination of the absorption behavior of the polymers in $CHCl_3$, $CHCl_3$/MeOH solutions, and in the solid state, corroborates the presence of solid-state aggregates in thin films. For instance, addition of a non-solvent (MeOH) to a $CHCl_3$ solution of **59** produces a new band centered at 365 nm, which corresponds to the strongest absorption observed in the solid state (Figure 34). Further support for the red-shift is provided by the observation of solid-state mercuriophilic interactions in the crystal structure of **61** (Figure 32). The long side chains in **58–60** should suppress aggregation or at least delay its onset, and indeed, changes in absorptions are observed for **59** only after addition of approximately 80% methanol. The behavior of **58–60** has an analogy in that of poly[(9,9-dialkyl)fluoren-2,7-yleneethynylene]s, 2,5-dialkylpoly(*p*-

phenyleneethynylenes) and polydiacetylenes where similar aggregate bands are described [162].

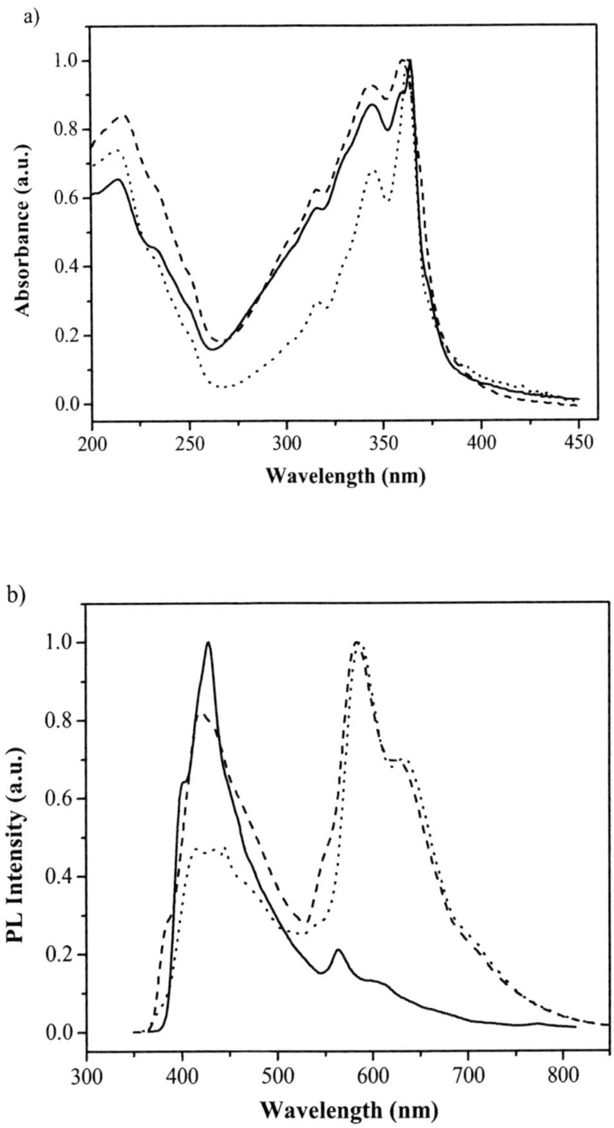

Figure 33. a) Absorption and b) emission spectra (11 K) of thin solid films of **58** (—), **59** (----) and **60** (····).

The PL spectra display a red-shift on going from binuclear to polynuclear structures (**61** *vs.* **59**), implying that the excited state is stabilized by the greater degree of delocalization in the polymers. In dilute fluid solutions, we observe an intense $^1(\pi\pi^*)$ emission peak near 400 nm for **58–60** due to fluorescence ($S_1 \rightarrow S_0$), which is in line with the small Stokes shift observed between the bands in the absorption and the emission spectra. Comparing with the solution PL data, the thin film singlet emissions in **58–60** appear broad at 290 K and are red-shifted which are consistent with their absorption features and are likely to be a result of interchain interactions due to aggregate formation. At 11 K, another well-resolved lower-lying emissions emerge around 570–590 nm for **58–60** and the large Stokes shifts of these peaks from the dipole-allowed absorptions (ca. 1.2–1.3 eV), plus the long emission lifetimes at 11 K (28 ± 0.2 (**58**), 35 ± 0.2 (**59**) and 42 ± 0.2 μs (**60**)) are indicative of their triplet parentage, and they are thus assigned to the $^3(\pi\pi^*)$ excited states of the diethynylfluorenes (Figure 33b). The triplet emission ($T_1 \rightarrow S_0$) bands increase in intensity with decreasing temperature for **58–60** and show vibronic structures at low temperatures (Figure 35), typical of the phosphorescence observed in related Pt(II) and Au(I) polyynes. The intensity of the phosphorescence relative to fluorescence increases as the value of *m* increases from **58** to **60**. To assess the relative ISC efficiency based on the peak height ratio from triplet emission to singlet emission at 11 K, we note that the order of S_1–T_1 crossover efficiency is **60** > **59** > **58**. The ISC rate is also higher in the polymers than in the monomers (e.g. **61** *vs.* **59**). Considering the spatial extent of the singlet and triplet excitons in the present system, values of $\Delta E(S_o–T_1)$ (energy gap between S_o and T_1) were found to be 2.10–2.23 eV for **58–60**, which can be compared to those observed in dialkoxy-substituted poly(*p*-phenyleneethynylene)s (PPEs, 1.90 eV) [95], [Cy_3PAu]-capped PPE (1.98–2.04 eV) [153-155] and Pt-containing poly[(9,9-dialkyl)fluoren-2,7-yleneethynylene]s (2.20 eV) [105]. The $\Delta E(S_1–T_1)$ values for **58–60** lie within the range of 0.73–0.79 eV and they match well the S_1–T_1 energy gap of 0.7 ± 0.1 eV for similar π-conjugated Pt and Au polyynes, and are close to the gaps estimated for a series of related organic polymers [163-165]. We attribute such a constant $\Delta E(S_1–T_1)$ value to the exchange energy and possibly some additional constant contribution due to the admixture of the Hg orbitals.

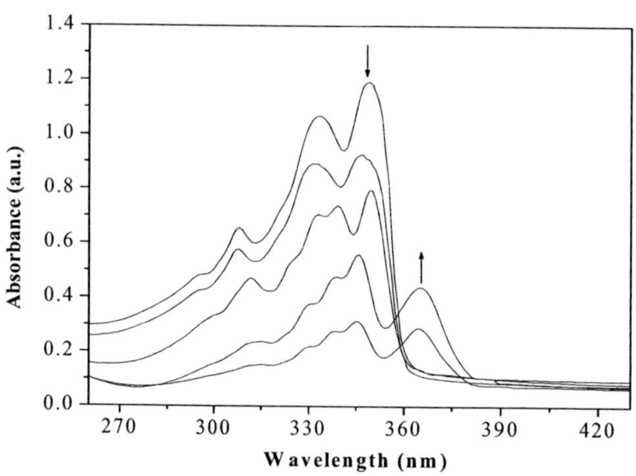

Figure 34. Absorption spectrum of **59** (CH_2Cl_2) upon the addition of MeOH (top to bottom: 0, 20, 40, 80, 90%).

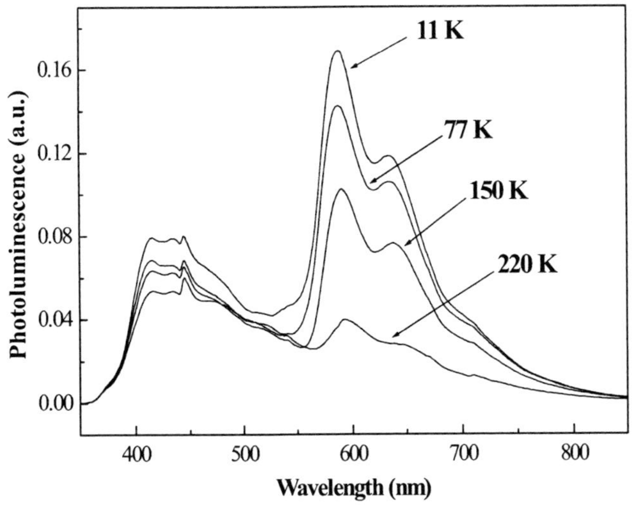

Figure 35. Temperature dependence of the PL of **60**.

2.7. Miscellaneous

Compared to the polyaromatic ligands, the fluorenyl group which can be considered as a dibenzocyclopentadienyl system has attracted relatively less attention from coordination chemists. The coordination chemistry of fluorene centered on mononuclear [166-171] and binuclear metal complexes [172,173] has been well-documented and will not be reported in this contribution. Recently, the interactions of 9-ethynylfluoren-9-ol with some bimetallic and multimetallic systems were examined. The thermal reaction of $[Ru_3(CO)_{12}]$ with 9-ethynylfluoren-9-ol in THF gave five new ruthenium fluorene clusters **62**–**66** in low yields which were fully characterized by IR, NMR, FABMS and X-ray crystallographic studies (Scheme 12) [174]. Most of them exhibit interesting structural features. The structure of **62** consists of a fluorene ring bound to the Ru_3 cluster unit via a typical μ_3-(η^2-$\perp$) coordination mode. Cluster **63** is based on a Ru_3 triangle containing a C_4 chain which arises from the head-to-tail coupling of two ligands with the formation of a ruthenole ring bonded in a novel μ_3-η^1:η^2:η^4 fashion. An interesting structural feature observed in **64** is the formation of a Ru_4O square pyramidal framework, with the four Ru atoms in a butterfly arrangement, and the hydroxy group originating from the ligand is bound to the cluster in a triply bridging manner. The bonding modes of the organic fragments in **65** and **66** are very similar. They both interact with the five Ru atoms but with a novel Ru_5 bent chain in **65** and a spiked butterfly in **66**. The synthesis and structural analysis of the corresponding dicobalt-alkyne complex derived from 9-ethynylfluoren-9-ol are also known [175].

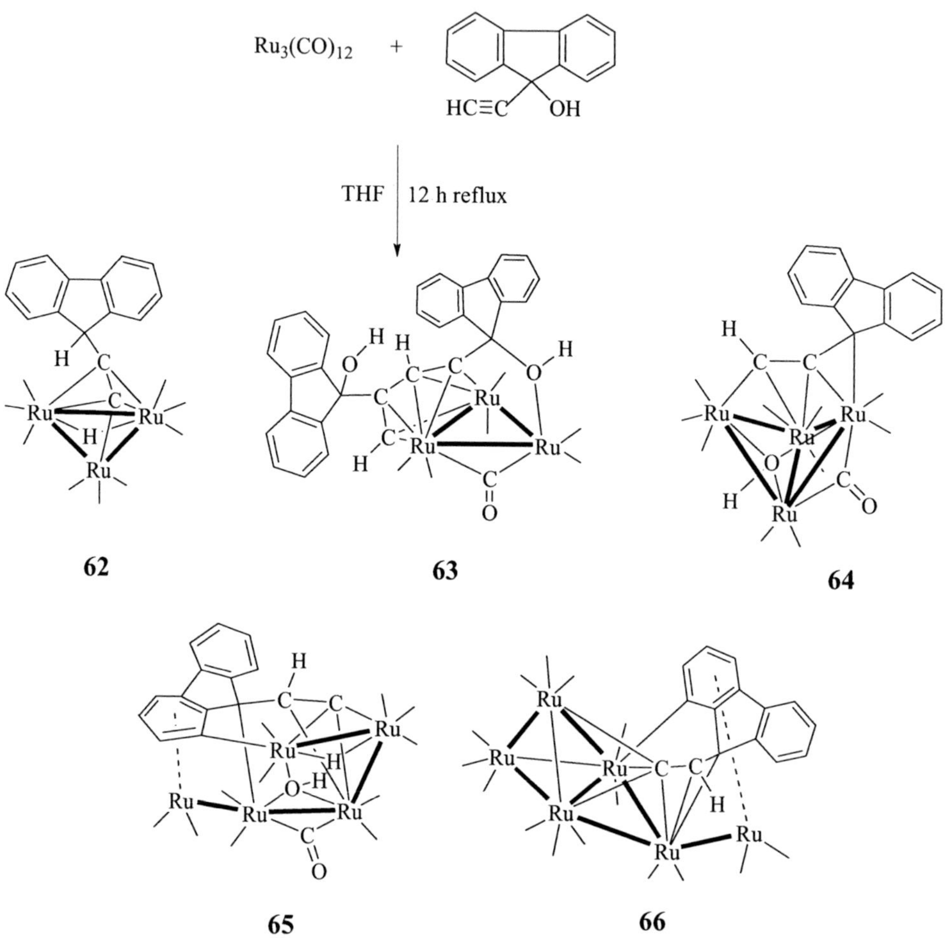

Scheme 12.

Fluorene- and carbazole-bridged diethynyl-pyridyl ligands are prepared as useful building blocks for the construction of soluble macrocyclic supramolecules with different geometries by self-assembly processes. They undergo facile ring closure reactions with $Re(CO)_5X$ (X = Cl, Br) to form molecular geometries of different dimensions. Fluorene-bridged ditopic ligands promote the formation of trimetallic triangles **67**, while the angular carbazole-bridged ligands favor the formation of cyclic dimers **68** [176]. These supramolecules display emission of different origins triggered by the bridging aromatic segment. Carbazole-based cyclic dimers emit green light that probably arises from a ligand-localized $^1(\pi\pi^*)$ excited state. However, emission is observed only at low temperature for the fluorene-containing triangles, and it shows $^3(\pi\pi^*)$ and ^{3}MLCT-based emissions with prolonged lifetimes.

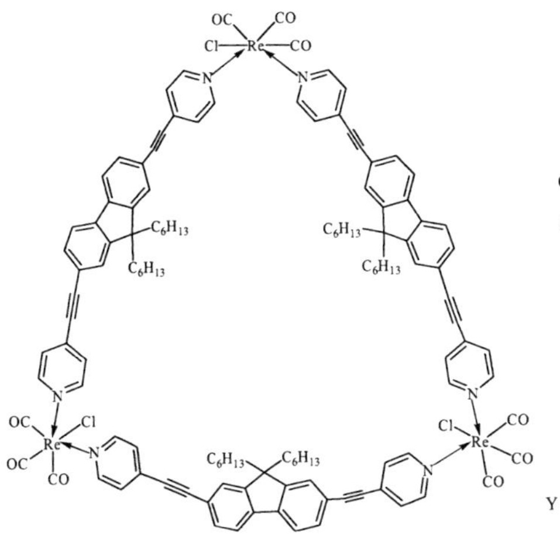

67

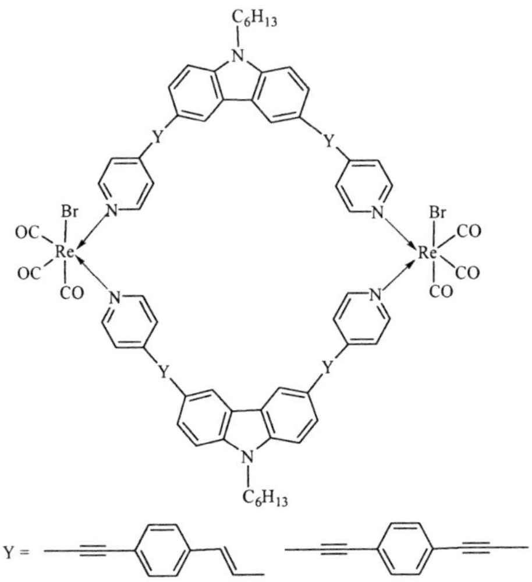

68

In a related context, the ligand $\mathbf{L_{10}}$ was shown to be capable of incorporating into infinite one-dimensional coordination polymers and square grid polymer structures. Treatment of $\mathbf{L_{10}}$ with copper nitrate in ethanol leads to exceptionally large, noninterpenetrating, square-grid polymers $[Cu(\mathbf{L_{10}})_2(NO_3)_2]$ with grid dimensions of 25 × 25 Å^2 [177]. The ABAB stacking of the layers generates reduced, yet still very large infinite channels of dimensions 16 × 16 Å^2. The solvent-accessible volume estimated is 54.1% of the volume of the unit cell. Bunz and co-workers also accomplished the first chiral noninterpenetrating square-grid coordination polymers $[Cu(\mathbf{L_{11}})_2(NO_3)_2]$ derived from a chiral fluorene-based ligand $\mathbf{L_{11}}$. The crystal of $[Cu(\mathbf{L_{11}})_2(NO_3)_2]$ crystallizes in the space group $P2_1$ and the grid dimension is still 25 × 25 Å^2. But, the ABCABC stacking significantly reduces the accessible channel dimensions to about 8 × 8 Å^2. Along the same line, two luminescent coordination polymers with a triple-helix structure $[HgX_2(L_{11})]_\infty$ (X = Cl, Br) **69** can be synthesized from HgX_2 and $\mathbf{L_{11}}$ [178]. The mercury atom is in a distorted tetrahedral environment. There is a large shift in the emission between the free ligand and the coordination polymers. The mercury atoms, which exert a strong electron-withdrawing effect as a result of their +2 charge and attached halides, have a significant influence on the fluorescent properties of the organic chromophore.

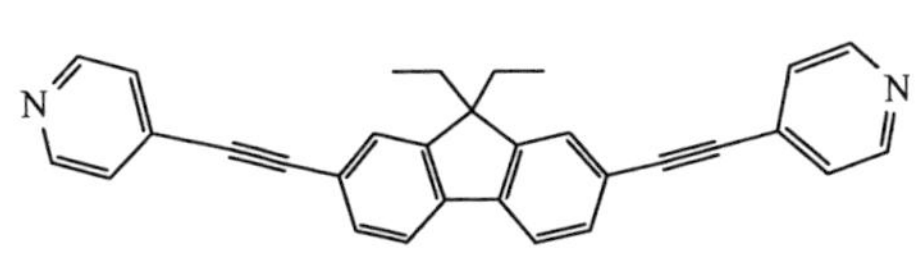

$\mathbf{L_{10}}$

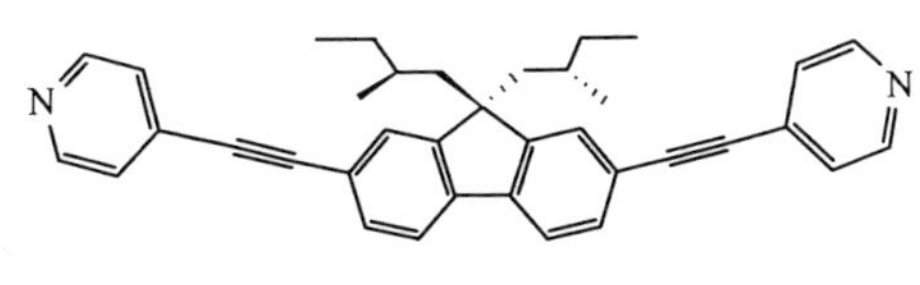

$\mathbf{L_{11}}$

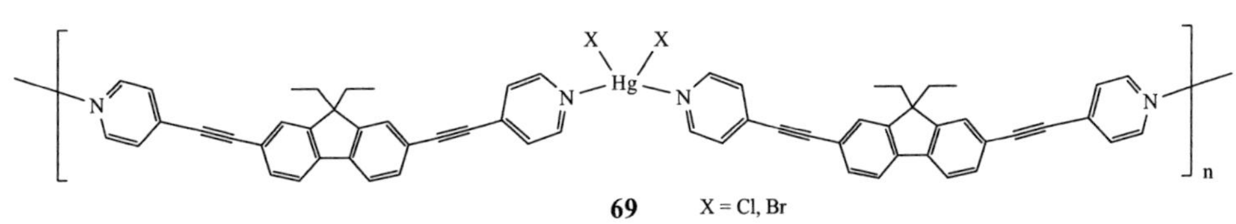

69 X = Cl, Br

The synthesis and characterization of carbazole-based ruthenium complexes linked by an acetylene spacer **70** and **71** were also reported [179]. The ground state electronic properties of these compounds were examined by CV technique and absorption spectra. The excited state behavior was investigated after pumping the carbazole moiety. The presence of Ru^{2+} unit strongly suppresses the fluorescence of the carbazole chromophore in both **70** and **71**. In the case of **70**, both Ru^{2+} and Ru^{3+} compounds (the latter being obtained by chemical or electrochemical oxidation) show a decrease of the fluorescence of carbazole, typical of other pentaammine complexes.

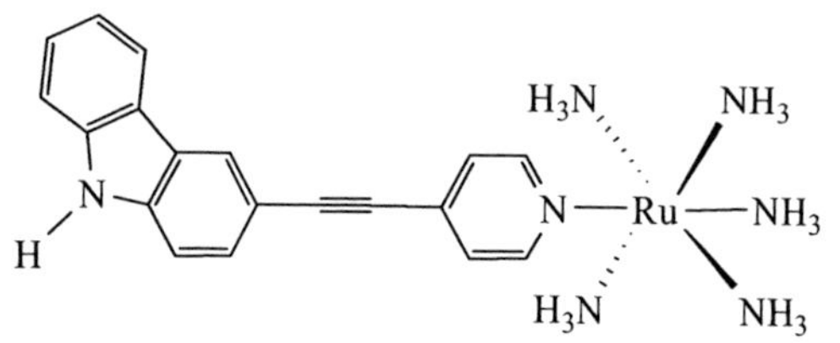

70

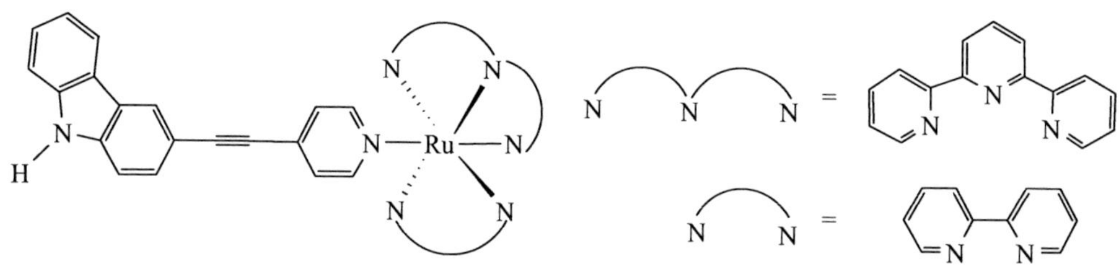

71

Fluorene-containing bidentate bipyridine ligands, which could serve as models for the polymeric analogues possessing alternating fluorene and bipyridine units, were prepared and used as complexing agents for a variety of transition metals [180]. New octahedral complexes **72**–**77** were thoroughly characterized by NMR, FABMS, electronic and fluorescence spectra as well as CV measurements. The ferrocene-tethered fluorenyl bipyridine affords novel heterobimetallic components. Except for the ferrocene-containing compounds, the metal complexes are luminescent at room temperature. The luminescence in the ruthenium complexes is possibly from a MLCT state, and the fluorescence due to the ligand is totally quenched. When compared to $[Ru(bipy)_3]^{2+}$, these ruthenium complexes exhibit longer lifetimes, which is probably due to the delocalization of charge into the extended fluorene conjugation.

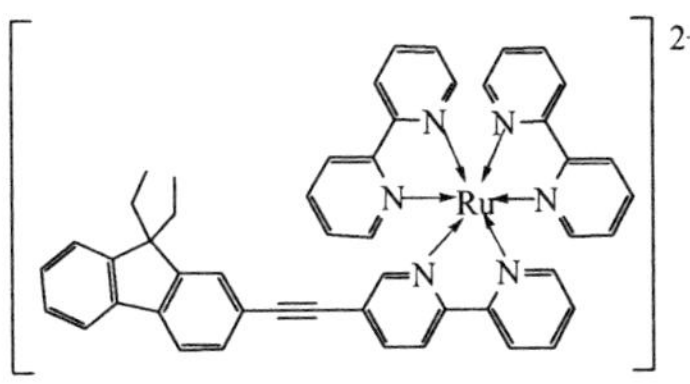

72

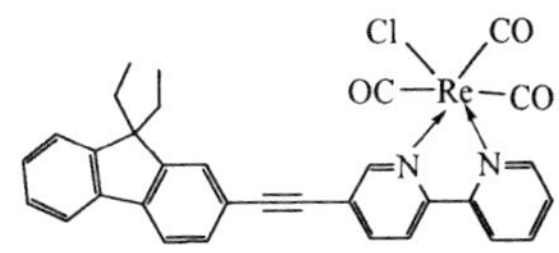

73

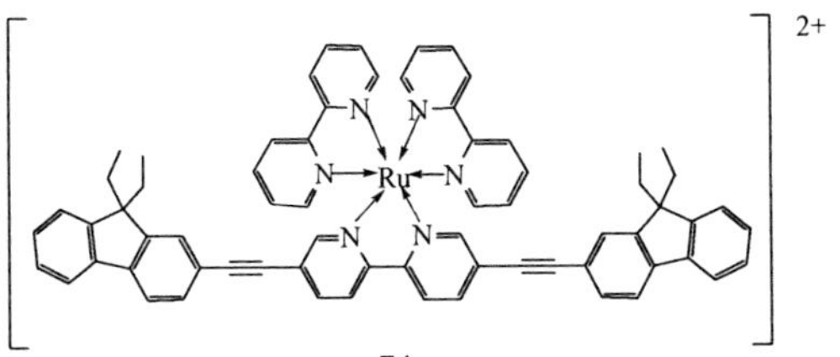

74

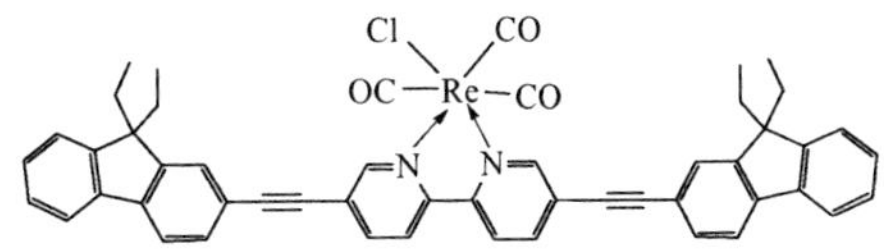

75

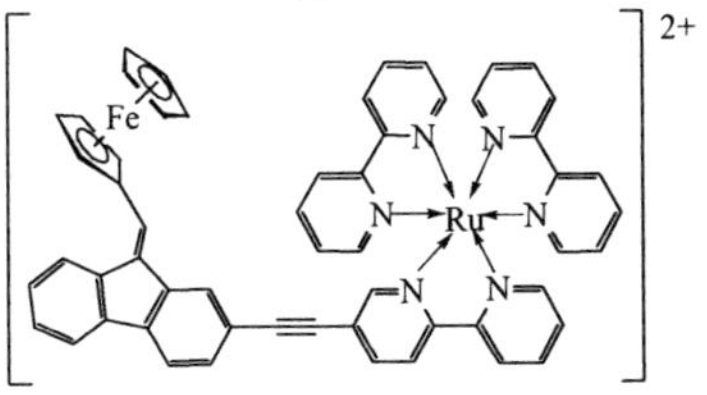

76

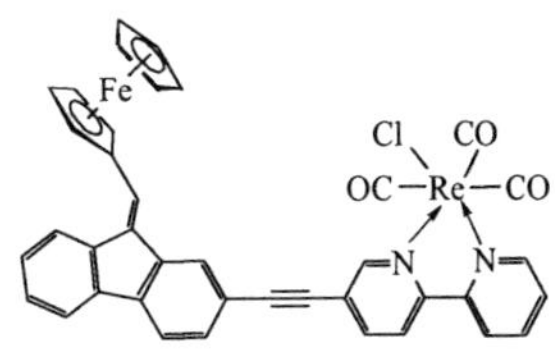

77

3. Conclusions and Outlook

It is clear from this chapter that a wide variety of metal alkynyl complexes containing the fluorene unit have been synthesized and characterized. The driving force for this work lies in the possibility of coupling the chemical, electronic and optical properties of metal complexes to those of the organic chromophore, thus accessing novel materials with new properties. These metal-based systems have demonstrated interesting reactivity, optical, luminescent and redox characteristics. In the majority of cases, the versatility of preparing new metal alkynyl complexes of fluorene derivatives has been demonstrated by insertion of different peripheral groups on the fluorene ring and variation of the 9-substituents can affect the optical and photophysical properties of these materials.

Some terminal ferrocenylacetylene precursors were successfully used as the synthons to afford a new class of carbon-rich organometallics with remarkable physical properties. We have been able to extend the dimensions of these rigid-rod-like complexes from the molecular scale to nanoscale. The production of these molecules provides us an opportunity to investigate novel multimetallic assemblies of nanosized dimensions which is essential for the further development of nanoscale molecular wires and related nanomaterials.

As a continuation of the studies on the nature of photoexcited states in organometallic conjugated polymers as a function of π-conjugation in the bridging ligand, a group of soluble, well-defined platinum-containing diyne and polyyne materials consisting of 9-functionalized fluorene and 9-butylcarbazole linking units was prepared and fully characterized. The polymers exhibit good thermal stability and photoconducting properties. A systematic correlation was made between the effective conjugation length (or conversely, bandgap) and the intersystem crossing rate in these polyynes. The larger the bandgap, the higher the efficiency of triplet emission. These polymers represent suitable model systems to investigate the relationship between chemical structure and the evolution of singlet and triplet excited states and the results contribute to our understanding of the structure-property relationship in this class of organometallic conjugated polymers. Similar to the Pt(II) polyynes, the use of mercury together with their associated high optical gaps can offer a good way to enhance the ISC rate and we have shown for the first time that the organic triplet emissions of fluorenyleneethynylenes have been "illuminated" by the heavy atom effect of mercury. The present investigation is desirable for applications that harvest the T_1 state for light emission and will provide impetus for future study. While the multidisciplinary nature of this field of molecular science continues to attract an abundance of high quality research, it seems likely that some fascinating chemistry and valuable applications can be developed.

4. Acknowledgment

This work was supported by CERG Grants from the Research Grants Council of the Hong Kong SAR, P.R. China (Project Nos. HKBU 2022/99P, 2048/01P and 2022/03P) and Faculty Research Grants from Hong Kong Baptist University (FRG/99-00/II-58 and 01-02/II-48). We also thank Dr. K.-W. Cheah at the Hong Kong Baptist University and Dr. W.-K. Chan at the University of Hong Kong for the access of their research facilities to perform the photoluminescence and photocurrent measurements, respectively. Dr. Z. Lin at the Hong Kong University of Science and Technology is also gratefully acknowledged for his contribution on molecular orbital calculations.

5. References

[1] Skotheim, T. A.; Elsenbaumer, R. L.; Reynolds, J. R. *Handbook of Conducting Polymers*; Marcel Dekker: New York, 1998.

[2] Nuguyen, P.; Gómez-Elipe, P.; Manners, I. *Chem. Rev.* 1999, *99*, 1515-1548.

[3] Manners, I. *Synthetic Metal-Containing Polymers*; Wiley: Weinheim, 2004, Chap. 5.

[4] Stott, T. L.; Wolf, M. O. *Coord. Chem. Rev.* 2003, *246*, 89-101.

[5] Kingsborough, R. P.; Swager, T. M. *Prog. Inorg. Chem.* 1999, *48*, 123-231.

[6] Long, N. J.; Williams, C. K. *Angew. Chem. Int. Ed.* 2003, *42*, 2586-2617.

[7] Neher, D. *Macromol. Rapid Commun.* 2001, *22*, 1365-1385.

[8] Tsuie, B.; Reddinger, J. L.; Sotzing, G. A.; Soloducho, J.; Katritzky, A. R.; Reynolds, J. R. *J. Mater. Chem.* 1999, *9*, 2189-2200.

[9] Wong, K.-T.; Chien, Y.-Y.; Chen, R.-T.; Wang, C.-F.; Lin, Y.-T.; Chiang, H.-H.; Hsieh, P.-Y.; Wu, C.-C.; Chou, C.-H.; Su, Y. O. *J. Am. Chem. Soc.* 2002, *124*, 11576-11577.

[10] Yu, W.-L.; Pei, J.; Huang, W.; Heeger, A. J. *Adv. Mater.* 2000, *12*, 828-831.

[11] Setayesh, S.; Grimsdale, A. C.; Weil, T.; Enkelmann, V.; Müllen, K.; Meghdadi, F.; List, E. J. W.; Leising, G. *J. Am. Chem. Soc.* 2001, *123*, 946-953.

[12] Marsitzky, D.; Vestberg, R.; Blainey, P.; Tang, B. T.; Hawker, C. J.; Carter, K. R. *J. Am. Chem. Soc.* 2001, *123*, 6965-6972.

[13] Janietz, S.; Bradley, D. D. C.; Grell, M.; Glebeler, C.; Inbasekaran, M.; Woo, E. P. *Appl. Phys. Lett.* 1998, *73*, 2453-2455.
[14] Yu, W.-L.; Pei, Y.; Cao, Y.; Huang, W.; Heeger, A. J. *Chem. Commun.* 1999, 1837-1838.
[15] Pschirer, N. G.; Bunz, U. H. F. *Macromolecules* 2000, *33*, 3961-3963.
[16] Scherf, U.; List, E. J. *Adv. Mater.* 2002, *14*, 477-487.
[17] Mysyk, D. D.; Perepichka, I. F.; Perepichka, D. F.; Bryce, M. R.; Popov, A. F.; Goldenberg, L. M.; Moore, A. J. *J. Org. Chem.* 1999, *64*, 6937-6950.
[18] Perepichka, I. F.; Popov, A. F.; Orekhova, T. V.; Bryce, M. R.; Andrievskii, A. M.; Batsanov, A. S.; Howard, J. A. K.; Sokolov, N. I. *J. Org. Chem.* 2000, *65*, 3053-3063.
[19] Paul, F.; Lapinte, C. *Coord. Chem. Rev.* 1998, *178-180*, 431-509.
[20] Yam, V. W.-W. *Chem. Commun.* 2001, 789-796.
[21] Sauvage, J.-P.; Collin, J.-P.; Chambron, J.-C.; Guillerez, S.; Coudret, C.; Balzani, V.; Barigelletti, F.; De Cola, L.; Flamigni, L. *Chem. Rev.* 1994, *94*, 993-1019.
[22] Ziessel, R.; Hissler, M.; El-Ghayoury, A.; Harriman, A. *Coord. Chem. Rev.* 1998, *178–180*, 1251-1298.
[23] Balzani, V.; Juris, A.; Venturi, M.; Campagna, S.; Serroni, S. *Chem. Rev.* 1996, *96*, 759-834.
[24] McCleverty, J. A.; Ward, M. D. *Acc. Chem. Res.* 1998, *31*, 842-851.
[25] Ward, M. D. *Chem. Soc. Rev.* 1995, 121-134.
[26] Lavastre, O.; Plass, J.; Bachmann, P.; Salaheddine, A.; Moinet, C.; Dixneuf, P. H. *Organometallics* 1997, *16*, 184-189.
[27] Schimanke, H.; Gleiter, R. *Organometallics* 1998, *17*, 275-277.
[28] Justin Thomas, K. R.; Lin, J. T.; Lin, K.-J. *Organometallics* 1999, *18*, 5285-5291.
[29] Hradsky, A.; Bildstein, B.; Schuler, N.; Schottenberger, H.; Jaitner, P.; Ongania, K.-H.; Wurst, K.; Launay, J.-P. *Organometallics* 1997, *16*, 392-402.
[30] Stang, S. L.; Paul, F.; Lapinte, C. *Organometallics* 2000, *19*, 1035-1043.
[31] Brady, M.; Weng, W.; Gladysz, J. A. *J. Chem. Soc. Chem. Commun.* 1994, 2655-2656.
[32] Justin Thomas, K. R.; Lin, J. T.; Wen, Y. S. *J. Organomet. Chem.* 1999, *575*, 301-309.
[33] Jones, N. D.; Wolf, M. O.; Giaquinta, D. M. *Organometallics* 1997, *16*, 1352-1354.
[34] Grosshenny, V.; Harriman, A.; Hissler, M.; Ziessel, R. *J. Chem. Soc. Faraday Trans.* 1996, 2223-2238.
[35] Thomas, J. A.; Jones, C. J.; McCleverty, J. A.; Hutchings, M. G. *Polyhedron* 1996, *15*, 1409-1414.
[36] Le Vanda, C.; Cowan, D. O.; Leitch, C.; Bechgaard, K. *J. Am. Chem. Soc.* 1974, *96*, 6788-6789.
[37] Peruga, A.; Mata, J. A.; Sainz, D. *J. Organomet. Chem.* 2001, *637-639*, 191-197.
[38] Justin Thomas, K. R.; Lin, J. T.; Wen, Y. S. *Organometallics* 2000, *19*, 1008-1012.
[39] Xue, W.-M.; Kühn, F. E.; Herdtweck, E.; Li, Q. *Eur. J. Inorg. Chem.* 2001, 213-221.
[40] Adams, R. D.; Kwon, O.-S.; Qu, B.; Smith, M. D. *Organometallics* 2001, *20*, 5225-5232.
[41] Adams, R. D.; Qu, B.; Smith, M. D.; Albright, T. A. *Organometallics* 2002, *21*, 2970-2978.
[42] Adams, R. D.; Qu, B.; Smith, M. D. *Organometallics* 2002, *21*, 3867-3872.
[43] Adams, R. D.; Qu, B.; Smith, M. D. *Organometallics* 2002, *21*, 4847-4852.

[44] Long, N. J.; Martin, A. J.; Vilar, R.; White, A. J. P.; Williams, D. J.; Younus, M. *Organometallics* 1999, *18*, 4261-4269.
[45] Colbert, M. C. B.; Lewis, J.; Long, N. J.; Raithby, P. R.; White, A. J. P.; Williams, D. J. *J. Chem. Soc. Dalton Trans.* 1997, 99-104.
[46] Justin Thomas, K. R.; Lin, J. T. *J. Organomet. Chem.* 2001, *637-639*, 139-144.
[47] Lavastre, O.; Even, E.; Dixneuf, P. H.; Pacreau, A.; Vairon, J.-P. *Organometallics* 1996, *15*, 1530-1531.
[48] Malaun, M.; Kowallick, R.; McDonagh, A. M.; Marcaccio, M.; Paul, R. L.; Asselberghs, I.; Clays, K.; Persoons, A.; Bildstein, B.; Fiorini, C.; Nunzi, J.-M.; Ward, M. D.; McCleverty, J. A. *J. Chem. Soc. Dalton Trans.* 2001, 3025-3038.
[49] Long, N. J.; Martin, A. J.; White, A. J. P.; Williams, D. J.; Fontani, M.; Laschi, F.; Zanello, P. *J. Chem. Soc. Dalton Trans.* 2000, 3387-3392.
[50] Wong, W.-Y.; Lu, G.-L.; Ng, K.-F.; Wong, C.-K.; Choi, K.-H. *J. Organomet. Chem.* 2001, *637-639*, 159-166.
[51] Wong, W.-Y.; Ho, K.-Y.; Choi, K.-H. *J. Organomet. Chem.* 2003, *670*, 17-26.
[52] Wong, W.-Y.; Lu, G.-L.; Ng, K.-F.; Choi, K.-H.; Lin, Z. *J. Chem. Soc. Dalton Trans.* 2001, 3250-3260.
[53] Osella, D.; Milone, L.; Nervi, C.; Ravera, M. *J. Organomet. Chem.* 1995, *488*, 1-3.
[54] Osella, D.; Gobetto, R.; Nervi, C.; Ravera, M.; D'Amato, R.; Russo, M. V. *Inorg. Chem. Commun.* 1998, *1*, 239-245.
[55] Wong, W.-Y.; Ho, K.-Y.; Ho, S.-L.; Lin, Z. *J. Organomet. Chem.* 2003, *683*, 341-353.
[56] Becke, A. D. *J. Chem. Phys.* 1993, *98*, 5648-5652.
[57] Miehlich, B.; Savin, A.; Stoll, H.; Preuss, H. *Chem. Phys. Lett.* 1989, *157*, 200-206.
[58] Lee, C.; Yang, W.; Parr, G. *Phys. Rev. B.* 1988, *37*, 785-789.
[59] Takahashi, S.; Kuroyama, Y.; Sonogashira, K.; Hagihara, N. *Synthesis* 1980, 627-630.
[60] Wong, W.-Y.; Lu, G.-L.; Choi, K.-H.; Lin, Z. *Eur. J. Org. Chem.* 2003, 365-373.
[61] Moore, A. J.; Chesney, A.; Bryce, M. R.; Batsanov, A. S.; Kelly, J. E.; Howard, J. A. K.; Perepichka, I. F.; Perepichka, D. F.; Meshulam, G.; Berkovic, G.; Kotler, Z.; Mazor, R.; Khodorkovsky, V. *Eur. J. Org. Chem.* 2001, 2671-2687.
[62] Lewis, J.; Raithby, P. R.; Wong, W.-Y. *J. Organomet. Chem.* 1998, *556*, 219-228.
[63] Wong, W.-Y.; Wong, C.-K.; Lu, G.-L. *J. Organomet. Chem.* 2003, *671*, 27-34.
[64] Younus, M.; Long, N. J.; Raithby, P. R.; Lewis, J. *J. Organomet. Chem.* 1998, *570*, 55-62.
[65] Brédas, J. L., Chance, R. R. (Eds.), *Conjugated Polymer Materials: Opportunities in Electronic, Optoelectronic and Molecular Electronics*; Kluwer Academic Publishers: Dordrecht, 1990.
[66] Sheets, J. E.; Carraher, C. E., Jr.; Pittman, C. U., Jr. (Eds.), *Metal-Containing Polymeric Systems*; Plenum: New York, 1985.
[67] Patai S. (Ed.), *The Chemistry of Carbon-Carbon Triple Bond*; Wiley-Interscience: Chichester, 1978.
[68] Diederich, F.; Rubin, Y. *Angew. Chem. Int. Ed. Engl.* 1992, *31*, 1101-1123.
[69] Rubin, Y.; Knobler, C. B.; Diederich, F. *J. Am. Chem. Soc.* 1990, *112*, 4966-4968.
[70] Haley, M. M.; Langsdorf, B. L. *Chem. Commun.* 1997, 1121-1122.
[71] Dickson, R. S.; Fraser, P. J. *Adv. Organomet. Chem.* 1974, *12*, 323-377.
[72] Pauson, P. L. *Tetrahedron* 1985, *41*, 5855-5860.
[73] Magnus, P.; Becker, D. P. *J. Chem. Soc. Chem. Commun.* 1985, 640-642.

[74] Agh-Atabay, W. E.; Lindsell, P. N.; Preston, P. J.; Tomb, A. D.; Lloyd, R.; Raugel-Rojo, G.; Spruce, B. S.; Wherrett, J. *J. Mater. Chem.* 1992, 2, 1241-1247.
[75] Geiger, W. E.; Connelly, N. G. *Adv. Organomet. Chem.* 1985, 24, 87-130.
[76] Robinson, B. H.; Simpson, J. *Paramagnetic Organometallic Species in Activation/Selectivity, Catalysis*; Chanon, M.; Juillard, M.; Poite, J. C. (Eds.), NATO ASI Series, Kluwer Academic Publishers: Dordrecht, 1989.
[77] Colbran, S. B.; Robinson, B. H.; Simpson, J. *J. Chem. Soc. Chem. Commun.* 1982, 1362-1364.
[78] Davidson, J. L. *Reactions of Coordinated Ligands*; Braterman, P. S. (Ed.), Plenum Press: New York, 1986, Vol. 1.
[79] Nicholas, K. M.; Nestle, M. O.; Seyferth, D. *Transition Metal Organometallics in Organic Synthesis*; Alper, H. (Ed.), Academic Press: New York, 1978, Vol. 2.
[80] Johnson, B. F. G.; Lewis, J.; Raithby, P. R.; Wilkinson, D. A. *J. Organomet. Chem.* 1991, *408*, C9-C12.
[81] Lewis, J.; Lin, B.; Khan, M. S.; Al-Mandhary, M. R. A.; Raithby, P. R. *J. Organomet. Chem.* 1994, *484*, 161-167.
[82] Arewgoda, M.; Rieger, P. H.; Robinson, B. H.; Simpson, J.; Visco, S. J. *J. Am. Chem. Soc.* 1982, *104*, 5633-5640.
[83] Wong, W.-Y.; Lam, H.-Y.; Lee, S.-M. *J. Organomet. Chem.* 2000, *595*, 70-80.
[84] Worth, G. H.; Robinson, B. H.; Simpson, J. *Organometallics* 1992, *11*, 3863-3874.
[85] Worth, G. H.; Robinson, B. H.; Simpson, J. *J. Organomet. Chem.* 1990, *387*, 337-355.
[86] Burroughes, J. H.; Bradley, D. D. C.; Brown, A. R.; Marks, R. N.; Mackay, K.; Friend, R. H.; Burn, P. L.; Holmes, A. B. *Nature*, 1990, *347*, 539-541.
[87] Tessler, N.; Denton, G. J.; Friend, R. H. *Nature*, 1996, *382*, 695-697.
[88] Halls, J. J. M.; Walsh, C. A.; Greenham, N. C.; Marseglia, E. A.; Friend, R. H.; Moratti, S. C.; Holmes, A. B. *Nature*, 1995, *376*, 498-500.
[89] Yu, G.; Gao, J.; Hummelen, J. C.; Wudl, F.; Heeger, A. J. *Science*, 1995, *270*, 1789-1791.
[90] Brown, A. R.; Bradley, D. D. C.; Burroughes, J. H.; Friend, R. H.; Greenham, N. C.; Burn, P. L.; Holmes, A. B.; Kraft, A. *Appl. Phys. Lett.* 1992, *61*, 2793-2795.
[91] Baldo, M. A.; Thompson, M. E.; Forrest, S. R. *Nature* 2000, *403*, 750-753.
[92] Baldo, M. A.; O'Brien, D. F.; You, Y.; Shoustikov, A.; Sibley, S.; Thompson, M. E.; Forrest, S. R. *Nature* 1998, *395*, 151-154.
[93] Friend, R. H.; Gymer, R. W.; Holmes, A. B.; Burroughes, J. H.; Marks, R. N.; Taliani, C.; Bradley, D. D. C.; dos Santos, D. A.; Brédas, J.-L.; Lögdlund, M.; Salaneck, W. R. *Nature* 1999, *397*, 121-128.
[94] Cleave, V.; Yahioglu, G.; Le Barny, P.; Friend, R. H.; Tessler, N. *Adv. Mater.* 1999, *11*, 285-288.
[95] Wilson, J. S.; Köhler, A.; Friend, R. H.; Al-Suti, M. K.; Al-Mandhary, M. R. A.; Khan, M. S.; Raithby, P. R. *J. Chem. Phys.* 2000, *113*, 7627-7634.
[96] Chawdhury, N.; Köhler, A.; Friend, R. H.; Wong, W.-Y.; Lewis, J.; Younus, M.; Raithby, P. R., Corcoran, T. C.; Al-Mandhary, M. R. A.; Khan, M. S. *J. Chem. Phys.* 1999, *110*, 4963-4970.
[97] Chawdhury, N.; Köhler, A.; Friend, R. H.; Younus, M.; Long, N. J.; Raithby, P. R.; Lewis, J. *Macromolecules*, 1998, *31*, 722-727.

[98] Wilson, J. S.; Chawdhury, N.; Al-Mandhary, M. R. A.; Younus, M.; Khan, M. S.; Raithby, P. R.; Köhler, A.; Friend, R. H. *J. Am. Chem. Soc.* 2001, *123*, 9412-9417.

[99] Wong, W.-Y.; Chan, S.-M.; Choi, K.-H.; Cheah, K.-W.; Chan, W.-K. *Macromol. Rapid Commun.* 2000, *21*, 453-457.

[100] Younus, M.; Köhler, A.; Cron, S.; Chawdhury, N.; Al-Mandhary, M. R. A.; Khan, M. S.; Lewis, J.; Long, N. J.; Friend, R. H.; Raithby, P. R. *Angew. Chem. Int. Ed.* 1998, *37*, 3036-3039.

[101] Wong, W.-Y.; Lu, G.-L.; Choi, K.-H.; Shi, J.-X. *Macromolecules* 2002, *35*, 3506-3513.

[102] Wong, W.-Y.; Choi, K.-H.; Lu, G.-H. *Macromol. Rapid Commun.* 2001, *22*, 461-465.

[103] Wong, W.-Y.; Wong, W.-K.; Raithby, P. R. *J. Chem. Soc. Dalton Trans.* 1998, 2761-2766.

[104] Khan, M. S.; Kakkar, A. K.; Long, N. J.; Lewis, J.; Raithby, P.; Nguyen, P.; Marder, T. B.; Wittmann, F.; Friend, R. H. *J. Mater. Chem.* 1994, *4*, 1227-1232.

[105] Khan, M. S.; Al-Mandhary, M. R. A.; Al-Suti, M. K.; Ahrens, B.; Mahon, M. F.; Male, L.; Raithby, P. R.; Boothby, C. E.; Köhler, A. *Dalton Trans.* 2003, 74-84.

[106] Wright, M. E.; Cochran, B. B. *Organometallics* 1993, *12*, 3873-3878.

[107] Zanello, P.; Opromolla, G.; de Biani, F. F.; Ceccanti, A.; Giorgi, G. *Inorg. Chim. Acta* 1997, *255*, 47-52.

[108] Chakraborty, A.K.; Bhattacharjee, A.; Mallik, B.; *Bull. Chem. Soc. Jpn.* 1994, *67*, 607-611.

[109] Puddephatt, R. J. *Chem. Commun.* 1998, 1055-1062.

[110] Vicente, J.; Chicote, M.-T.; Abrisqueta, M.-D. *Organometallics* 1997, *16*, 5628-5636.

[111] Yam, V. W.-W.; Choi, S. W.-K.; Cheung, K.-K. *Organometallics* 1996, *15*, 1734-1739.

[112] Payne, N. C.; Ramachandran, R.; Puddephatt, R. J. *Can. J. Chem.* 1995, *73*, 6-11.

[113] Payne, N. C.; Ramachandran, R.; Treurnicht, I.; Puddephatt, R. J. *Organometallics* 1990, *9*, 880-882.

[114] Jia, G.; Puddephatt, R. J.; Scott, J.; Vittal, J. J. *Organometallics* 1993, *12*, 3565-3574.

[115] Vicente, J.; Chicote, M.-T.; Abrisqueta, M.-D.; Jones, P. G. *Organometallics* 2000, *19*, 2629-2632.

[116] Müller, T. E.; Choi, S. W.-K.; Mingos, D. M. P.; Murphy, D.; Williams, D. J.; Yam, V. W.-W. *J. Organomet. Chem.* 1994, *484*, 209-224.

[117] Yam, V. W.-W.; Choi, S. W.-K. *J. Chem. Soc. Dalton Trans.* 1996, 4227-4232.

[118] Irwin, M. J.; Vittal, J. J.; Puddephatt, R. J. *Organometallics* 1997, *16*, 3541-3547.

[119] Hunks, W. J.; MacDonald, M.-A.; Jennings, M. C.; Puddephatt, R. J. *Organometallics* 2000, *19*, 5063-5070.

[120] McArdle, C. P.; Irwin, M. J.; Jennings, M. C.; Puddephatt, R. J. *Angew. Chem. Int. Ed.* 1999, *38*, 3376-3378.

[121] Alejos, P.; Coco, S.; Espinet, P. *New J. Chem.* 1995, *19*, 799-805.

[122] Altmann, M.; Bunz, U. H. F. *Angew. Chem. Int. Ed.* 1995, *34*, 569-571.

[123] Irwin, M. J.; Jia, G.; Payne, N. C.; Puddephatt, R. J. *Organometallics* 1996, *15*, 51-57.

[124] Whittall, I. R.; Humphrey, M. G.; Samoc, M.; Luther-Davies, B. *Angew. Chem. Int. Ed. Engl.* 1997, *36*, 370-371.

[125] Vicente, J.; Chicote, M.-T.; Abrisqueta, M.-D.; Ramírez de Arellano, M. C. *Organometallics*, 2000, *19*, 2968-2974.

[126] Forward, J. M.; Fackler, J. P., Jr.; Assefa, Z. *Optoelectronic Properties of Inorganic Compounds*; Roundhill, D. M., Fackler, J. P., Jr. (Eds.); Plenum Press: New York, 1999; pp. 195–239.
[127] Yam, V. W.-W.; Lo, K. K.-W. *Chem. Soc. Rev.* 1999, *28*, 323-334.
[128] Hong, X.; Cheung, K.-K.; Che, C.-M. *J. Chem. Soc. Dalton Trans.* 1994, 1867-1871.
[129] Yam, V. W.-W.; Choi, S. W.-K.; Cheung, K.-K. *J. Chem. Soc. Dalton Trans.* 1996, 3411-3415.
[130] Tzeng, B.-C.; Lo, W.-C.; Che, C.-M.; Peng, S.-M. *J. Chem. Soc. Chem. Commun.* 1996, 181-182.
[131] Xiao, H.; Weng, Y.-X.; Peng, S.-M.; Che, C.-M. *J. Chem. Soc. Dalton Trans.* 1996, 3155-3157.
[132] Yam, V. W.-W.; Lo, K. K.-W.; Wong, K. M.-C. *J. Organomet. Chem.* 1999, *578*, 3-30.
[133] Whittall, I. R.; Humphrey, M. G.; Houbrechts, S.; Persoons, A. *Organometallics* 1996, *15*, 5738-5745.
[134] Naulty, R. H.; Cifuentes, M. P.; Humphrey, M. G.; Houbrechts, S.; Boutton, C.; Persoons, A.; Heath, G. A.; Hockless, D. C. R.; Luther-Davies, B.; Samoc, M. *J. Chem. Soc. Dalton Trans.* 1997, 4167-4174.
[135] Tullius, T. D. *Comprehensive Supramolecular Chemistry*; Atwood, J. L.; Davies, J. E. D.; MacNicol, D. D.; Vögtle, F.; Suslick, K. S. (Eds.), Pergamon Press: Oxford, 1996.
[136] Onitsuka, K.; Yamamoto, S.; Takahashi, S. *Angew. Chem. Int. Ed.* 1999, *38*, 174-176.
[137] Wade, K. *Transition Metal Clusters*; Johnson, B. F. G. (Ed.), Wiley: New York, 1980, Chap. 3.
[138] Evans, D. G.; Mingos, D. M. P. *J. Organomet. Chem.* 1982, *232*, 171-190.
[139] Hoffmann, R. *Angew. Chem. Int. Ed. Engl.* 1982, *21*, 711-724.
[140] Miessler, G. L.; Tarr, D. A. *Inorganic Chemistry*; Prentice Hall: New Jersey, 2004, Chap. 15.
[141] Wong, W.-Y.; Choi, K.-H.; Lu, G.-L.; Lin, Z. *Organometallics* 2002, *21*, 4475-4481.
[142] Coates, G. E.; Parkin, C. *J. Chem. Soc.* 1962, 3220-3226.
[143] Bruce, M. I.; Horn, M. I.; Matisons, J. G.; Snow, M. R. *Aust. J. Chem.* 1984, *37*, 1163-1170.
[144] Cross, R. J.; Davidson, M. F.; McLennan, A. J. *J. Organomet. Chem.* 1984, *265*, C37-C39.
[145] Wong, W.-Y.; Choi, K.-H.; Lu, G.-L.; Shi, J.-X.; Lai, P.-Y.; Chan, S.-M. *Organometallics* 2001, *20*, 5446-5454.
[146] Fabbri, D.; Lombardo, M.; Trombini, C.; Vassura, I. *Appl. Organomet. Chem.* 1995, *9*, 713-718.
[147] Li, P.; Ahrens, B.; Choi, K.-H.; Khan, M. S.; Raithby, P. R.; Wilson, P. J.; Wong, W.-Y. *CrystEngComm.* 2002, *4*, 405-412.
[148] Rais, D.; Mingos, D. M. P.; Vilar, R.; White, A. J. P.; Williams, D. J. *Organometallics* 2000, *19*, 5209-5217.
[149] Gutierrez-Puebla, E.; Vegas, A.; Garcia-Blanco, S. *Acta Crystallogr. B* 1978, *34*, 3382-3384.
[150] Hoskins, B. F.; Robson, R.; Sutherland, E. E. *J. Organomet. Chem.* 1996, *515*, 259-260.
[151] Hartbaum, C.; Roth, G.; Fischer, H. *Eur. J. Inorg. Chem.* 1998, 191-202.
[152] Ghosh, I.; Mishra, R.; Poddar, D.; Mukherjee, A. K. *Chem. Commun.* 1996, 435-436.

[153] Chao, H.-Y.; Lu, W.; Li, Y.; Chan, M. C.-W.; Che, C.-M.; Cheung, K.-K.; Zhu, N. *J. Am. Chem. Soc.* 2002, *124*, 14696-14706.
[154] Lu, W.; Xiang, H.-F.; Zhu, N.; Che, C.-M. *Organometallics*, 2002, *21*, 2343-2346.
[155] Che, C.-M.; Chao, H.-Y.; Miskowski, V. M.; Li, Y.; Cheung, K.-K. *J. Am. Chem. Soc.* 2001, *123*, 4985-4991.
[156] Wong, W.-Y.; Liu, L.; Shi, J.-X. *Angew. Chem. Int. Ed.* 2003, *42*, 4064-4068.
[157] Sebald, A.; Wrackmeyer, B. *Spectrochim. Acta, Part A* 1982, *38*, 163-173.
[158] Faville, S. J.; Henderson, W.; Mathieson, T. J.; Nicholson, B. K. *J. Organomet. Chem.* 1999, *580*, 363-369.
[159] Pyykkö, P. *Chem. Rev.* 1997, *97*, 597-636.
[160] Batsanov, S. S. *J. Chem. Soc. Dalton Trans.* 1998, 1541-1546.
[161] Wong, W.-Y.; Lu, G.-L.; Liu, L.; Shi, J.-X.; Lin, Z. *Eur. J. Inorg. Chem.* 2004, 2066-2077.
[162] Bunz, U. H. F. *Chem. Rev.* 2000, *100*, 1605-1644.
[163] Hertel, D.; Setayesh, S.; Nothofer, H. G.; Scherf, U.; Mullen, K.; Bässler, H. *Adv. Mater.* 2001, *13*, 65-70.
[164] Monkman, A. P.; Burrows, H. D.; Hartwell, L. J.; Horsburgh, L. E.; Hamblett, I.; Navaratnam, S. *Phys. Rev. Lett.* 2001, *86*, 1358-1361.
[165] Romanovskii, Y. V.; Gerhard, A.; Schweitzer, B.; Scherf, U.; Personov, R. I.; Bassler, H. *Phys. Rev. Lett.* 2000, *84*, 1027-1030.
[166] Andell, O.; Goddard, R.; Holle, S.; Jolly, P. W.; Krüger, C.; Tsay, Y. H. *Polyhedron* 1989, *8*, 203-209.
[167] Fischer, B.; Boersma, J.; Koten, G. V.; Smeets, W. J. J.; Spek, A. L. *Organometallics* 1989, *8*, 667-672.
[168] Channareddy, S.; Linti, G.; Nöth, H. *Angew. Chem. Int. Ed. Engl.* 1990, *29*, 199-201.
[169] Helm, S. W.; Linti, G.; Noth, H.; Channareddy, S.; Hofmann, P. *Chem. Ber.* 1992, *125*, 73-86.
[170] Suravajjala, S.; Polam, J. R.; Porter, L. C. *Organometallics* 1994, *13*, 37-42.
[171] Hon, Z.; Fujita, A.; Yamazaki, H.; Wakatsuki, Y. *Chem. Commun.* 1998, 669-670.
[172] Curtis, M. D.; Messerle, L.; D'Errico, J. J.; Solis, H. E.; Barcelo, I. D.; Butler, W. M. *J. Am. Chem. Soc.* 1987, *109*, 3603-3616.
[173] D'Errico, J. J.; Curtis, M. D. *J. Am. Chem. Soc.* 1983, *105*, 4479-4480.
[174] Lau, C. S.-W.; Wong, W.-T. *J. Chem. Soc. Dalton Trans.* 1999, 607-613.
[175] Schottenberger, H.; Wurst, K.; Buchmeiser, M. R. *J. Organomet. Chem.* 1999, *584*, 301-309.
[176] Justin Thomas, K. R.; Lin, J. T.; Lin, Y.-Y.; Tsai, C.; Sun, S.-S. *Organometallics* 2001, *20*, 2262-2269.
[177] Pschirer, N. G.; Ciurtin, D. M.; Smith, M. D.; Bunz, U. H. F.; zur Loye, H.-C. *Angew. Chem. Int. Ed.* 2002, *41*, 583-585.
[178] Ciurtin, D. M.; Pschirer, N. G.; Smith, M. D.; Bunz, U. H. F.; zur Loye, H.-C. *Chem. Mater.* 2001, *13*, 2743-2745.
[179] Ribou, A.-C.; Wada, T.; Sasabe, H. *Inorg. Chim. Acta* 1999, *288*, 134-141.
[180] Justin Thomas, K. R.; Lin, J. T.; Lin, H.-M.; Chang, C.-P.; Chuen, C.-H. *Organometallics* 2001, *20*, 557-563.

In: Focus on Organometallic Chemistry Research
Editor: Marin A. Cato, pp. 53-78
ISBN 1-59454-495-6

Chapter 2

Organometallic Intermediates Formed in the Reaction of Ethers with Potassium Anions

Andrzej Stolarzewicz and Zbigniew Grobelny
Institute of Materials Science, University of Silesia, 40-007 Katowice, Poland

Adalbert Maercker
Institut für Organische Chemie, Universität Siegen, D-57068 Siegen, Germany

Abstract

Ethers can be easily cleaved with various electron-transfer reagents, i.e. alkali metals, alkali metal aromatic radical anions, alkalides, and low-valent transition metal compounds. Alkalides (M^-, M^+C) are salts containing metal anions and complexed metal cations (M: Na, K, Rb, and Cs; C: crown ether or cryptand). Some of them, e.g. K^-, K^+(18-crown-6) **1** or K^-, $K^+(15\text{-crown-}5)_2$ **2**, are recently used for the reductive C-O bond cleavage in linear ethers, oxiranes and other cyclic ethers, as oxetane and tetrahydrofuran. Moreover, crown ethers, which are necessary to prepare concentrated solutions of alkalides are also destroyed by metal anions during the self-decomposition process.

Organometallic compounds being intermediate products of these reactions are generally unstable at ambient temperature. They react rapidly with crown ether complexing the metal cation and/or with ether substrate; some of them eliminate unsaturated hydrocarbon, i.e. ethylene or cyclohexene. The first type of the reaction leads to the crown ring cleavage. Potassium tetraethylene glycoxide vinyl ether is its product if 15-crown-5 is used as a ligand.

In the system containing oxiranes the organometallic intermediate can also eliminate potassium alkoxide or potassium phenoxide. Particularly interesting is the reaction of **2** with (benzyloxymethyl)oxirane. The selective scission of the linear ether bond occurs in the latter without the opening of its oxirane ring. Benzylpotassium formed as the intermediate cleaves immediately the same kind of the ether bond in the next (benzyloxymethyl)oxirane molecule. Bibenzyl and potassium glycidoxide are the final reaction produts. The alkoxide obtained can be applied to the synthesis of glycidyl ethers,

glycidyl esters, crown-like oligomers with alkoxide groups as well as star-shaped and hyperbranched polymers.
An exception to the rule of instability of the organometallic compounds makes triphenylmethylpotassium formed by the linear ether bond cleavage in (triphenylmethoxymethyl)oxirane. It does not take part in any further processes.
The appropriate reaction mechanisms of the formation and decay of organometallic intermediates in the systems containing ethers and potassium alkalides are discussed in this chapter. A qualitative method of identification of the short-lived species is proposed.

1. Introduction

Among organometallic compounds basing on main groups of metals, organoalkali metal ones are especially interesting for chemists working not only in science but also in industry, because many products are still made using cheap alkali metals.

Organoalkali metal compounds can be formed by the usual insertion of a metal into a carbon-halogen bond as well as by the cleavage of the bonds between carbon and other leaving groups, even of less reactive bonds like those of amines and ethers [1]. These processes are easily mediated with various electron-transfer reagents. They involve alkali metals and low-valent transition metal compounds, e.g. $TiCp_2Cl$ [2-7]. The former are applied in the form of a dispersion (as M^0) [8-16] or in the ionic form, as alkali metal aromatic radical anions ($Ar^{\bullet -}$, M^+) [17-20] or alkalides (M^-, M^+C), i.e. salts possessing alkali metal anions and cations complexed by a ligand C [21-24].

The study of the reduction of aryl methyl ethers by Li, Na, K, and Cs in tetrahydrofuran or 1,2-dimethoxyethane confirms that radical anions are obtained in such systems. Then, they undergo unimolecular fragmentation, which can take place both by Ar-O or CH_3-O bond scission [25]. In the case of 1,2,3-trimethoxybenzene the methoxy group in the 2-position can be regioselectively removed by electron-transfer from alkali metals and replaced with a variety of electrophiles [14].

The reaction of oxiranes with alkali metals in protic solvents, e.g. ammonia or amines leads to the formation of simple alkoxides [26,27]. 2-Metalloalkoxides are obtained in this case as intermediates. Most of them are stable only at low temperatures. If the reaction is performed in aprotic solvents, such as tetrahydrofuran, various products are formed depending on the nature of the metal and the structure of the oxirane [28]. Deoxygenation to olefins is the main reaction in the case of lithium. The rearrangement to carbonyl compounds, reduction to alcohols, and formation of dimeric products occur when oxiranes are treated with sodium.

Bartmann [20] synthesized some stable 2-metalloalkoxides using salts of aromatic radical anions, e.g. lithium biphenylide or lithium naphthalenide as well as their potassium or magnesium analogues. It was possible to convert them into appropriate derivatives in the reaction with some electrophiles in relatively high yields. Interesting reactions of oxiranes with lithium in the presence of a catalytic amount of 4,4'-di-*tert*-butylbiphenylide have been presented by Yus [29-32] in several review articles. All the reactions could be rationalized by a mechanism involving an initial single electron transfer.

Metal anions of alkalides have been for the first time identified unequivocally with the NMR technique [33] in 1975. They were then successfully applied to the reduction of alkenes

[34-37], alkines [38], alkyl halides [39,40], silyl halides [41], ketones [42,43], esters [44-48] and alcohols [49]. Alkali metal anions behave as two-electron-transfer reagents. The reaction occurs in two steps. At first M^- transfers one electron to an acceptor molecule. Then, the resulting M^0 transfers the second electron and becomes M^+ [50].

The present chapter deals with the organometallic compounds formed as intermediate products in the reaction of alkalides with ethers. Two alkalides, i.e. K^-, K^+(18-crown-6) **1** and K^-, K^+(15-crown-5)$_2$ **2** were selected for the study. The structure of **1** differs from that of **2** [51]. 18-Crown-6 (18C6) forms a flat 1 : 1 complex with the potassium cation whereas 15-crown-5 (15C5) a sandwich 2 : 1 one does. It strongly influences the stability of these alkalides as well as of their solutions. Tetrahydrofuran solution of **2** is a particularly stable system from the midst of known alkalide solutions. Its ^{39}K NMR spectrum consists of two signals; the wide signal derives from K^+(15C5)$_2$ and the narrow one from K^- [52]. The concentration of metal anions in this solution at 20 °C was found to be constant for 1 h, followed by [K^-] decrease during 24 h. On the contrary, the decomposition of potassium solution in tetrahydrofuran containing 18-crown-6 took place gradually within about 1.5 h at 25 °C [53].

The alkalides were prepared by dissolution of metallic potassium in tetrahydrofuran containing the appropriate crown ether. In most of the cited in this chapter works, the alkalide solution was then introduced into the substrate solution in the same solvent at ambient temperature while mixing. The process was conducted in the apparatus presented in Ref. [40]. The ether substrate was used in the excess in relation to alkalide. In other cases, i.e. at the excess of alkalide or during a simple titration of alkalide with an ether solution until discoloration of the reaction mixture, which usually undergoes at the molar ratio of reagents close to 1 : 1, may provide wrong conclusions. The metal anions usually react then with products obtained initially [49]. The latter can disappear and new ones are formed. It makes difficulties by the explanation of the real reaction mechanism. However, in all the cases a characteristic for potassium anions deep blue colour vanishes immediately at the end of reaction. It indicates that all K^- are converted into K^+ [54].

In separate experiments the alkalide solution was allowed to stay until its discoloration. In that system potassium anions reacted slowly with crown ether complexing the potassium cation and with ethereal solvent.

2. Organopotassium Compounds Formed from Linear Ethers

The process starts with the transfer of one electron from K^- of alkalide to the ether molecule (Scheme 1). The potassium atom (K^0) and ether radical anion are the initial reaction products. The charged species is unstable and undergoes unimolecular fragmentation giving a radical and an anion by mesolytic cleavage of the C-O bond in the homolytic or heterolytic mode [55,56]. In the next step the second electron is transferred from K^0 to the radical. It results in an unstable intermediate organopotassium compound with more or less ionic character. Then, this intermediate takes part in various very rapid reactions. Potassium alkoxide and a hydrocarbon are the end products of the process.

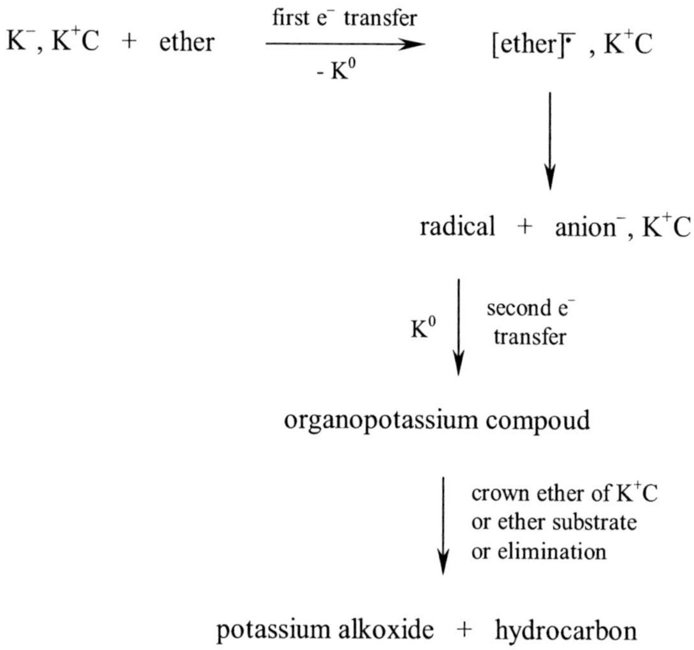

Scheme 1.

The first report on the reaction of linear ethers with alkalides concerned the regioselectivity of the C-O bond cleavage in aromatic ethers by **1** [48]. A problem of organometallic intermediates was not discussed in that case. Alkalide **2** in tetrahydrofuran solution was then mainly used in other studies involving reactions of ethers because, as mentioned, this system has been found to be significantly more stable than **1** [52].

Table 1 shows formulae of the organopotassium compounds formed from various linear ethers by K^- of **2** as well as the direction of their further reactions.

Most of them react with one of two crown ether molecules complexing the potassium cation of the second reaction product. For example, butylpotassium being intermediate product of the ether bond cleavage in butyl vinyl ether opens the 15-crown-5 ring of potassium acetaldehyde enolate (Scheme 2) [58]. Potassium tetraethylene glycoxide vinyl ether and butane are products of the reaction. Surprisingly, this reaction is rather slow in comparison to most of the potassium anion reactions known till now.

Benzyl phenyl ether decomposes in the presence of K^- by CH_2-O bond scission (Scheme 3) [60]. It results in potassium phenoxide and benzyl radical. The latter does not dimerize but receives one electron from K^0. Benzylpotassium formed reacts then with the ether substrate being in the excess. Bibenzyl and potassium phenoxide are obtained in this case.

Table 1. Organopotassium intermediates formed in the reaction of linear ethers with 2. $[Ether]_o > [2]_o$. The direction of the C-O bond cleavage with K^- of 2 is marked by .⟶

No.	Linear ether substrate	Organopotassium intermediate product	Intermediate reacts with	Ref.
1	Ph−O↑Me	Me^-, K^+	15C5	[24,57]
2	Ph−O↑Et	Et^-, K^+	15C5	[24,57]
3	Ph−O↑$CH=CH_2$	$CH_2=CH^-$, K^+	15C5	[58]
4	Ph−O↑$CH_2CH=CH_2$	$CH_2=CHCH_2^-$, K^+	15C5	[59]
5	Ph−O↑CH_2Ph	$PhCH_2^-$, K^+	ether substrate	[60]
6	$CH_2=CHCH_2$−O↑CH_2Ph	$PhCH_2^-$, K^+ [1/]	ether substrate	[61]
7	$CH_2=CHCH_2$↑O−CH_2Ph	$CH_2=CHCH_2^-$, K^+	ether substrate	[61]
8	$CH_2=CH$−O↑Bu	Bu^-, K^+	15C5	[58]
9	$CH_2=CH$↑O−CH_2CH_2OBu	$CH_2=CH^-$, K^+	15C5	[58]
10	$CH_2=CH$−O↑CH_2CH_2OBu	$BuOCH_2CH_2^-$, K^+	eliminates ethylene	[58]
11	$CH_2=CH$↑O−$CH_2CH_2O(CH_2CH$:	$CH_2=CH^-$, K^+	15C5	[58]
12	$CH_2=CH$−O↑$CH_2CH_2O(CH_2CH$:	$Me(OCH_2CH_2)_2OCH_2CH_2^-$, K^+	eliminates ethylene	[58]

[1/] the main reaction product

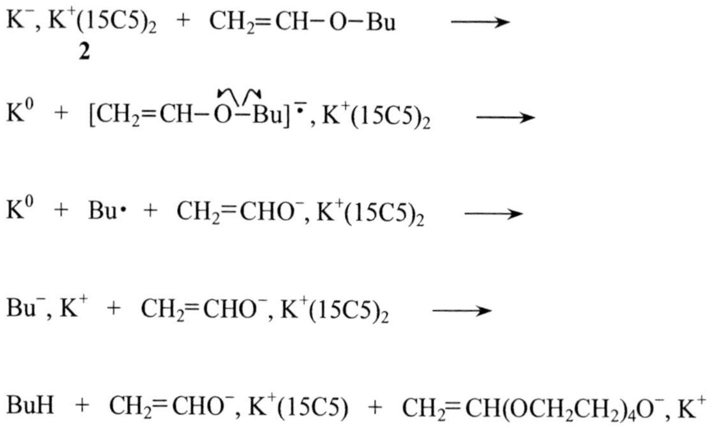

Scheme 2.

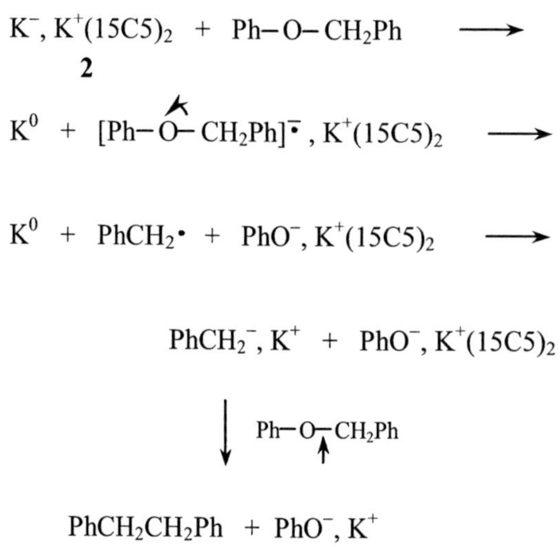

Scheme 3.

Various products of allyl ethers cleavage were presented in the literature so far. Angelo [17] found that the reaction of sodium naphthalenide, i.e. the single-electron-transfer reagent, with allyl ethyl ether resulted in biallyl (in 30 % yield) as the main product. Ethanol was not isolated. Biallyl (40 %) and phenol (60 %) were found in the case of allyl phenyl ether. It was suggested that the cleavage of the ether bond between the allyl carbon and the oxygen atom took place in the first step of the process and then allyl radicals formed underwent dimerization giving biallyl. Biallyl was also obtained by Weber [62] in the reaction of allyl methyl ether with sodium, being another single-electron-transfer reagent [13], in the presence of trimethylchlorsilane. The mechanism of this reaction was not proposed. On the other hand, Eisch [63] obtained allyllithium from allyl phenyl ether using the excess of lithium in tetrahydrofuran at –10 °C. However, Crandall [64] reported that allyl benzyl ether under the influence of lithium in tetrahydrofuran was cleaved to benzyllithium and lithium allyloxide. Allyllithium was not observed in the reaction mixture.

Products of the reaction of **2** with allyl phenyl ether indicate that allyl radicals formed initially do not dimerize but rather receive the second electron from K^0 giving allylpotassium [59]. The latter reacts then with crown ether similarly to other alkylpotassium compounds formed from selected phenyl ethers (Table 1). It seems that the rate of electron-transfer reaction is much faster then the rate of dimerization. Biallyl was not found in the reaction mixture.

It is interesting that the ether bond is opened in both the possible positions in the reaction of K^- with allyl benzyl ether. The cleavage occurs between the benzyl group and oxygen atom as well as between the allyl group and oxygen (Scheme 4) [61]. It results in the formation of two organometallic compounds, namely benzylpotassium and allylpotassium. They react then immediately with the ether substrate. In this case the C-O bond is cleaved only in one position, i.e. between the benzyl group and oxygen. The final reaction products involve potassium allyloxide, potassium benzyloxide, bibenzyl and 3-butenylbenzene. The crown ring opening was not observed.

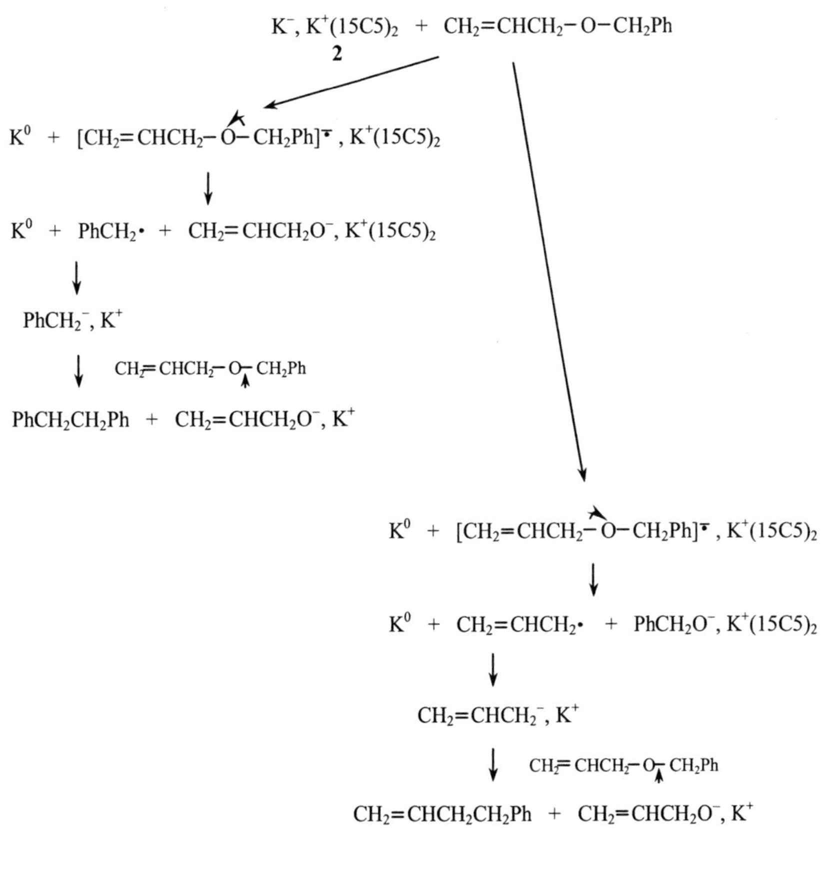

Scheme 4.

The reaction mechanism concerning ethers with one or more CH_2CH_2O units is also a complex one [58]. Organopotassium intermediates containing the CH_2CH_2O group or groups eliminate ethylene forming potassium alkoxides. Potassium enolate of acetaldehyde is one of the final products of this reaction. It undergoes exclusively *O*-benzylation after treating with benzyl bromide, i.e. it behaves rather as an alkoxide than as organopotassium compound. Scheme 5 shows the reaction of **2** with ethylene glycol butyl vinyl ether.

In general, all these processes with only one exception occurred very fast. It was impossible to isolate the corresponding organometallic compounds from the reaction mixture. Their direct identification was found to be unsuccessful. Attempts to obtain their volatile derivatives, e.g. by methylation or benzylation of the reaction mixture proved usually abortive [23]. Therefore, their intermediate existence was proven by the analysis of the end products forming with the participation of crown ether or a substrate as well as by elimination of ethylene. It was assumed that they could not be obtained otherwise.

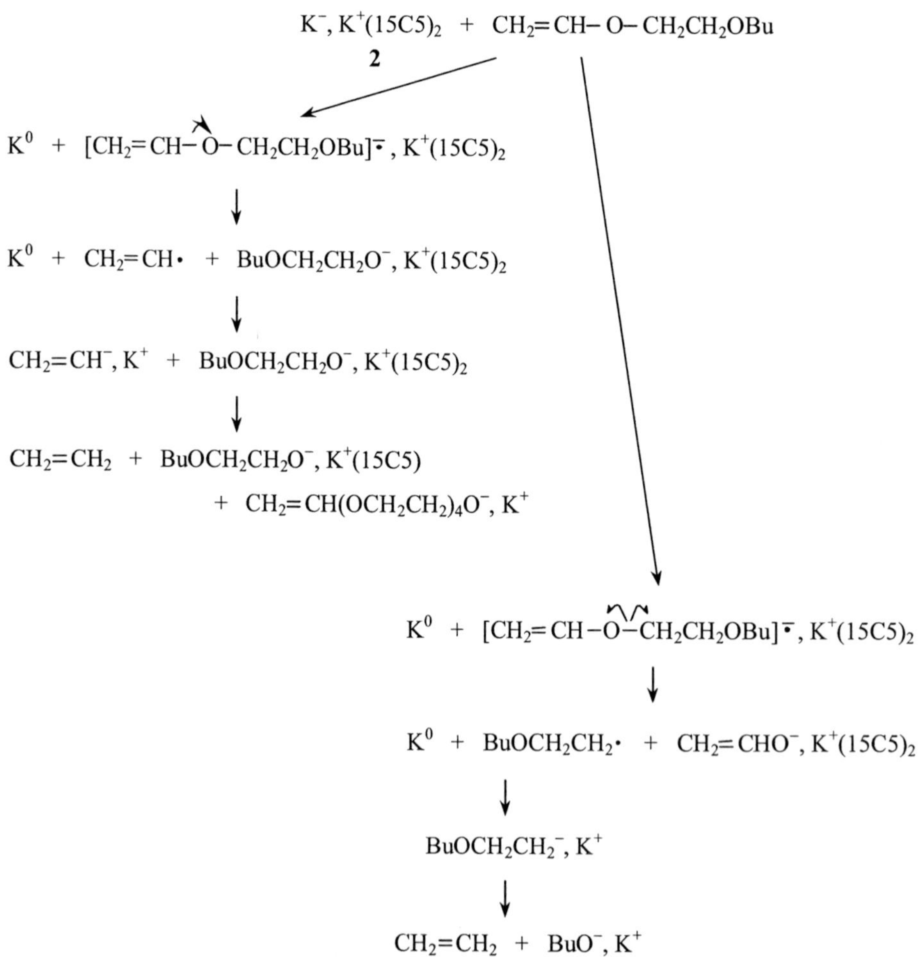

Scheme 5.

3. Organopotassium Intermediates Derived from Monosubstituted Oxiranes

The structure of organopotassium compounds formed during the reaction of **2** with selected oxiranes is presented in Table 2. The greater part of them is a product of the oxirane ring opening. It can occur in the β-position (**3**, **5** – **7**, **10**) or in the α-position (**4**, **8**, **11**) or in both the positions with the prevailing β-opening (**13** – **16**). The resulting potassium 2-potassioalkoxides possess the potassium cation complexed with two 15-crown-5 molecules. On the other hand, **9**, **12** and **17** obtained by the cleavage of the linear ether bond present in the oxirane molecule are not alkoxides and their potassium cations are not complexed. However, both the cyclic and linear ether bonds are opened by heterolytic mode in all the studied systems.

Table 2. Organopotassium intermediates formed in the reaction of monosubstituted oxiranes with 2. [Oxirane]$_o$ > [2]$_o$

No.	Oxirane substrate	Organopotassium intermediate product	Intermediate reacts with	Ref.
1	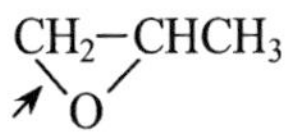	K^+, $^-CH_2CH(CH_3)O^-$, $K^+(15C5)_2$ **3**	15C5	[40]
2		K^+, $^-CH(CH{=}CH_2)CH_2O^-$, $K^+(15C5)_2$ **4**	15C5	[65]
3	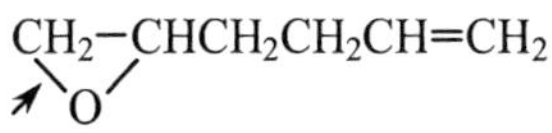	K^+, $^-CH_2CH(CH_2CH_2CH{=}CH_2)O^-$, $K^+(15C5)_2$ **5**	15C5	[65]
4	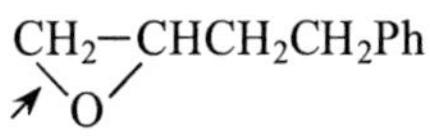	K^+, $^-CH_2CH(CH_2CH_2Ph)O^-$, $K^+(15C5)_2$ **6**	15C5	[66]
5	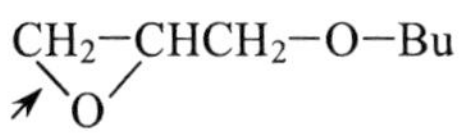	K^+, $^-CH_2CH(CH_2OBu)O^-$, $K^+(15C5)_2$ **7**	15C5	[59]
6	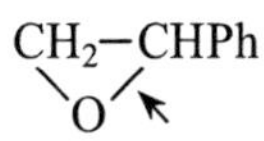	K^+, $^-CH(Ph)CH_2O^-$, $K^+(15C5)_2$ **8**	oxirane substrate	[66]
7	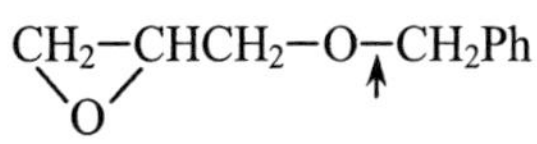	$PhCH_2^-$, K^+ **9**	oxirane substrate	[60]
8	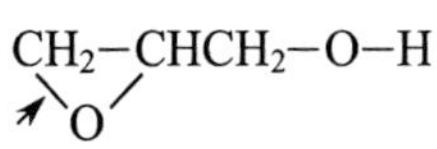	K^+, $^-CH_2CH(CH_2OH)O^-$, $K^+(15C5)_2$ **10**	oxirane substrate	[49]

Table 2. Continued

No.	Oxirane substrate	Organopotassium intermediate product	Intermediate reacts with	Ref.
9	CH_2-CHCH_2Ph O	CH_2Ph $K^+, {}^-CHCH_2O^-, K^+(15C5)_2$ **11**	15C5 and oxirane substrate	[66]
10	$CH_2-CHCH_2-O-CH_2CH=CH_2$ O	$CH_2=CHCH_2^-, K^+$ **12**	15C5 [1] and oxirane substrate	[59]
11	$CH_2-CHCH_2-O-CH_2CH_2Cb$ [2] O	$CH_2OCH_2CH_2Cb$ $K^+, {}^-CH_2CHO^-, K^+(15C5)_2$ **13** [3]	15C5 and oxirane substrate	[67]
12	$CH_2-CHCH_2-O-CH_2CH_2Cb$ O	$CH_2OCH_2CH_2Cb$ $K^+, {}^-CHCH_2O^-, K^+(15C5)_2$ **14**	undergoes β-elimination	[67]
13	CH_2-CHCH_2-O-Ph O	CH_2OPh $K^+, {}^-CH_2CHO^-, K^+(15C5)_2$ **15** [3]	undergoes γ-elimination	[68]
14	CH_2-CHCH_2-O-Ph O	CH_2OPh $K^+, {}^-CHCH_2O^-, K^+(15C5)_2$ **16**	undergoes β-elimination	[68]
15	$CH_2-CHCH_2-O-C(Ph)_2-Ph$ O	$Ph-C^-(Ph)_2, K^+$ **17**	is inert	[60]

[1] 15C5 complexes the potassium cation of the second reaction product, i.e. potassium glycidoxide
[2] Cb denotes the carbazolyl group
[3] the main reaction product

All organopotassium intermediates, except **17**, undergo further rapid reactions. Carbanions of compounds **3** – **7** react exclusively with one of two 15-crown-5 molecules complexing their own counter-ion. Potassium tetraethylene glycoxide vinyl ether and an appropriate alkoxide, i.e. potassium isopropoxide [40], potassium 3-butene-1-oxide [65], potassium 5-hexene-2-oxide [65], potassium 4-phenylbutane-2-oxide [66], and potassium 3-butoxypropane-2-oxide [59] are the reaction products, respectively. Their formation from organopotassium compounds **3**, **5** – **7**, i.e. those obtained by the oxirane ring opening in the β-position, is shown in Scheme 6. The presented path concerns the intramolecular reaction, however, an intermolecular one is also possible.

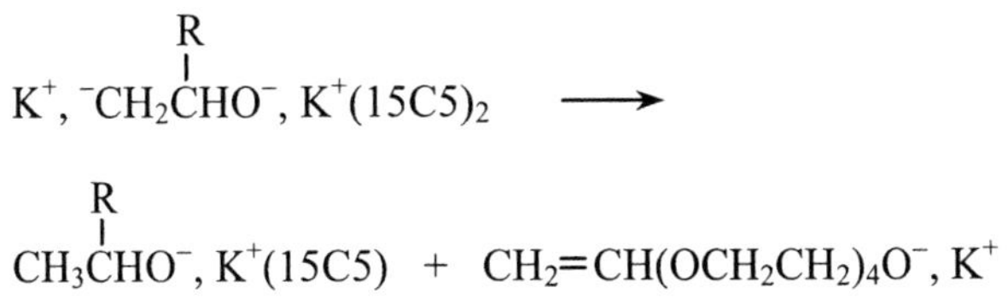

where R:
CH_3 (**3**), $CH_2CH_2CH{=}CH_2$ (**5**), CH_2CH_2Ph (**6**), CH_2OBu (**7**)

Scheme 6.

The opening of 15-crown-5 ring resulting in potassium tetraethylene glycoxide vinyl ether was reported for the first time in the reaction of **2** with methyloxirane or methyl iodide [40], and then with phenoxymethyloxirane [68] and oxetane [22]. In all these processes organopotassium intermediates attacked the crown ether molecule. For example, potassium 2-potassio-1-methylethoxide was one of them in the case of methyloxirane [40]. The mechanism of its formation is presented in Scheme 7.

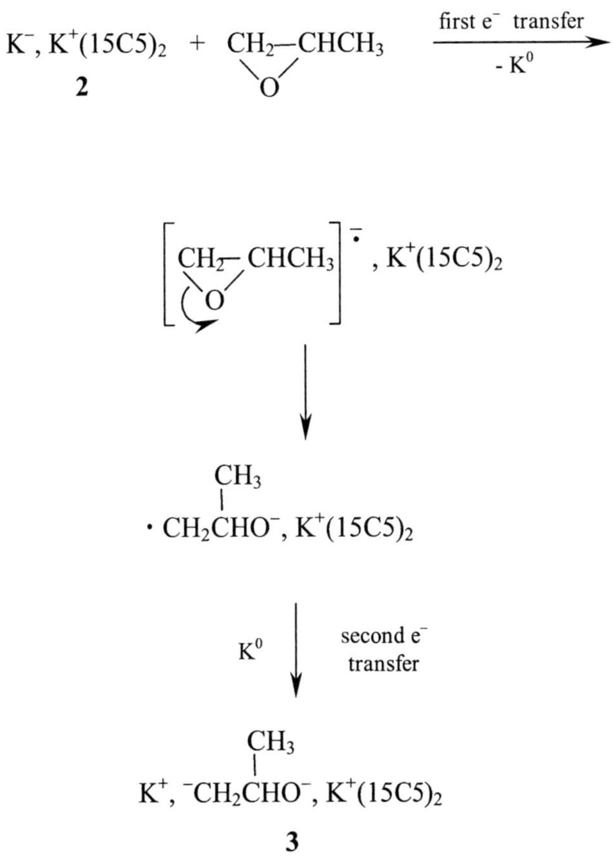

Scheme 7.

Radical anion of methyloxirane and K^0 are formed after the transfer of one electron from K^- of **2** to the oxirane molecule. This radical anion is unstable and decomposes by mesolytic cleavage of the cyclic C-O bond. The resulting radical receives the second electron from K^0 giving the organopotassium product.

The mechanism different from that one was presented for the reaction of **1** with methyloxirane [69] as well as styrene [34], methyl methacrylate [35], and isoprene [36]. The proton exchange was assumed to occur between the organometallic intermediate and tetrahydrofuran used as the solvent. However, no evidence was given for such a hypothesis as well as for the role of the second product expected to form in this case. Thus, it was not clear what a compound really took part in the reaction with the organometallic intermediate in those systems.

A more detailed study of this problem shows that 18-crown-6 is opened during the reaction of **1** with methyloxirane [23]. Potassium isopropoxide and potassium pentaethylene glycoxide vinyl ether are found as the reaction products. Their methylated or benzylated derivatives are obtained when the reaction mixture is treated with methyl iodide or benzyl bromide, respectively. The ring opening of 18-crown-6 is also observed in the reaction of potassium anions of **1** with oxetane and methyl iodide. It allows to state that organopotassium intermediates are present in these systems, too, and they easily opened up the crown ring. Moreover, it would indicate that potassium oligoethylene glycoxides vinyl ethers could be obtained in such a way from unsubstituted crown ethers. However, further experiments showed that this is not a rule.

The next three organopotassium compounds, i.e. **8** - **10** react exclusively with the appropriate oxirane substrate. The mechanism of this process depends on the structure of the oxirane molecule. 8 opens the oxirane ring of phenyloxirane in the β-position which leads to a dimeric product, i.e. dipotassium 1,3-diphenylbutane-1,4-dioxide (Scheme 8) [69]. It is known that 2 as well as other electron-transfer reagents, e.g. metallic sodium [28], lithium naphthalenide or potassium biphenylide [20,66] caused the α-scission in phenyloxirane. Only one exception to this rule was described in the literature [69], i.e. both the α- and the β-ring opening was reported in phenyloxirane when using **1**. However, such a result might be due to the method of substrate delivery. Phenyloxirane was introduced gradually into the blue alkalide solution until its discoloration and this could cause side reactions between potassium anions and primary reaction products [49,60,66]. Experiments performed with an excess of phenyloxirane, i.e. using the method presented in Ref. [66], showed that only the α-opening occurs in the presence of **1** [24]; the final reaction products were the same as in the system containing **2**.

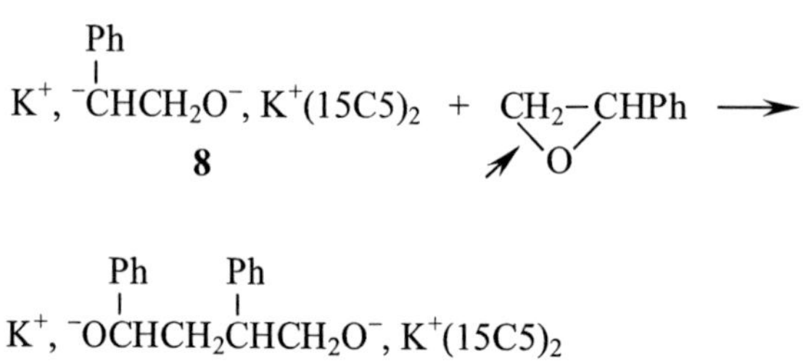

Scheme 8.

Intermediate **9** cleaves unexpectedly one of the linear ether bonds of (benzyloxymethyl)oxirane. It results in potassium glycidoxide and bibenzyl [60]. The cited work demonstrates for the first time the possibility of the selective scission of the linear ether bond by K^- (Scheme 9) or by organopotassium compound (Scheme 10) without the opening of the oxirane ring present in the same molecule. Crown ether does not take part in this process as the reagent. Its role is limited to the complexation of the potassium cation in spite of the formation of an organometallic intermediate. Moreover, the formation of potassium glycidoxide under such conditions was earlier not described in the literature. This alkoxide can be used for the synthesis of some glycidyl ethers or glycidyl esters. Its reaction with aliphatic or aromatic halides in tetrahydrofuran occurs very fast at ambient temperature. Potassium glycidoxide is also an inimer [70], i.e. it can both initiate the polymerization as well as take part in this process as a monomer. Its crown-like oligomers with alkoxide active centers are multifunctional initiators. Their use leads to new star-shaped and hyperbranched polymers with a cyclic core [71,72].

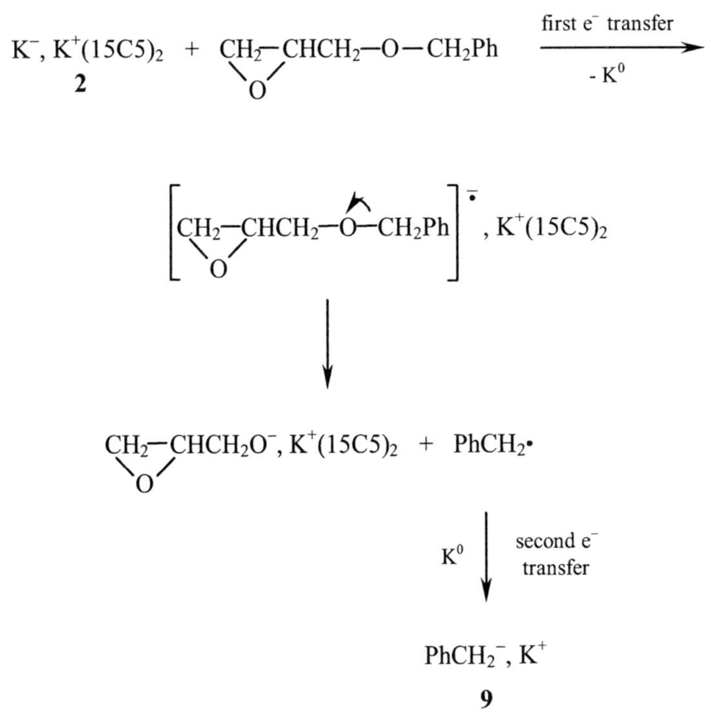

Scheme 9.

Compound **10** reacts with the hydroxyl group of glycidol giving two alkoxides, i.e. potassium 1-hydroxypropane-2-oxide and potassium glycidoxide (Scheme 11) [49]. It is worth noting that in this system **2** opens both the oxirane ring and the O-H bond of glycidol. The rule presented earlier in the literature [73] that the activity of one of those groups was connected with the inactivity of another, is not valid in the studied system.

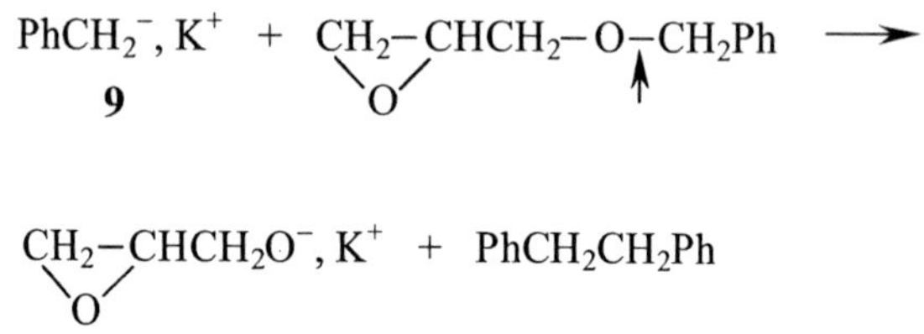

Scheme 10.

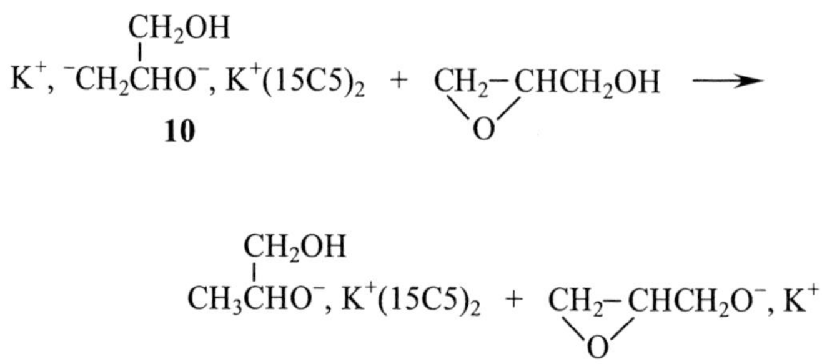

Scheme 11.

Compounds **11** – **13** react both with 15-crown-5 and with the appropriate oxirane. The first reaction results in the crown ring opening, i.e. in potassium tetraethylene glycoxide vinyl ether. The mechanism of the second one depends on the oxirane structure. The reaction of **11** with (phenylmethyl)oxirane [66] is not a simple addition to the oxirane ring known in the literature [74]. It seems to be a proton exchange process similar to the chain transfer reaction to the monomer observed during the anionic polymerization of methyloxirane [75,76]. Two alkoxides, i.e. potassium 3-phenylpropoxide and potassium 3-phenylallyloxide are the end products of this reaction (Scheme 12).

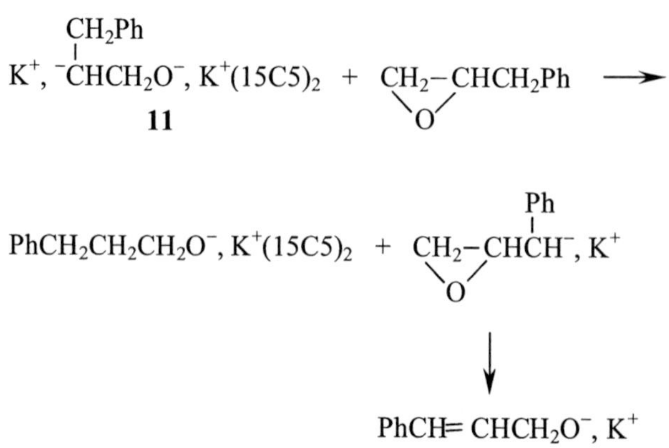

Scheme 12.

The oxirane ring opening in the β-position occurs in the reaction of **12** with (allyloxymethyl)oxirane. It results in potassium alkoxide with two double bonds (Scheme 13) [59]. It is noteworthy that **12** was obtained by the cleavage of the linear ether bond in (allyloxymethyl)oxirane, i.e. in a similar process as observed for (benzyloxymethyl)oxirane. Potassium glycidoxide was the second reaction product in both the systems.

The behaviour of **13** is more intricate. β-Elimination takes place during its reaction with (2-(9-carbazolyl)ethoxymethyl)oxirane giving potassium 3-(9-carbazolyl)ethoxy-2-oxide, potassium glycidoxide and 9-vinylcarbazole (Scheme 14) [67].

Compound **14** does not react with the oxirane substrate but undergoes β-elimination to potassium allyloxide and potassium 2-carbazolylethoxide (Scheme 15) [67]. Similarly, γ-elimination of **15** leads to potassium cyclopropoxide and potassium phenoxide (Scheme 16) [68]. **16** decomposes by β-elimination to potassium allyloxide and potassium phenoxide (Scheme 17). The elimination reaction is preferred in this case probably due to the formation of the phenoxide anion stabilized by resonance.

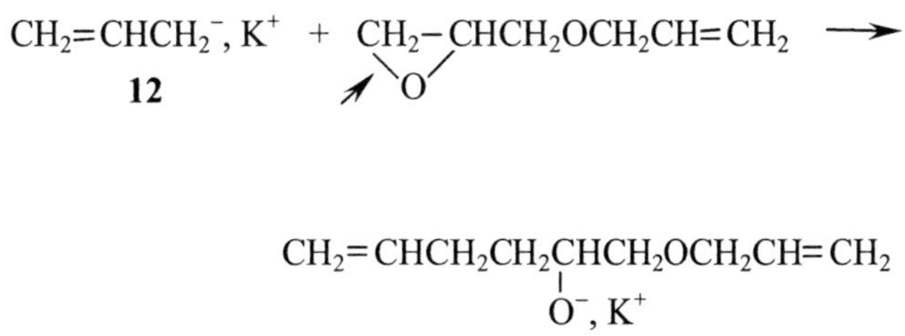

Scheme 13.

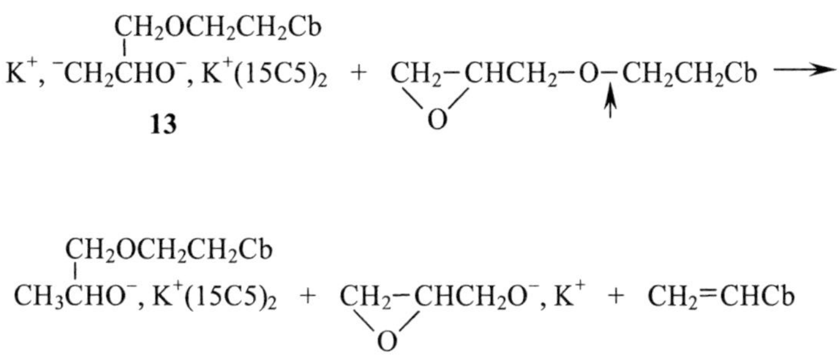

Scheme 14.

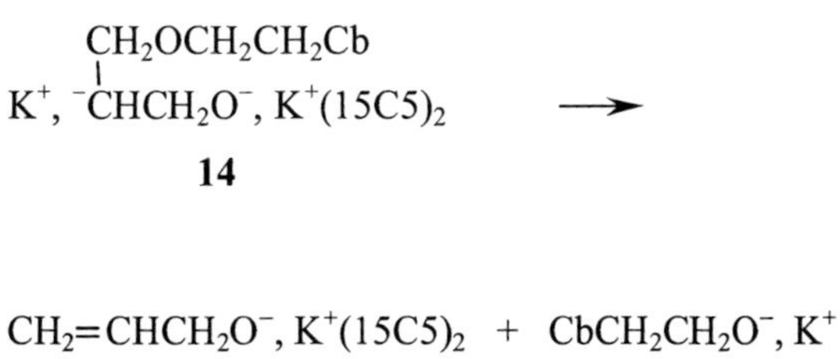

Scheme 15.

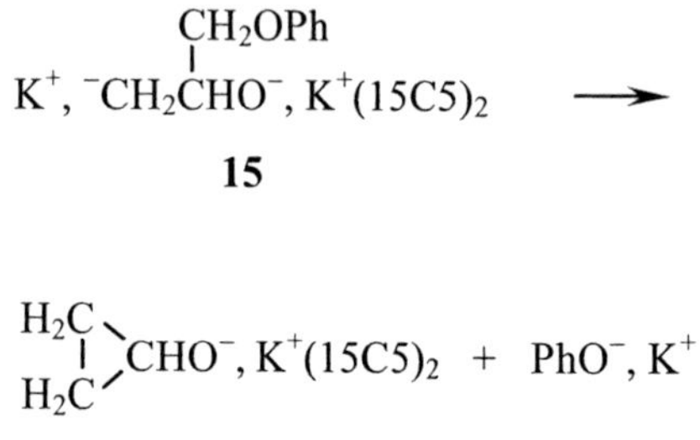

Scheme 16.

$$K^+, {}^-CH(CH_2OPh)CH_2O^-, K^+(15C5)_2 \longrightarrow$$

16

$$CH_2{=}CHCH_2O^-, K^+(15C5)_2 + PhO^-, K^+$$

Scheme 17.

As mentioned above, from the midst of organopotassium intermediates only **17** was found to be stable one because of a high resonance stability of the triphenylmethyl anion [60]. This anion is formed after the second electron transfer, i.e. from K^0 to the triphenylmethyl radical. It is worth noting that the latter obtained in the reaction of metallic zinc with triphenylmethyl chloride has been found to exist in an equilibrium with a dimeric form, which even prevailed in that system [77,78]. Such a phenomenon was not observed by the use of **2**.

The proton exchange between the 2-potassioalkoxide anion and crown ether being in the complex with the counter-ion is much faster in the greater part of the systems under study than the addition of alkoxide to the oxirane substrate. In other systems the proton exchange with the oxirane molecule or the resonance stability of the reaction product is the main factor responsible for the behaviour of organopotassium compounds.

4. Organopotassium Compounds Obtained from Ethers of a Larger Cycle

Cyclic ethers with a ring larger than that of the oxirane ring, e.g. oxetane, tetrahydrofuran and 2-methyltetrahydrofuran also undergo the ring opening reaction under the influence of potassium anions of **2**. It leads to various potassium potassioalkoxides (Table 3). They are unstable and react exclusively with crown ether causing its ring opening. The final reaction products involve potassium tetraethylene glycoxide vinyl ether and appropriate potassium alkoxide.

Table 3. Organopotassium intermediates formed in the reaction of cyclic four- and five-membered ethers with 2. $[Ether]_o > [2]_o$

No.	Cyclic ether substrate	Intermediate organopotassium product	Intermediate reacts with	Ref.
1	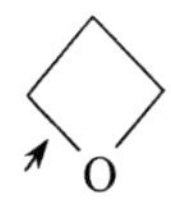 1/	$K^+, {}^-CH_2O^-, K^+(15C5)_2$	15C5	[22]
		$K^+, {}^-CH_2CH_2CH_2O^-, K^+(15C5)_2$	15C5	
2	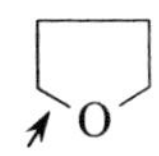 2/	$K^+, {}^-CH_2CH_2CH_2CH_2O^-, K^+(15C5)_2$	15C5	[81]
3	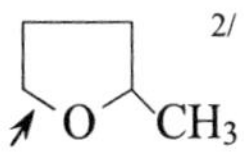 2/	$K^+, {}^-CH_2CH(CH_3)O^-, K^+(15C5)_2$	15C5	[81]

1/ two organopotassium intermediates are formed
2/ used as the solvent of **2**

Potassium 3-potassiopropoxide was the main intermediate in the case of oxetane (Scheme 18) [22]. The same results were observed both at +25 °C and at –20 °C. In the contrary, it has been earlier reported that this organopotassium alkoxide obtained in the reaction of oxetane with another alkalide, i.e. with **1**, was stable at –20 °C and it could be protonated or methylated [79]. No reaction with crown ether was postulated in that system.

Attention was separately paid to tetrahydrofuran used in the studied processes as a solvent. That cyclic ether had been known to decompose to ethylene and potassium acetaldehyde enolate under the influence of butylpotassium [80]. However, none of its expected decomposition products were found in each reaction mixture conducted in tetrahydrofuran solution [23]. It means that tetrahydrofuran was stable in the presence of organometallic compounds in all the systems containing a linear ether substrate or an oxirane or oxetane one. An explanation of this phenomenon is possible for a higher reactivity of crown ether in comparison with tetrahydrofuran. It may originate from a specific activating effect of several oxygen atoms in the polydentate ligand as well as from the fact that crown ether takes part in the reaction in the form of a complex with the potassium cation. The complexation of the cation by crown molecules would tend to make the C-H bond more acidic [40]. Summarising, the reaction with crown ether and other reactions occurring with the participation of organometallic intermediates are faster than that with the solvent.

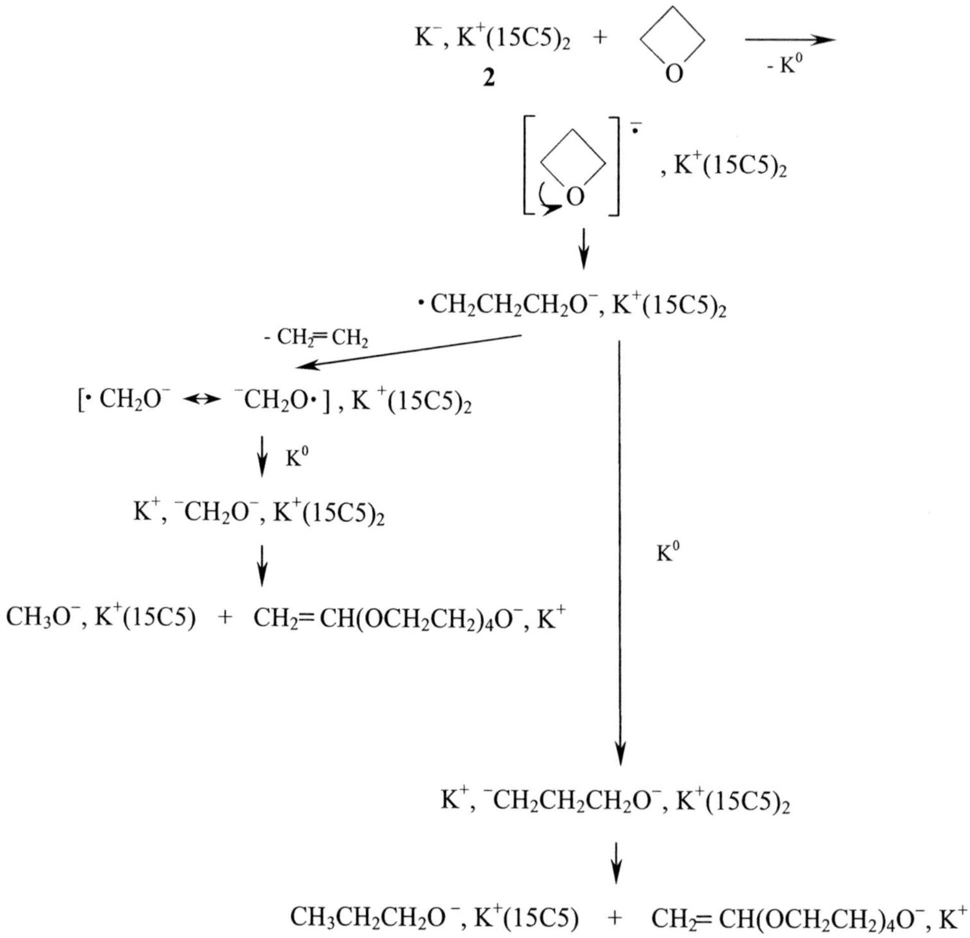

Scheme 18.

However, this situation changes when the blue tetrahydrofuran solution of **1** or **2** is left until its discoloration connected with the self-decomposition of these systems. The presence of benzyl butyl ether in benzylated samples of discoloured solutions shows that the potassium anion cleaves the solvent molecule indeed [81]. The reaction of K^- with tetrahydrofuran leads to an intermediate product, i.e. potassium 4-potassiobutoxide (Scheme 19). The latter opens the crown ether ring forming potassium butoxide and potassium penta- or tetraethylene glycoxide vinyl ether from 18-crown-6 or 15-crown-5, respectively.

The blue solution of **2** was then prepared using 2-methyltetrahydrofuran. In such a case benzyl isopropyl ether was detected after benzylation of the colourless mixture instead of expected benzyl 1-methylbutyl ether. It could indicate that potassium 2-potassio-1-methylethoxide is formed initially in that process with the elimination of ethylene (Scheme 20). Then, the crown ether molecule is opened, which results in potassium isopropoxide and potassium tetraethylene glycoxide vinyl ether.

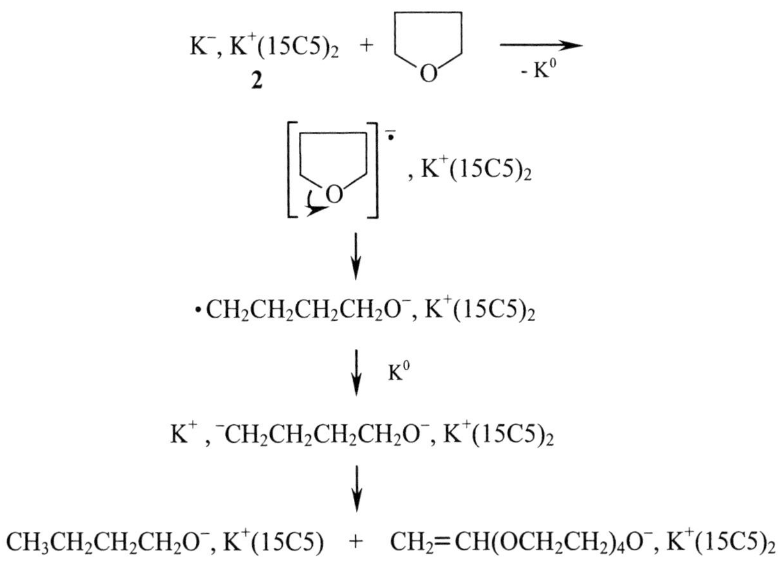

Scheme 19.

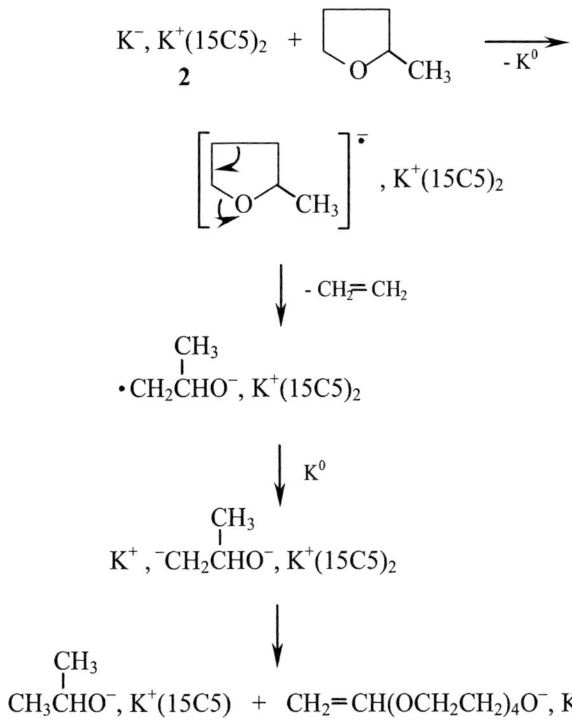

Scheme 20.

Crown ethers, which are necessary to prepare concentrated alkalide solutions, also generate organopotassium alkoxides under the influence of K^-. They are formed during the self-decomposition of alkalide solutions in the first step of the process. Then, they immediately decayed by elimination of an unsaturated hydrocarbon, i.e. ethylene or cyclohexene (Table 4). It results in respective dipotassium glycoxides. However, in some cases further reactions are possible.

Table 4. Organopotassium intermediates formed from crown ethers during the self-decomposition of alkalide tetrahydrofuran solution obtained with the appropriate ligand

No.	Crown ether substrate	Direction of the C-O bond scission with K^-	Intermediate organopotassium product	Intermediate decomposes by elimination of	Ref.
1	15C5	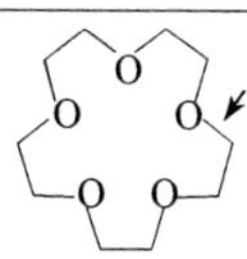	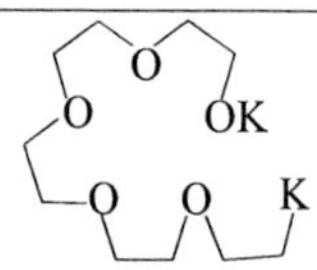	ethylene	[81]
2	CH15C5	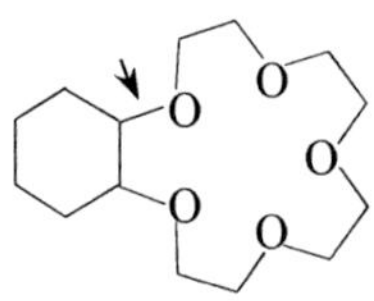	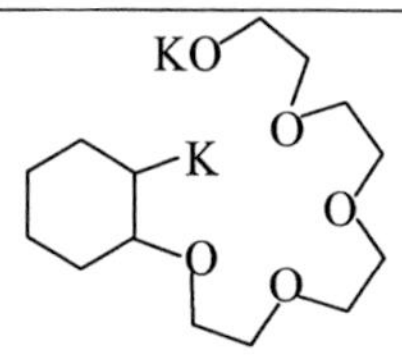	cyclohexene	[82]
3		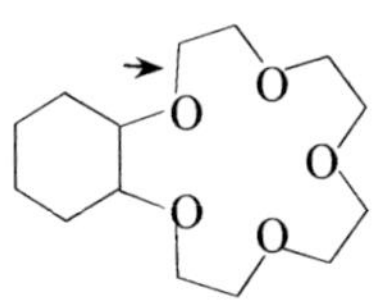	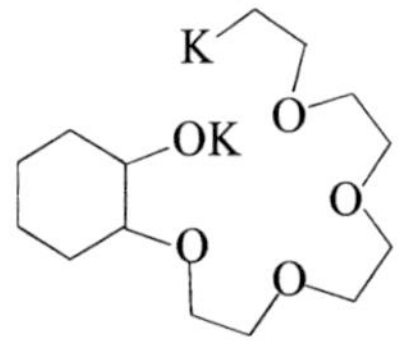	ethylene	[82]
4		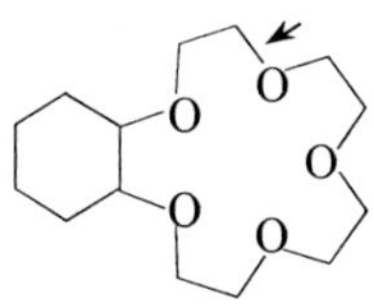	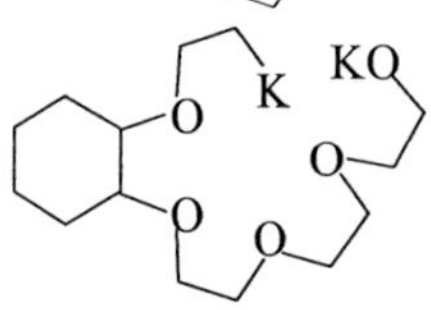	ethylene	[82]
5		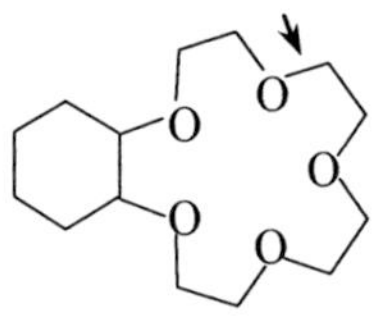	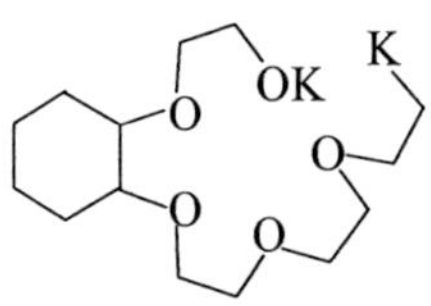	ethylene	[82]
6		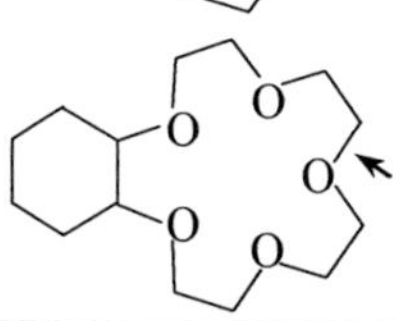	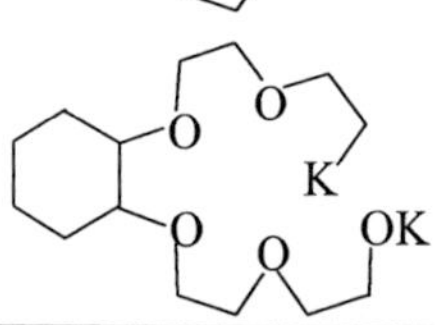	ethylene	[82]
7	DCH15C5	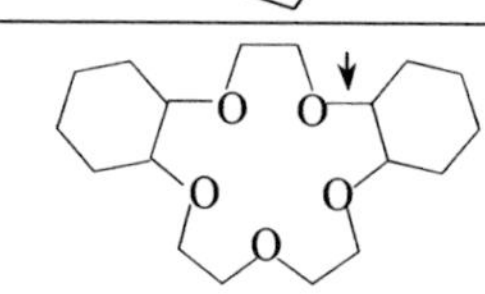	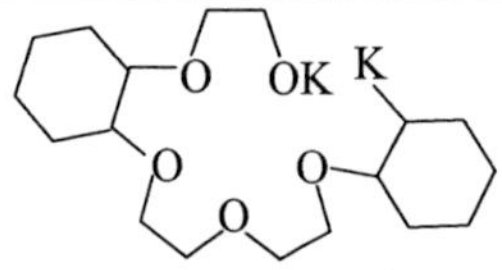	cyclohexene	[82]
8		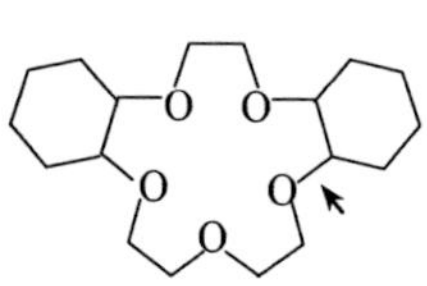	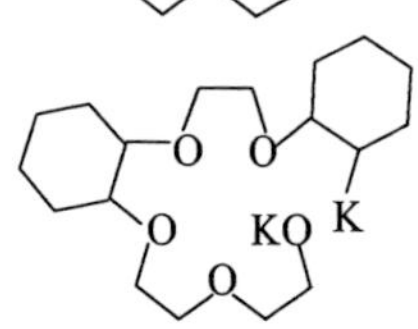	cyclohexene	[82]

Table 4. Continued

No.	Crown ether substrate	Direction of the C-O bond scission with K^-	Intermediate organopotassium product	Intermediate decomposes by elimination of	Ref.
9	18C6	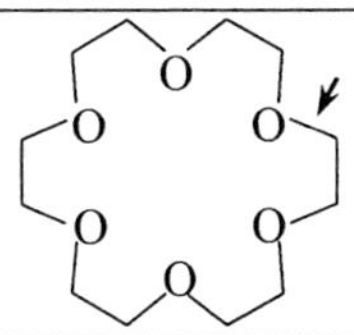	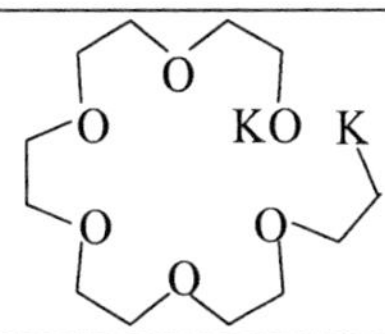	ethylene	[81]
10	CH18C6	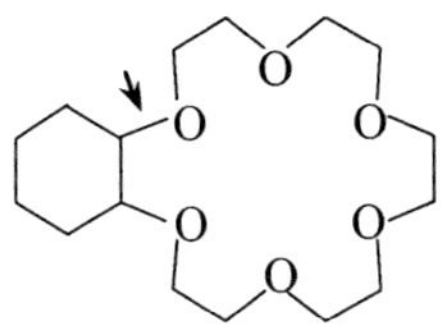	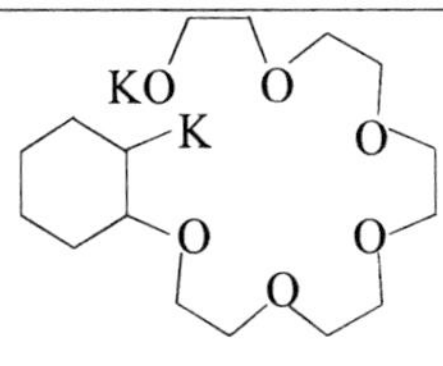	cyclohexene	[82]
11		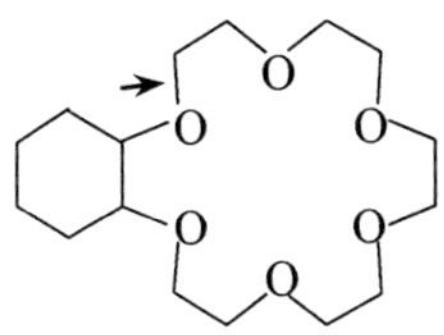	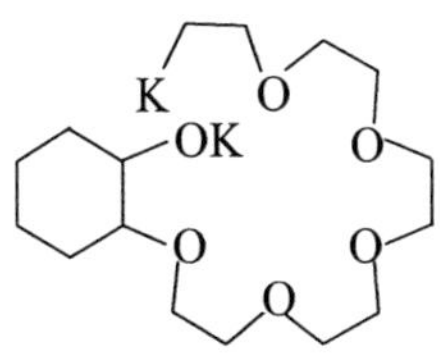	ethylene	[82]
12		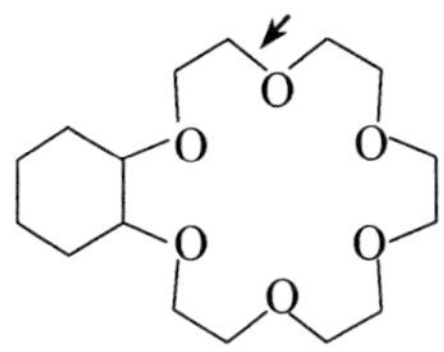	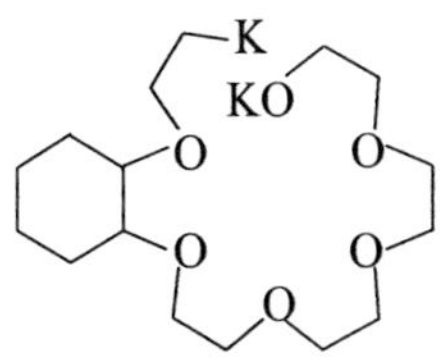	ethylene	[82]
13		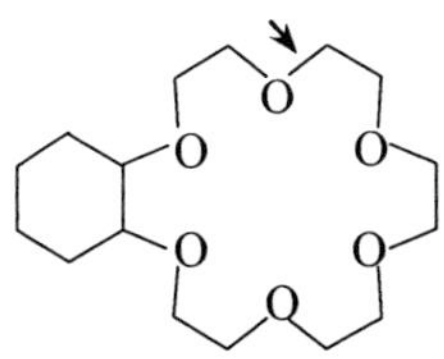	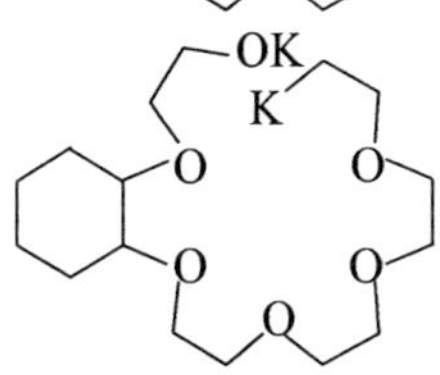	ethylene	[82]
14		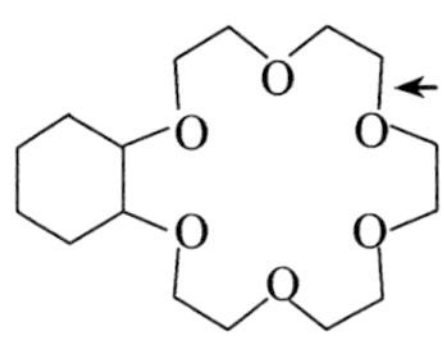	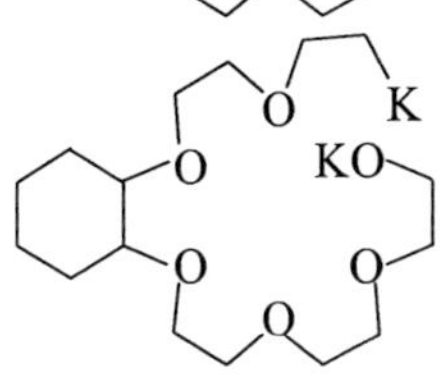	ethylene	[82]
15		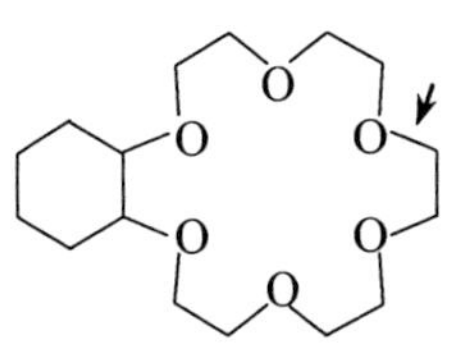	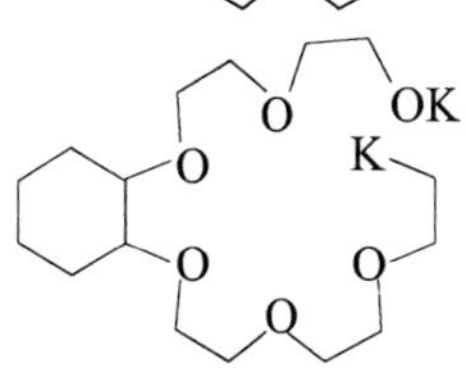	ethylene	[82]
16	DCH18C6	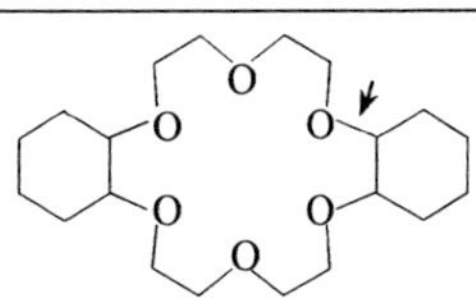	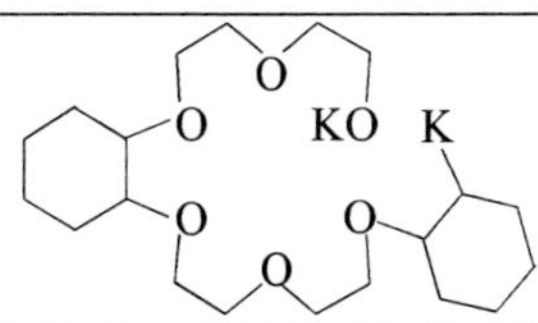	cyclohexene	[82]

For example, the cleavage of the C-O bond in the 15-crown-5 molecule by K^- gives an unstable potassium organometallic alkoxide (Scheme 21) [81]. Dipotassium tetraethylene glycoxide is obtained after the reductive elimination of ethylene. The former can partly decompose under the influence of K^- to dipotassium triethylene glycoxide and potassium 2-potassioethoxide (Scheme 22). This organometallic compound reacts rapidly with crown ether causing its ring opening. The process results in one potassium alkoxide and two glycoxides.

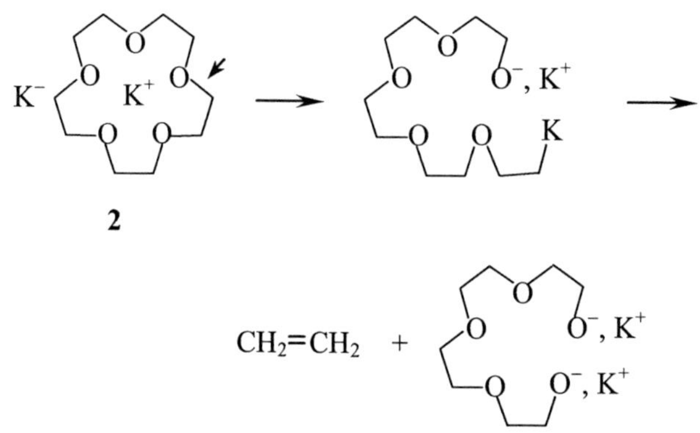

(the second 15C5 molecule was omitted)

Scheme 21.

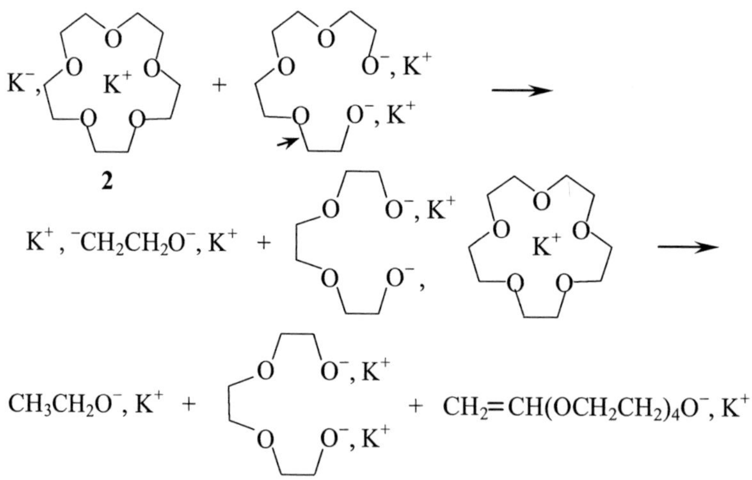

Scheme 22.

Cleavage to different degrees takes place at each of the C-O bonds in cyclohexano-15-crown-5 (CH15C5) of K^-, $K^+(CH15C5)_2$ [82]. Five different potassium potassioalkoxides are intermediate products in this case. However, elimination of ethylene and cyclohexene gives finally only three dipotassium glycoxides. A similar situation was observed for decomposition of K^-, $K^+(CH18C6)$, where CH18C6 denotes the cyclohexano-18-crown-6 molecule. All of the C-O bonds were sensitive for cleavage. Unexpectedly, for analogous dicyclohexano-substituted crown ethers, K^- ruptured only the C-O bonds that link the cyclohexano substituent to the crown ether framework. Indeed, two kinds of potassium potassioalkoxides could be formed from DCH15C5 resulting in the same alkoxide after elimination of cyclohexene. Ethylene was not found as a reaction product in both the systems.

The mechanism of 18-crown-6 decomposition by K^- is until now not enough clear. It seems that the reaction starts with the crown ring opening, which results in potassium pentaethylene glycoxide 2-potassioethyl ether (Table 4, No. 9). It eliminates ethylene giving dipotassium pentaethylene glycoxide [53]. The formation of four other dipotassium glycoxides identified in this system might also occur via organopotassium intermediates [81]. However, no evidence for this supposition was found until now in the literature.

5. Conclusions

Organopotassium compounds formed from the linear and cyclic ethers in the presence of K^- are very unstable. They react immediately mainly with crown ether, which was used for the preparation of alkalide solution. Crown ethers known till now as stable compounds and widely applied to enhance the reaction rate in many chemical processes unexpectedly can act both as activators and as reagents. Their oxacyclic ring is opened with organopotassium compounds. It results in potassium tetraethylene glycoxide vinyl ether for 15-crown-5 and in potassium pentaethylene glycoxide vinyl ether in the case of 18-crown-6. This reaction might have applications as a method helpful for checking of the formation of short-lived organoalkali metal intermediates, i.e. for their qualitative analysis. The proposed method bases on the addition of one of these crown ethers to the reaction mixture and identification of its ring opening product. We have found that it works in 17 processes from 27 described in the literature for 15-crown-5. This makes 63 % probability that the test is successfully conducted and the organopotassium compound is identified as a short-lived intermediate reaction product. However, the negative result of such experiment, i.e. the absence of the expected alkoxide can not be a proof that the process occurs without the participation of organopotassium compounds. In some cases the latter decompose in the reaction with the ether substrate or by elimination.

Organopotassium intermediates formed in the studied systems are precursors of many potassium alkoxides. The latter are known as effective initiators of the anionic polymerization of various cyclic and vinyl monomers, especially in the presence of crown ether. Potassium glycidoxide is particularly interesting among them because it can be used as the reagent in various organic syntheses, e.g. for the preparation of star-shaped polyethers.

References

[1] Houben-Weyl: *Methoden der Organischen Chemie*; Band XIII/1; Thieme-Verlag; Stuttgart, 1970; *ibid.* Band XIII/2a; Thieme-Verlag; Stuttgart, 1973.

[2] Nugent, W.A.; RajanBabu, T.V. *J. Am. Chem. Soc.* 1988, *110*, 8561.

[3] RajanBabu, T.V.; Nugent, W.A.; Beattie, M.S. *J. Am. Chem. Soc.* 1990, *112*, 6408.

[4] RajanBabu, T.V.; Nugent, W.A. *J. Am. Chem. Soc.* 1994, *116*, 986.

[5] Gansäuer, A.; Bluhm, H. *Chem. Rev.* 2000, *100*, 2771.
[6] Gansäuer, A.; Rinker, B. *Tetrahedron* 2002, *58*, 7017.
[7] Chakraborty, T.K.; Das, S. *Tetrahedron Lett.* 2002, *43*, 2313.
[8] Azzena, U.; Denurra, T.; Fenude, G.; Melloni, G.; Rassu, G. *Synthesis* 1989, 28.
[9] Thornton, T.A.; Woolsley, N.F.; Bartak, D.E. *J. Am. Chem. Soc.* 1986, *108*, 6497.
[10] Koppang, M.D.; Woolsley, N.F.; Bartak, D.E. *J. Am. Chem. Soc.* 1984, *106*, 2799.
[11] Patel, K.M.; Baltisberger, R.J.; Stenberg, V.I.; Woolsley, N.F. *J. Org. Chem.* 1982, *47*, 4250.
[12] Bhatt, M.V.; Kulkarni, S.A. *Synthesis* 1983, 249.
[13] Maercker A., *Angew. Chem. Int. Ed. Engl.* 1987, *26*, 972.
[14] Azzena, U.; Denurra, T.; Melloni, G.; Piroddi, A.M. *J. Org. Chem.* 1990, *55*, 5386.
[15] Azzena, U.; Denurra, T.; Melloni, G.; Fenude, E.; Rassu, G. *J. Org. Chem.* 1992, *57*, 1444.
[16] Bowers, K.W. In *Radical Ions*, Kaiser, E.T.; Kevan, L., Eds.; John Wiley & Sons, New York, 1968, p. 232.
[17] Angelo, B. *Bull. Chim. Soc. Fr.* 1966, 1091.
[18] Maslak, P.; Guthrie, R.D. *J. Am. Chem. Soc.* 1986, *108*, 2628 and 2637.
[19] Dorigo, A.; Houk, K.N.; Cohen, T. *J. Am. Chem. Soc.* 1989, *111*, 8976.
[20] Bartmann, E. *Angew. Chem. Int. Ed. Engl.* 1986, *25*, 653.
[21] Jedliński, Z. *Acc. Chem. Res.* 1998, *31*, 55.
[22] Grobelny, Z.; Stolarzewicz, A.; Maercker, A. *J. Organomet. Chem.* 2000, *604*, 283.
[23] Grobelny, Z. *J. Organomet. Chem.* 2003, *677*, 118.
[24] Grobelny, Z. *Eur. J. Org. Chem.* 2004, 2973.
[25] Lazana, M.C.R.L.R.; Franco, M.L.T.M.B.; Herold, B.J. *J. Am. Chem. Soc.* 1989, *111*, 8640.
[26] Smith, J.G. *Synthesis* 1984, 629.
[27] Kaiser, E.M.; Edmonds, C.G.; Grubb, S.D.; Smith, J.W.; Tramp, D. *J. Org. Chem.* 1971, *36*, 330.
[28] Gurudutt, K.N.; Pasha, M.A.; Ravindranath, B.; Srinivas, P. *Tetrahedron* 1983, *40*, 1629.
[29] Yus, M. *Pure Appl. Chem.* 2003, *75*, 1453.
[30] Nájera, C. Yus, M. *Curr. Org. Chem.* 2003, *7*, 1.
[31] Yus, M. *Synlett* 2001, *8*, 1197.
[32] Ramon, D.J.; Yus, M. *Eur. J. Org. Chem.* 2000, 225.
[33] Dye, J.L.; Andrews, C.W.; Ceraso, J.M. *J. Phys. Chem.* 1975, *26*, 3076.
[34] Szwarc, M.; Janeczek, H.; Jedliński, Z. *Macromolecules* 1997, *30*, 4498.
[35] Jedliński, Z.; Janeczek, H.; Bosek, I. *J. Org. Chem.* 1999, *64*, 4956.
[36] Janeczek, H.; Jedliński, Z. *Polymer* 2002, *43*, 7219.
[37] Stolarzewicz, A.; Morejko-Buż, B.; Grobelny, Z.; Neugebauer, D. *Macromol. Symp.* 2002, *184*, 325.
[38] Mathre, D.J.; Guida, W.C. *Tetrahedron Lett.* 1980, *21*, 4773.
[39] Jedliński, Z.; Janeczek, H.; Matuszowicz, A.; Grobelny, Z. *Macromol. Rapid Commun.* 1997, *18*, 529.
[40] Grobelny, Z.; Stolarzewicz, A.; Czaja, M.; Demuth, W.; Maercker, A. *J. Org. Chem.* 1999, *64*, 8990.
[41] Jedliński, Z.; Kurcok, P.; Nozirow, F. *Macromol. Rapid Commun.* 1997, *18*, 483.

[42] Jedliński, Z.; Misiołek, A.; Główkowski, W. *Synlett* 1990, 213.
[43] Jedliński, Z.; Misiołek, A.; Główkowski, W.; Janeczek, H. *Bull. Polish Acad. Sci. Chem.* 1991, 1513.
[44] Barett, A.G.M.; Godfrey, Ch.R.A.; Hollinshead, D.M.; Prokopiou, P.A.; Barton, D.H.R.; Boar, R.B.; Joukhadar, L.; McGhie, J.F.; Misra, S.C. *J. Chem. Soc. Perkin I* 1981, 1501.
[45] Barett, A.G.M.; Prokopiou, P.A.; Barton, D.H.; Boar, J.B.; McGhie, J.F. *J. Chem. Soc. Chem Commun.* 1979, 1173.
[46] Jedliński, Z.; Misiołek, A.; Kurcok, P. *J. Org. Chem.* 1989, *54*, 3242.
[47] Jedliński, Z.; Kowalczuk, M.; Kurcok, P.; Grzegorzek, M.; Ermel, J. *J. Org. Chem.* 1987, *52*, 4501.
[48] Fish, R.H.; Dupon, J.W. *J. Org. Chem.* 1988, *53*, 5230.
[49] Grobelny, Z.; Stolarzewicz, A.; Maercker, A.; Krompiec, S.; Kasperczyk, J.; Rzepa, J.; Frey, H. *J. Organomet. Chem.* 2004, *689*, 2361.
[50] Perrin, C.L.; Wang, J.; Szwarc, M.; *J. Am. Chem. Soc.* 2000, *122*, 4569.
[51] Dye, J.L. *J. Phys. IV* 1991, *1*, 259.
[52] Grobelny, Z.; Stolarzewicz, A.; Sokół, M.; Grobelny, J.; Janeczek, H. *J. Phys. Chem.* 1992, *96*, 5193.
[53] Jedliński, Z.; Stolarzewicz, A.; Grobelny, Z. *Makromol. Chem.* 1986, *187*, 795.
[54] Stolarzewicz, A.; Grobelny, Z; Grobelny, J. *Spectrochim. Acta Part A* 2000, *56*, 1257.
[55] Casado, F.; Pisano, L.; Farriol, M.; Gallards, I.; Marquet, J.; Melloni, G. *J. Org. Chem.* 2000, *65*, 322.
[56] Pisano, L.; Farriol, M.; Asansio, X.; Gallardo, I.; Gonzáles-Lafont, A.; Lluch, J.M.; Marquet, J. *J. Am. Chem. Soc.* 2002, *124*, 4708.
[57] Stolarzewicz, A.; Grobelny, Z.; Kowalczuk, M. *J. Organomet. Chem.* 1995, *492*, 111.
[58] Grobelny, Z.; Stolarzewicz, A.; Maercker, A.; Krompiec, S.; Kasperczyk, J.; Rzepa, J. *J. Organomet. Chem.* 2004, *689*, 1580.
[59] Grobelny, Z.; Stolarzewicz, A.; Maercker, A.; Krompiec, S.; Bieg, T. *J. Organomet. Chem.* 2002, *660*, 133.
[60] Grobelny, Z.; Stolarzewicz, A.; Morejko-Buż, B.; Maercker, A. *J. Organomet. Chem.* 2002, *660*, 6.
[61] Grobelny, Z. unpublished data.
[62] Tzeng, D.; Weber, W.P. *J. Org. Chem.* 1981, *46*, 265.
[63] Eisch, J.J.; Jacobs, A.M. *J. Org. Chem.* 1963, *28*, 2145.
[64] Crandall, J.K.; Clark, A.C. *J. Org. Chem.* 1972, *37*, 4236.
[65] Grobelny, Z.; Stolarzewicz, A.; Morejko-Buż, B.; Maercker, A. *Eur. Polym. J.* 2004, *40*, 409.
[66] Grobelny, Z.; Stolarzewicz, A.; Morejko-Buż, B.; Maercker, A.; Krompiec, S.; Bieg, T. *J. Organomet. Chem.* 2003, *672*, 43.
[67] Grobelny, Z.; Stolarzewicz, A.; Morejko-Buż, B.; Buika, G.; Gražulevičius, J.; Maercker, A. *Eur. Polym. J.* 2002, *38*, 2359.
[68] Grobelny, Z.; Stolarzewicz, A.; Maercker, A.; Demuth, W. *J. Organomet. Chem.* 1999, *590*, 153.
[69] Janeczek, H.; Jedliński, Z. *Pol. J. Appl. Chem.* 1997, *46*, 377.
[70] Müller, A.H.E.; Yan, D.; Wulkow, M. *Macromolecules* 1997, *30*, 7015.

[71] Stolarzewicz, A.; Morejko-Buż, B.; Grobelny, Z.; Pisarski, W.; Lanzendörfer, Ch.; Müller, A. *Rapid Commun. Mass Spectrom.* 2004, *18*, 716.
[72] Stolarzewicz, A.; Morejko-Buż, B.; Grobelny, Z.; Pisarski, W.; Frey, H. *Polymer* 2004, *45*, 7047.
[73] Weigert, W.M.; Kleeman, A.; Schreyer, G. *Chemiker Ztg.* 1975, *99*, 19.
[74] Dworak, A.; Jedliński, Z. *Polymer* 1980, *1*, 93.
[75] Steiner, E.C.; Pelletier, R.R.; Trucks, R.O. *J. Am. Chem. Soc.* 1964, *86*, 4678.
[76] Grobelny, Z.; Stolarzewicz, A.; Neugebauer, D.; Morejko-Buż, B. *Eur. Polym. J.* 2002, *38*, 1065.
[77] Sholle, V.D.; Rozantsev, G.G. *Russ. Chem. Rev.* 1973, *42*, 1011.
[78] McBride, J.M. *Tetrahedron* 1974, *30*, 2009.
[79] Jedliński, Z.; Misiołek, A.; Jankowski, A.; Janeczek, H. *J. Chem. Soc., Chem. Commun.* 1991, 1513.
[80] Lehman, R.; Schlosser, M. *Tetrahedron Lett.* 1984, *25*, 745.
[81] Grobelny, Z.; Stolarzewicz, A.; Maercker, A. *J. Organomet. Chem.* 2001, *628*, 65.
[82] Grobelny, Z.; Stolarzewicz, A.; Morejko-Buż, B.; Bartsch, R.A.; Yamato, K.; Fernandez, F.A.; Maercker, A. *J. Org. Chem.* 2002, *67*, 7807.

In: Focus on Organometallic Chemistry Research
Editor: Marin A. Cato, pp. 79-109
ISBN 1-59454-495-6

Chapter 3

ORGANOMETALLIC COMPOUNDS CONTAINING TELLURIUM LIGANDS: SYNTHESIS REACTIVITY AND STRUCTURES

Ajai K. Singh
Department of Chemistry, Indian Institute of Technology, New Delhi 110016, India

ABSTRACT

Tellurium ligands of different types viz. RR'Te, RTe^-, Te^{2-}, halotellurium ligands, polytelluride ligands and tellurium containing macrocycles have attracted inorganic chemists during last three decades. Many organometallic compounds (mononuclear as well as polynuclear) containing such ligands have been designed. They include metal carbonyls, half sandwich compounds and metal alkyl based derivatives. Tellurium ligands have been found very good building blocks for designing metal carbonyl clusters. Unique structural features observed in solid state as well as in solutions for many such compounds are attractive. The secondary Te····Cl interactions are capable of generating supramolecular assemblies. In this paper the chemistry of organometallic compounds containing Te ligands is reviewed. In general metal tellurium bonds are not very stable as tellurium is noblest metalloid. Some metallic centers like mercury(II) promote the cleavage of Te- C bonds when organotellurium ligands coordinate with them. The greater emphasis in this review is laid on compounds for which single crystal structure determination by X-ray diffraction or detailed NMR studies have been reported. The trends in the reactivity due to the presence of Te ligands are highlighted. The possibilities of employing Te ligands for the further growth of organometallic chemistry are envisaged.

1. INTRODUCTION

The great deal of organometallic chemistry is developed by using phosphines as co-ligands of the organic ones. The known organometallic compounds containing other 'Soft' donors like arsines, thioethers, selenoethers, etc are much less in number. Organometallic compounds containing tellurium ligands were scantly known before 1980, but in last nearly two and a half decade they have become more visible in the literature. However, their total count is still very small. In fact the ligand chemistry of tellurium was known to limited extent up to late seventies probably due to commercial non-availability of a wide variety of tellurium / organotellurium ligands and some misconception that they are air sensitive, foul smelling and toxic. However, during the last 3 decades many publications devoted to these ligands have appeared in the literature which indicates considerable and continuing interest in them. With modern FT-NMR instruments there is a good possibility to use ^{125}Te-NMR to understand the solution behavior of metal–tellurium bond-containing systems. Such an option does not exist for sulfur ligands. This coupled with the fact that metal complexes of organotellurium ligands may be used as single source precursors for MOCVD of II–VI semiconductors, has enhanced interest in the ligand chemistry of tellurium. The other reasons for the currently growing interest are the increasing evidence of enhanced ligating properties of telluroether compared to thioethers and the availability of standardized synthetic routes to these ligands at present. So far no attempt is made to review critically an overall status of organometallic compounds containing tellurium ligands. Some reviews are available which deal with chemistry of some specific type of Te ligands. The ligand chemistry of the polytelluride ion [1] was reviewed in 1994 covering literature up to early 1993. Adamantane like tellurolate bridged cages of zinc group metals were reviewed in 1992 by Dean and Vittal [2]. The chemistry of inorganic halotellurium ligands, e.g. TeX_3^- was compiled by Beck [3] and of tellurometalates $[M_xTe_y]^{2-}$ by Ansari et al. [4]. Similarly ligation of telluroethers and multidentate organotellurium ligands have been reviewed [5-7] and few general reviews on tellurium ligands have appeared [8-11]. Metal carbonyls and cyclopentadienyl carbonyls [12-13] containing tellurium ligands (including bridging ones) and metal tellurolates [14] have also been reviewed. The focus of the present review is to sketch the current overall status of organometallic compounds like half sandwich compounds, carbonyl clusters and metal alkyl / aryls containing tellurium ligands.

2. Tellurium Ligands

The following are the important type of tellurium containing moieties used as ligands.

2.1. Telluroethers

The monodentate acyclic telluroethers like R_2Te, RR'Te, Ar_2Te or ArAr'Te are among the tellurium ligands which have been more extensively coordinated with organometallic species. The bidentate and tridentate telluroethers such as $ArTeCH_2TeAr$, $RTe(CH_2)_3TeR$, *o*-$C_6H_4(TeMe)_2$, *o*-$C_6H_4(CH_2TeR)_2$, $MeC(CH_2TeR)_3$, and $Te(CH_2CH_2CH_2TeR)_2$ are among the tellurium ligands of current interest. The telluroethers containing other donor sites behave as hybrid ligands and are called as hybrid telluroethers. Singh and his coworkers [5, 8, 15] have synthesized and explored many such derivatives. The tellurium analogue of 1,2-bisdiphenylphosphinoethane is yet unknown but closely related hybrid systems $RTeCH_2CH_2X$, where X = SR, NR_2, NH_2 or other organic group which can provide a donor site (suitable for coordination of the ligand in chelating mode) are synthesized and used as ligands. When X has ether oxygen as donor site such systems behave as hemilabile ligands [16-17]. Large number of hybrid organotellurium ligands are synthesized by following two reactions.

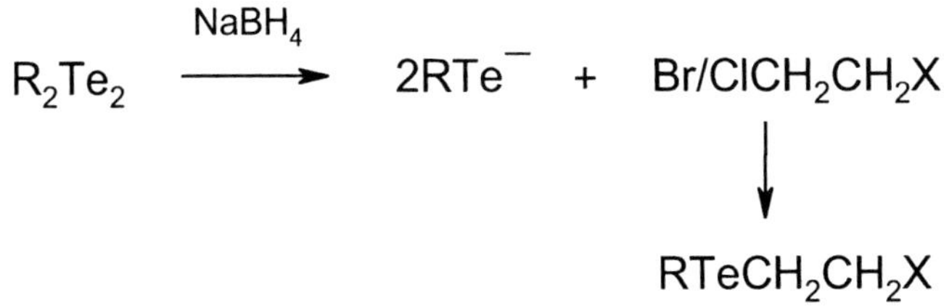

1

2

2.2. Anionic Organotellurium Ligands

The best known example of this class of ligands present in several organometallic compounds is RTe^-. It has been generated in situ from R_2Te_2 or some other organotellurium compound [8]. A bidentate dianionic organotellurium ligand **1** has been generated and identified in its complex [18-19].

1

2.3. Inorganic Tellurium Ligands

The organometallic compounds containing halotellurium ligands like TeX_2 and $[TeX_3]^-$ (Where X = Cl, Br or I), Te^{2-}, polytellurides (e.g. Te_2^{2-} and Te_4^{2-}) and mixed tellurium containing chalcogenides (e.g. $Te_2Se_2^{2-}$) have been reported in the last one and a half decade [8].

Apart from these three major classes of tellurium ligands, diorganoditelluride, cyclic and macrocyclic tellurium donor site containing ligands have also been explored for their ligating properties with organometallic species. The metal-Te bond formed when Te-ligands coordinate with metal center (soft) is generally not very strong. Some centers particularly mercury(II) promotes the cleavage of Te-C bonds when organotellurium ligands coordinate with them. For example complexation of 1-ethylthio-2-[2-thienyltelluro]ethane with mercury(II) bromide results in an unusual cleavage of the ligand and formation of a novel tetrameric bromo bridged mercury complex [(EtS(CH_2)$_2$SEt)HgBr(μ-Br)Hg(Br)(μ-Br)$_2$Hg(Br)(μ-Br)BrHg(EtS(CH_2)$_2$SEt)].2$HgBr_2$ [20a] having weak Hg···Br interactions (Fig 1)

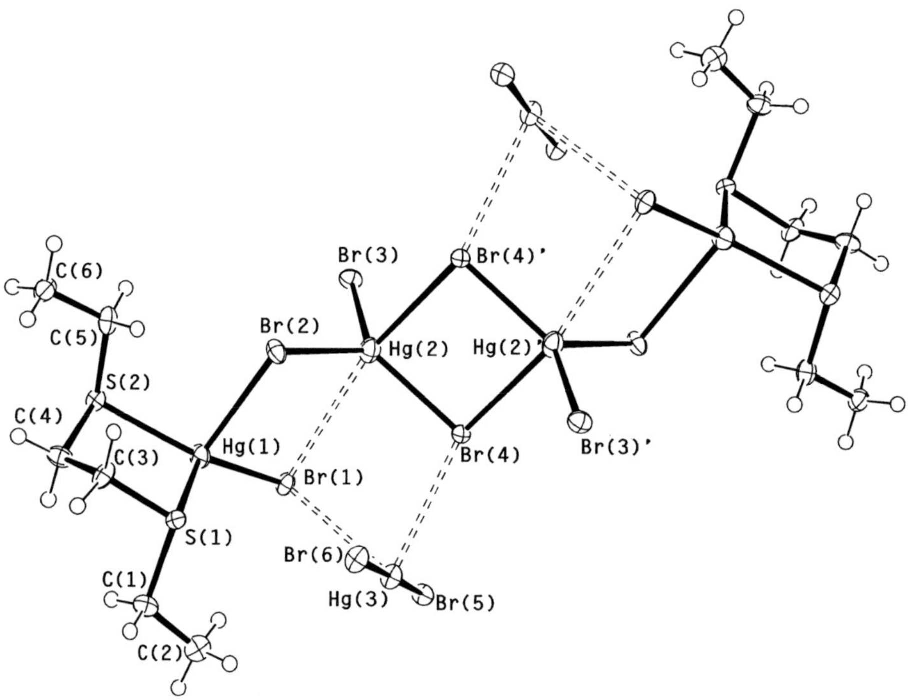

Figure 1. ORTEP Diagram of [(EtS(CH_2)$_2$SEt)HgBr(μ-Br)Hg(Br)(μ-Br)$_2$Hg(Br)(μ-Br)BrHg(EtS(CH_2)$_2$SEt)].2$HgBr_2$

3. Mono- and Di-Nuclear Metal Carbonyls

Metal carbonyls generally offer 'Soft' metal centers, which are preferred by tellurium ligands. Therefore metal carbonyls or their fragments have been reacted with a variety of tellurium ligands. The early reports on tellurium ligand containing mononuclear carbonyls were made in 1962 by Hieber and Kruck [20b]. [$Fe(CO)_4TePh_2$] and [$Mn(CO)_3(TePh_2)_2X$] were synthesized by reacting appropriate metal carbonyl fragments with $TePh_2$ [20b]. In some such reactions $PhTe^-$ was generated and gave species like [$Mn(CO)_4$(μ-TePh)]$_2$. [$Mn(CO)_3L_2Cl$] (where L = phenoxtellurin) was also reported by Hieber and Kruck [20]. In Table 1 the various mononuclear metal carbonyls containing tellurium ligand synthesized upto 1984 are summarized. Metal carbonyl fragments or intermediates containing them (synthesized generally by activation of metal carbonyls e.g. by photon) were generally reacted with the Te ligands for these preparations. The single crystal structure of [Cr(CO)$_5$(tellurourea)] was determined. Cr has octahedral geometry with Cr-Te bond distance 2.765(4) Å.

Table 1. Some mono- and di-nuclear metal carbonyls containing Te ligands reported upto 1984

Metal carbonyl complex	L	Ref.
[Cr(CO)$_5$L] (**2**)	TeEt$_2$	21
[Cr(CO)$_3$$\eta^5$-L] (**3**)	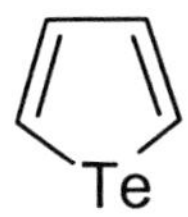	22
[Rh(CO)ClL$_2$] (**4**)	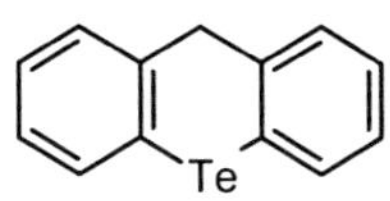 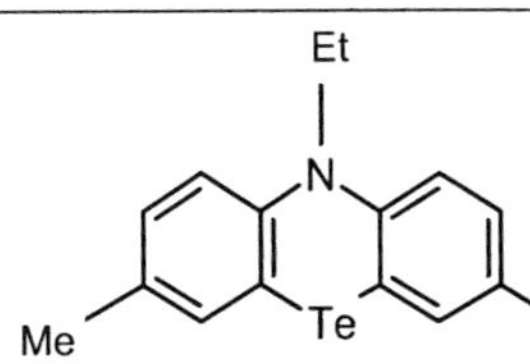	23
[Cr(CO)$_5$L] (**5**)	Te(EMe$_3$)$_2$ E = Ge, Sn Pb	24
[M(CO)$_5$L] (**6**); M= Cr, Mo or W	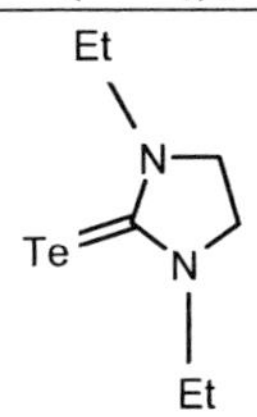 (Tellurourea)	25
[W(CO)$_5$L] (**7**)	Te=CPh$_2$	26
[Cr(CO)$_4$(CNEt$_2$)(L)] (**8**)	PhTe$^-$ (used as Li salt)	27
[Os(CO)$_2$(PPh$_3$)$_2$(Me)(L) (**9**)	MeTe$^-$	28
PPN[Cr(CO)$_5$L] (**10**)	TeH$^-$ (generated from Na$_2$Te)	21
[(CO)$_3$M(μ-Br)$_2$(μ-L)M(CO)$_3$ M = Mn (**11**) or Re (**12**)	Te$_2$Ph$_2$	29, 30
[(CO)$_4$W(μ-L)$_2$W(CO)$_4$](**13a**) [M(CO)$_3$TePh]$_2$ (**13b**) (M = Ru or Fe) [Ru(CO)$_2$(EPh)$_2$]n	PhTe$^-$	31a 31b
[(CO)$_4$Re(μ-L)$_2$Re(CO)$_4$] (**14**)	[TeSnMe$_3$]$^-$	32
[Ph$_4$As][(CO)$_5$W(μ-L)W(CO)$_5$] (**15**)	TeH$^-$	21
Os(CO)$_2$(PPh$_3$)$_2$(η^2-CH$_2$Te) (**15a**)	CH$_2$Te	28

For the synthesis of metal carbonyls containing RTe$^-$ ligands, the oxidative addition of Te species (e.g. R$_2$Te$_2$) to an appropriate metal carbonyl complex or reaction of RTeLi/Na (generated in situ by Te insertion in PhLi / Na or reduction of R$_2$Te$_2$ by sodium borohydride) with the carbonyl complex has been used in most of the cases. Spectroscopic data and properties of such compounds synthesized in seventies and early eighties (Table 1) are extensively reviewed by Gysling [19-10]. Some of the complexes given in Table 1 are unstable e.g. species $[Cr(CO)_5Te(EMe_3)_2]$ (E = Ge, Sn Pb) are not only air and moisture sensitive but decompose at 0 °C in solid state. The terminal TeH^- can be alkylated in $PPN[Cr(CO)_5TeH]$ to the neutral dialkyl telluride analogue (equation 3) in ethanol medium

$$PPN[Cr(CO)_5TeH] + [Et_3O]BF_4 \rightarrow [(OC)_5CrTeEt_2]\mathbf{3}$$

In the last two decades also the interest in complexation reactions of metal carbonyl fragments with tellurium ligands has continued. Photolysis of $Et_4N[V(CO)_6]$ and $TePh_2$ in THF gives $Et_4N[V(CO)_5TePh_2]$. Mono and di-nuclear metal carbonyls containing tellurium ligands reported in literature during last twenty years are summarized in Table 2 with reactants used for synthesis. The crystal structure of **25** shows it be a *fac* isomer in which Fe has nearly octahedral geometry. There is a Te···I weak secondary interaction (Distance 3.423(2)–3.486(2) Å; I---Fe---I angle 83.55(7)°). IR spectrum of **25** has three (CO) bands at 2087, 2042 and 2025 cm^{-1} concurring with its unsymmetrical structure. The **25** is significant in view of the fact that the complexes of telluroethers with Fe(II) were reported earlier [8] to have inherent instability and non-crystalline nature. The compound very similar to **25**, $[(MeTePh)Fe(CO)_3I_2]$ has been prepared by the sequence of reactions given in equation 4.

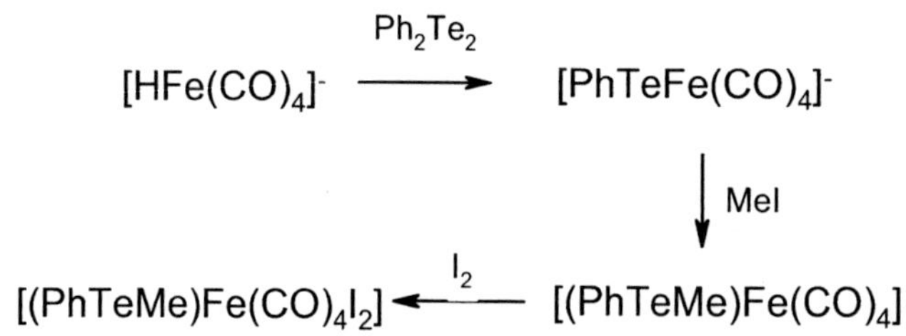

4

The crystal structures of **26** and **27** are solved and like Fe in **25**, W in **26** and Re in **27** have octahedral geometry. On reacting dialkyltelluride with $RMn(CO)_5$, a CO ligand is inserted into the R---Mn bond, resulting in $RCOMn(CO)_4(TeR_2)$ (R=Me, Et) [41]. When R = benzyl (Bz) this product is unstable and undergoes decarbonylation resulting in $BzMn(CO)_4(TeR_2)$. Only spectral data of these complexes are reported in the support of octahedral geometry of Mn and its *cis* stereochemistry. A novel diorganotelluride $[\eta^1$-$\{(Me_3Si)_3C_5H_2\}]_2Te$ stabilized in its adduct $[\eta^1$-$\{(Me_3Si)_3C_5H_2\}]_2TeW(CO)_5]$ [42] was synthesized by reacting bis(diethyldithiocarbamate)tellurium(II) with lithium tris(trimethylsilyl)cyclopentadienide. The 28 is structurally characterized and has octahedral Re. Oxidative addition of Ph_2Te_2 to $[Mn(CO)_5]^-$ gives *cis*-$[Mn(CO)_4(TePh)_2]^-$ which may be crystallized as *cis*-$[Na(18\text{-crown-}6)(THF)]_2[Mn(CO)_4(TePh)_2]$ (**35**) The anion of **35** undergoes a variety of reactions shown in Scheme 1 [43]. The metal ions have a geometry very close to octahedral.

Table 2. Some Mono- and di-nuclear metal carbonyls containing Te ligands reported in last two decades

Synthesis route	Metal carbonyl complex	Ref.
mer-$[RhCl_3L_3\}$ + CO L = Te heterocycle	$[RhCl_3(CO)L_2]$ (**16**)	33
$Ar_2TeO + Cr(CO)_6$ CO_2 is formed	$[Cr(CO)_5TeAr_2]$ (**17**); Ar = 4-$MeOC_6H_4$	34
$[Cr(CO)_5THF] + TeMe_2$	$[(Cr(CO)_5)_2$-μ-$TeMe_2]$ (**18**)	35
[4+2] cycloaddition of cyclopentadiene and 2,3-dimethyl-1,3-butadiene with $[(CO)_5W(Te{=}CPh_2)]$	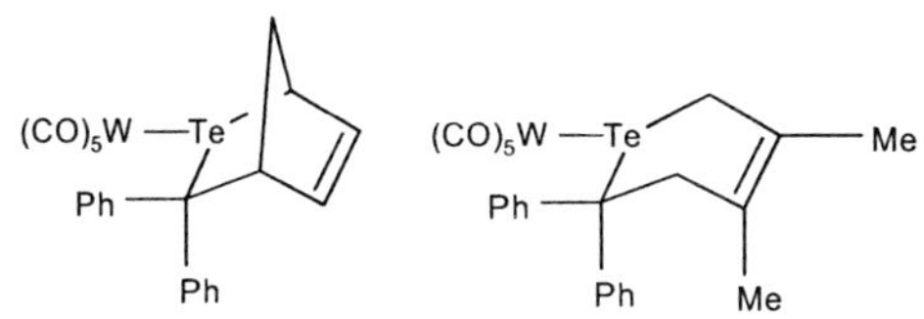 (**19**) (**20**)	36
$M_2(CO)_{10} + Me_2TeI_2$ M = Mn or Re	*Cis*-$[M(CO)_4(TeMe_2)I]$ (**21**)	37
$Co_2(CO)_8 + Me_2TeI_2$	$[Co(CO)_5TeMe_2]n$ (**22**)	37
$Na[Co(CO)_4] + Me_2TeI_2$	$[(Co(CO)_4)_2$ (μ-$TeMe_2)]$ (**23**)	37
$Na[Mn(CO)_5] + Me_2TeI_2$	$[(Mn(CO)_5)_2$ (μ-$TeMe_2)]$ (**24**)	37
$Fe(CO)_5 + Ph_2TeI_2$	$[Ph_2TeFe(CO)_5I_2]$ (**25**)	38
$[W(CO)_5(THF)] + Ph_2Te$	$[Ph_2TeW(CO)_5]$ (**26**)	38
$[PhTeRe(CO)_5 + [Me_3O]BF_4$	$[(PhTeMe)Re(CO)_5]BF_4$ (**27**)	39
$[Re(CO)_5Cl]$ + NaTePh	$[PhTeRe(CO)_5]$ (**28**)	39
$[M(CO)_6]$ + (*p*-$MeOC_6H_4)_2TeO$	$[M(CO)_5Te$ (*p*-$MeOC_6H_4)_2]$ (**29**) 0M= Mo or W	40
[PPN] $[HFe(CO)_4] + Ph_2Te_2$	[PPN] $[Fe(CO)_4(TePh)]$(**30**); [PPN] $[Fe(CO)_3(TePh)_3]$ (**31**)	47, 48
[PPN] $[Fe(CO)_3(TePh)_3] + NOBF_4$	$[Fe_2$(μ-$TePh)_2(NO)_4]$ (**32**); $[Fe_2$(μ-$TePh)_2(CO)_6]$ (**33**)	49
[PPN] $[Fe(CO)_4(TePh)] + Ph_2Te_2$ / Ph_2Se_2	$[PPN][Fe(CO)_3(TePh)_n(SePh)_{3-n}]$ (**34**)	50

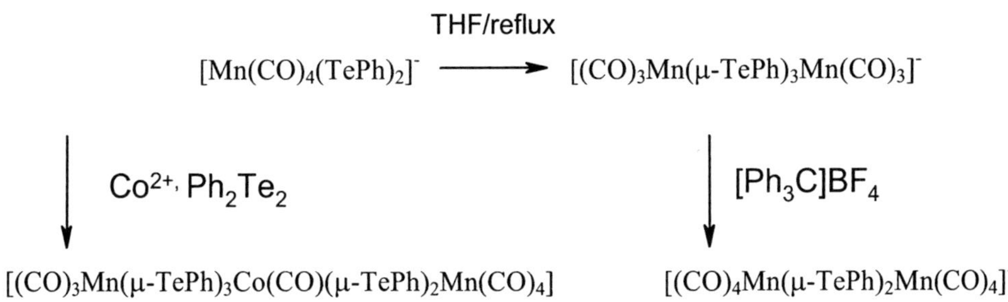

Scheme 1.

The *cis*-$[Mn(CO)_4(TePh)_2]^-$ has been further explored as metalloligand. On reaction with $BrMn(CO)_5$ it gives $[(CO)_3BrMn(\mu\text{-}TePh)_2Mn(CO)_4]^-$ and $[Mn_2(\mu\text{-}TePh)_2(CO)_8]$ [44]. In the presence of $[Mn(CO)_4(SePh)_2]^-$ and $[Mn(CO)_4(TePh)_2]^-$ oxidation of Co^{2+} to Co^{3+} by Ph_2Se_2 results in a novel complex $[(CO)_4Mn(\mu\text{-}TePh)_2Co(CO)(\mu\text{-}SePh)_3Mn(CO)_3]$ (**30**) [45]. The cation $Me_2TeTeMe^+$ also acts as a source of $MeTe^-$ and its reaction with $Mn(CO)_5^-$ results first in [$MeTeMn(CO)_5$] [44] which dimerizes to [$Mn_2(\mu\text{-}TeMe)_2(CO)_8$]. Arnold [14] has extensively reviewed metal-tellurolates including of organometallic species. The 1,3-dihydrobenzo[*c*]tellurophene (L) behaves as a simple Te-donor ligand in [$Mo(CO)_5(L)$], *cis*-[$Mo(CO)_4(L)_2$] and *fac*-[$Mo(CO)_3(L)_3$] as confirmed by multinuclear NMR studies and the crystal structure of the tetracarbonyl (Fig 1a) [46]

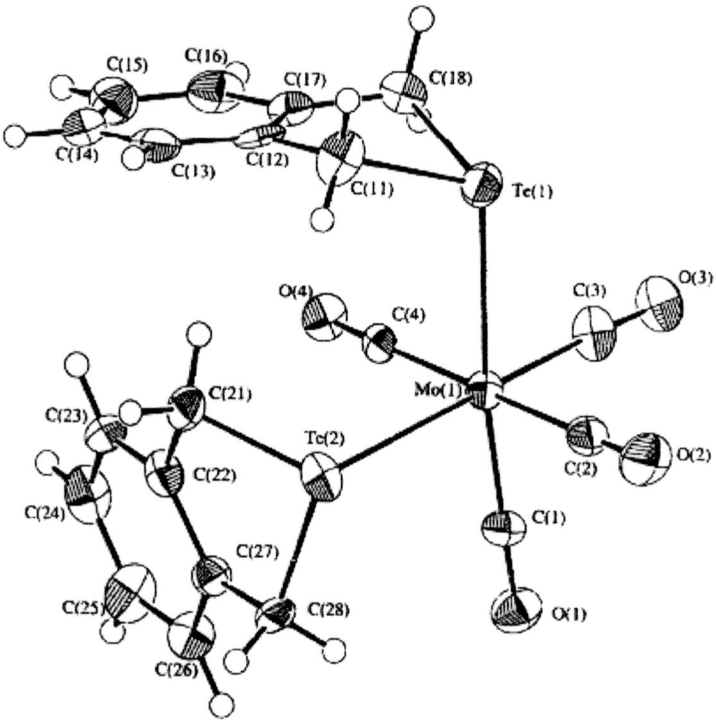

Figure 1a. Molecular structure of *cis*-[$Mo(CO)_4$ (1,3-dihydrobenzo[*c*]tellurophene)$_2$]

A series of compounds having stoichiometry [$M(CO)_4$(L-L)] (M = Cr, Mo or W; L-L = $MeTe(CH_2)_3TeMe$, $PhTe(CH_2)_3TePh$, *o*-$C_6H_4(TeMe)_2$) have been prepared [51] from [$Cr(CO)_4$(norbornadiene)], [$Mo(CO)_4$(norbornadiene)] and [$W(CO)_4(Me_2N(CH_2)_3NMe_2)$], respectively. On the basis of IR and multinuclear NMR (^1H, $^{13}C\{^1H\}$, ^{77}Se, ^{125}Te and ^{95}Mo) studies on these systems and their S and Se analogues it appears that σ-donation to the metal increases down the Group 16. 1,2-Bis(methyltelluromethyl)benzene, forms [$W(CO)_4$(L-L)] (36) and [$Mo(CO)_3Cl$(L-L)] (37) The structures of both the complexes were determined by X-ray crystallography [52]. A 7-membered chelate ring is formed (See structures in Fig. 2). The *fac*-[$M(CO)_3$(tripod)] (M = Mo or W, tripod = $MeC(CH_2TeMe)_3$) made from [$M(CO)_3(MeCN)_3$] and the tripod ligand, was unstable in solution decomposing to [$M(CO)_4$(tripod)] and other products. However, [$Cr(CO)_3$(tripod)] made in solution was identified spectroscopically [53]. The tritelluroether $\{MeTe(CH_2)_3\}_2Te$ forms *fac*-[$Mo(CO)_3\{MeTe(CH_2)_3\}_2Te\}$], but this is less stable than the tripod telluroether analogue and decomposes very rapidly in solution [54]. The reactions of $OTe(4\text{-}MeOC_6H_4)_2$ with [$M_2(CO)_{10}$] (M = Mn or Re) results in the formation of [$M_2(CO)_8(TeR_2)_2$] (R = 4-$MeOC_6H_4$)

via oxygen atom transfer, which on the basis of IR, ^{1}H- and ^{13}C-NMR spectra were described as the diaxial isomers[55].

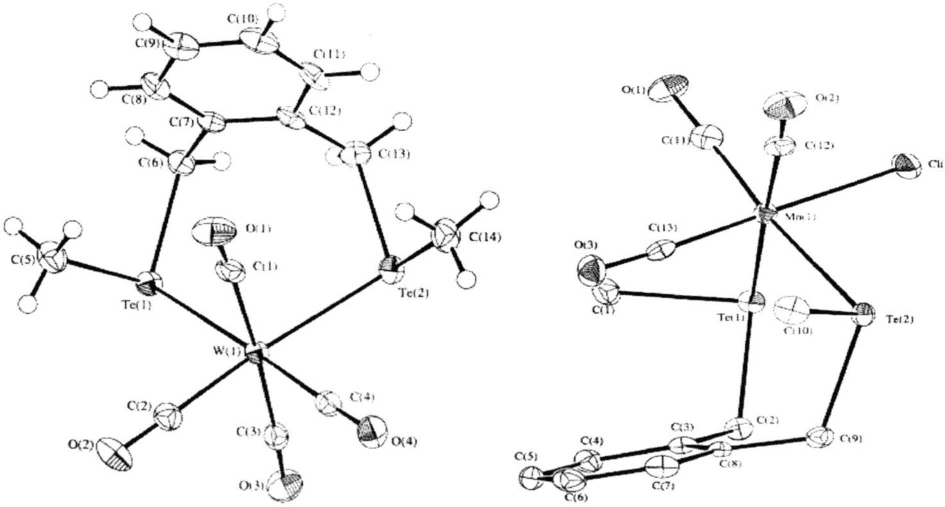

3637

Figure 2. Molecular structures of **36** and **37**

Manganese halocarbonyl derivatives $[Mn(CO)_4Br(TeR_2)]$ (R=Me or Et) $[Mn(CO)_3Br(TeEt_2)_2]$ [41] $[Mn(CO)_3Cl(TeMe_2)_2]$ [56] and $[Mn(CO)_3Cl\{$1,3-dihydrobenzo(*c*)tellurophene$\}]$ [46] have been characterized by spectroscopic methods. The $[Mn(CO)_3X(L)_2]$ complexes have *fac* geometries, except for $[Mn(CO)_3Cl(TeMe_2)_2]$ for which ^{55}Mn- and ^{125}Te-NMR spectra showed two species present, the second most probably *mer*, *trans* isomer [56]. The complexes *fac*-[M $(CO)_3X$(L---L)] (M = Mn or Re, X = Cl, Br or I, L---L = $RTe(CH_2)_3TeR$ (R = Me or Ph), and *o*-$C_6H_4(TeMe)_2$ have been reported [56]. The $[Mn(CO)_3X\{PhTe(CH_2)_3TePh\}]$ is unstable in solution, but the other complexes are reasonably stable as solids, although both light and oxygen sensitive in solution. The X-ray structure of $[Mn(CO)_3Cl\{$*meso*-2-*o*-$C_6H_4(CH_2TeMe)_2\}]$ shows significantly longer Mn---Te bonds (Fig.3). The crystal structures of $[Re(CO)_3Cl\{$*meso*-2-*o*-$C_6H_4(TeMe)_2\}]$ and $[Re(CO)_3Br\{$*meso*-2-$PhTe(CH_2)_3TePh\}]$ are also reported. A series of cationic *fac*-$[Mn(CO)_3(L)]CF_3SO_3$ (L = $MeC(CH_2TeMe)_3$, $MeC(CH_2TePh)_3$) have been synthesised from $[Mn(CO)_3(Me_2CO)_3]^+$ and the corresponding ligands [57-58]. X-ray structure is available for $[Mn(CO)_3\{MeC(CH_2TePh)_3\}]^+$ (Fig 4). Analysis of the spectroscopic and structural data suggest that, $MeC(CH_2TePh)_3$ is a weaker σ-donor than $MeC(CH_2TeMe)_3$ as expected. The *fac*-$[Mn(CO)_3(RTe(CH_2)_3)_2Te)]CF_3SO_3$ has been prepared for R=Me or Ph) and contain a mixture of invertomers [54].

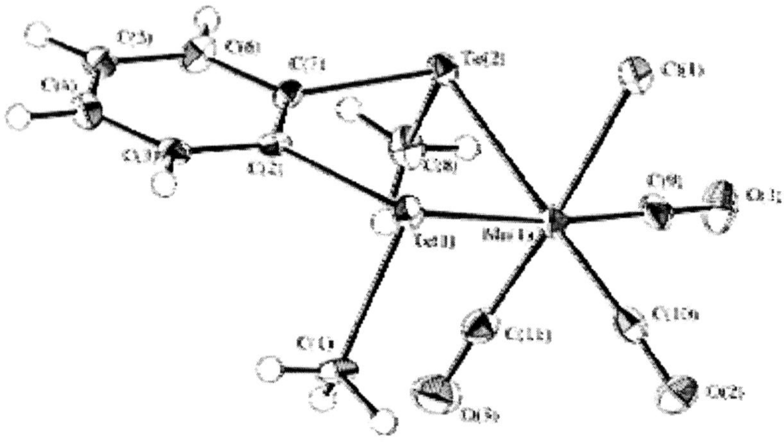

Figure 3. Molecular structure of [$Mn(CO)_3Cl$\{*meso*-2-*o*-$C_6H_4(CH_2TeMe)_2$\}]

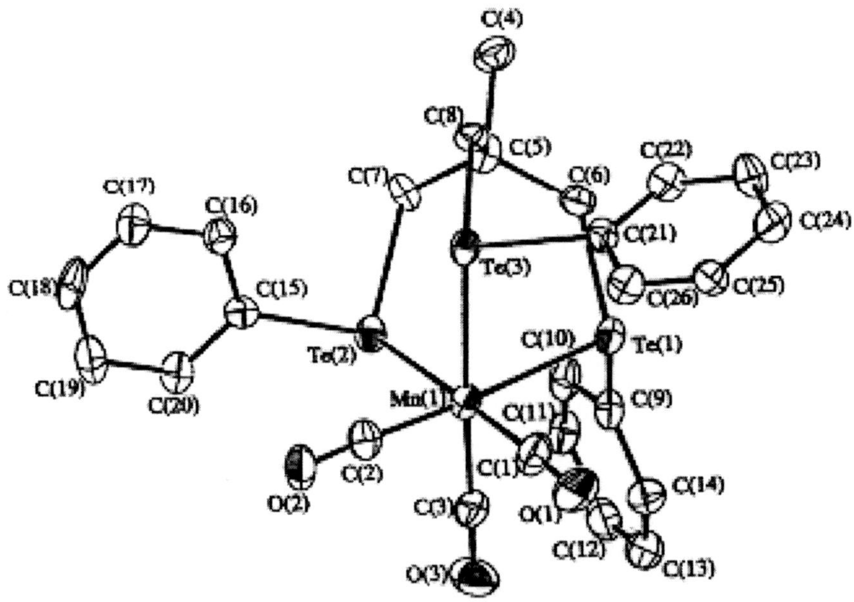

Figure 4. Molecular structure of [$Mn(CO)_3$\{$MeC(CH_2TePh)_3$\}]$^+$

The hybrid tridentate (2-$Me_2NCH_2C_6H_4)_2$Te (L) coordinates only via Te in [$Cr(CO)_5$(L)] [59]. The [$Ru(CO)_2Cl_2(TePh_2)_2$] has been prepared in improved yield from [\{$Ru(CO)_3Cl_2$\}$_2$] and the ligand as a single isomer, shown by an X-ray crystal structure to contain octahedral Ru with *trans* $TePh_2$, *cis* CO and *cis* Cl [60]. The reaction of [\{$Ru(CO)_3Cl_2$\}$_2$] with (4-$EtOC_6H_4Te)_2CH_2$, various hybrid telluroethers, including $PhTeCH_2CH_2SMe$ and \{2-(4-$MeOC_6H_4Te)_2C_2H_4$\}$_2$NH, has been reported. Various oligomeric structures are suggested, but none of the complexes gives crystals suitable for by X-ray crystallography [61]. The vanadium-51 shielding in the complexes [$V(CO)_5L$]$^-$ and [V(NO)$(CO)_4$L] (L = $TePh_2$) shows expected normal trend, i.e. increase on going down chalcogen group [73]. This reflects a decrease of the paramagnetic contribution to the local terms of the overall shielding. The

$[Co(CO)_3(NO)]$ reacted with $TeMe_2$ or $TeEt_2$ in the absence of solvent to form $[Co(CO)_2(NO)(TeR_2)]$ [62] . The dinuclear complexes containing bridging Te^{2-} and Te_2^{2-} as ligands are also reported. The μ_2-Te$[V(CO)_3(dppe)]_2$ and $(OC)_3Fe(\mu_2-\eta^2-Te_2)Fe(CO)_3$ are two such examples [63]. The latter one, reacts efficiently with $Pt(PPh_3)_2C_2H_4$ giving heterometallic clusters, $Fe_2Pt(\mu_3-Te)_2(CO)_6(PPh_3)_2$. This reaction formally involves the homolytic cleavage of the Te-Te bond in the μ_2-Te_2 precursor complexes [64]. Recently organometallic compounds of type $Mn_2(CO)_8E(CF_3)_2E'R$, in which the groups $E(CF_3)_2$ and E'R act as bridging ligands, are prepared [69] either by direct reactions of $Mn_2(CO)_{10}$ with $(F_3C)_2EE'R$ (E = P, As; E' = S, Se, Te) or by substitution of the iodine bridge in $Mn_2(CO)_8E(CF_3)_2I$ (E = P, As) with Hg compounds $Hg(E'R)_2$. The pyramidal $Mn_2E'R$ fragment shows dynamic behavior in solution. via inversion between two identical structures. The reaction of $[Mo(CO)_3(MeCN)_3]$ and Te_2Ph_2 yields the edge-sharing bioctahedral, metal-metal bonded Mo(I) product $[Mo(CO)_3(MeCN)(\mu-EPh)]_2$ characterized structurally [70]. The five-coordinate sixteen-electron Mn(I) complex $[Mn(CO)_3(-TeC_6H_4-o-E'-)]^-$ (E' = NH) is obtained by oxidative addition of 2-aminophenyl dichalcogenides to anionic Mn(0)-carbonyl [71]. It has distorted trigonal bipyramidal structure and its reactions with electrophiles occur at the more electron-rich amide site which indicates that the more electron-rich amide donor of the chelating 2-tellurolatophenyl-amido ligand occupies an equatorial site as indicated by a shorter Mn-N bond length of the $[Mn(CO)_3(-TeC_6H_4-o-NH-)]^-$. The reaction of $[IrCl(CO)(PPh_3)_2]$ with $TeCl_4$ yielded $[Ir(TeCl_3)Cl_2(CO)(PPh_3)_2]$, which was characterized by X-ray crystallography [72]. The geometry around iridium atom is essentially octahedral with *cis* chloro and *trans* phosphine ligands. These complexes are among the first mononuclear transition metal complexes ligated by the trichlorotelluronium group. The syntheses and structures of $[K(2,2\text{-crypt})]_2[Cr(CO_5)Te_3]\cdot(0.5en)$ and $[K(2,2\text{-crypt})]_2[\{Cr(CO)_5\}_2Te_2]$ were reported. In the first case a Te_3 chain is bonded through one end and in the second two $Cr(CO)_5$ units are bridged by Te_2^{2-} [85]. The iron pentacarbonyl reacts with $As_2Te_6^{2-}$ (prepared by reducing As_2Te_3 with K in dry DMF) resulting in $[Fe_2(AsTe_4)(CO)_4]^{2-}$, which is, characterized structurally [86]. The Te---Fe bond lengths are in the range 2.570(7)–2.6226(8) Å. The possibilities of using metal carbonyls containing Te ligands as precursors for material development have been explored. Pyrolysis of $[RCOMn(CO)_4(TeR'_2)]$ in hydrogen produced films containing both Te and Mn, although pre-dissociation of the complexes was problematic [41]. In the case of $[Co(CO)_2(NO)(TeMe_2)]$, the thermal decomposition of the vapour in hydrogen gave a cobalt film and $TeMe_2$, although the complex decomposes at too low a temperature to form $CoTe_x$ films [62].

POLYNUCLEAR METAL CARBONYLS AND CLUSTERS

Telluride, polytellurides and related species are among the most abundant tellurium ligands present in polynuclear metal carbonyls and clusters. The $Fe_3(\mu_3-Te)_2(CO)_9$ and (**38**) was synthesized as early as 1958 [66] by adding TeO_3^- to aqueous alkaline $Fe(CO)_4^-$, $Fe_2(CO)_8^-$, or $Fe_3(CO)_{11}^-$ solutions and acidifying the reaction mixture. The $Fe_3(CO)_{12}$ was also a product. The important polynuclear metal carbonyls and clusters reported up to 1990 are summarized in Table 3. In **38** the μ_3-Te moieties are strained to acute approximately 60° Fe-Te-Fe angles which are fixed by Fe-Fe bonding [68]. The CO groups in **38** have been substituted to get **39** and **41** (Table 3). In **40** L is bound to an Fe center and not to the Te atoms. The structure consists of an isosceles triangle of Fe atoms tethered by capping μ_3-Te atoms. The basal $Fe_2(CO)_6$ unit contains a Fe-Fe bond while the apical $Fe(CO)_3L$ (L = PPh_3) is octahedral, two coordination sites being occupied by the bridging Te atoms. The **38** is a good precursor for generating Te containing metal carbonyl clusters which occur via intermediate $Fe_2(\mu\text{-}Te_2)(CO)_6$ which has been isolated and characterized. Gysling has described the more details of the compound **38-52** in the review published in 1986 [10]. Iridium clusters $[PPh_4][Ir_6(CO)_{12}(\mu\text{-}CO)_2(\mu\text{-}TePh)]$ **(53)** and $[Ir_6(CO)_{12}(\mu\text{-}CO)(\mu\text{-}TePh)_2]$ (**54**) were synthesized by refluxing $[Ir_6(CO)_{15}]^{2-}$ with Ph_2Te_2 in THF. In these clusters there is an octahedral Ir_6 core. Each Ir has two terminal CO groups. The structure of **54** is similar to that of **53**, except one bridging CO is replaced by a PhTe bridging group. The coordination around the Te atoms is pyramidal. On the basis of cyclic voltammetric studies it is shown that addition of $PhTe^-$ occurs upon oxidation of the parent cluster and the mechanism of reaction of Ph_2Te_2 is through the monoanion $[Ir_6(CO)_{15}]^-$, in contrast to that of Ph_2S_2 which is said to involve $[Ir_6(CO)_{15}]$. An unusual carbonyl cluster anion is $[PPN][(TeMe_2)Mn(CO)_4(\mu_5\text{-}Te)(\mu_4\text{-}Te)Mn_4(CO)_{12}]$ (**55**) made by alkylation of $[(\mu_4\text{-}Te)_2Mn_4(CO)_{12}]^-$ with $MeSO_3CF_3$. The structure reveals a $(TeMe_2)Mn(CO)_4$ fragment bonded via Mn to μ_5-Te group in the Mn_4Te_2 cluster [81]. The reaction of the mixed metal clusters $[HRuCo_3(CO)_{11}(L)]$ (L= $TeMe_2$) with PMe_2Ph resulted in simple replacement of the Ru-bound chalcogen with the phosphine. In contrast, $[HRuCo_3(CO)_{11}(PMe_2Ph)]$ with $TeMe_2$ produced $[HRuCo(CO)_{10}(PMe_2Ph)(TeMe_2)]$ [82]. The $MeTe^-$ group is generated in a methanothermal reaction between $M_3(CO)_{12}$ (M = Fe, Ru) and Na_2Te_2 carried out in 0.5 ml of CH_3OH in the presence of Me_4NBr in a sealed Pyrex tube, and gives the cluster $[M_4(Te_2)_2(Te)_2(TeMe)_2(CO)_8]^{2-}$ (**56**) [83]. Mixed-metal carbonyl telluride clusters $(Ph_4P)_2[Fe_2MTe_3(CO)_{11}]$ (M = Mo (**57**), W (**58**)) and $Na(Et_4N)_2[Fe_3W_2Te_8(TeMe)(CO)_{12}]\cdot 2H_2O$ (**59**) were synthesized methanothermally at 80 °C using mixtures of $Fe_3(CO)_{12}$, $M(CO)_6$ (M = Mo, W), and Na_2Te_2 with Ph_4PBr, or $Et_4NCl\cdot H_2O$, in a sealed Pyrex tube and characterized by single crystal X-ray diffraction [84]. In the anionic $[Fe_2Mo(Te_2)(Te)(CO)_{11}]^{2-}$ cluster, a $[Fe_2(Te_2)(Te)(CO)_6]^{2-}$ unit is joined to an $Mo(CO)_5$ moiety by its Te_2^{2-} ligand. In the $[Fe_3W_2(Te_2)_3(Te)_2(TeMe)(CO)_{12}]^{3-}$ a W-W unit is connected to the rest of the cluster through telluride ligands to form an unsymmetrical cage like structure. ^{125}Te NMR spectra of **57** and **58** and $(Ph_4P)_2[Fe_2(Te_2)(Te)(CO)_6]$ display three-lines, while seven singlets were observed in the case of **59**, corresponding to seven different environments for the tellurium atoms in it. $(Me_4N)_2[Fe_4(Te_2)_2(Te)_2(TeMe)_2(CO)_8]$ (**60**) was synthesized [84] by methanothermal technique at 110 °C using a mixture. of $Fe_3(CO)_{12}$, $5Na_2Te_2$ and $6Me_4NBr$ in sealed Pyrex tube. $(Ph_4P)_2[Ru_4(Te_2)_2(Te)_2(TeMe)_2(CO)_8]$ (**61**) was prepared similarly at 80-130 °C using a mixture of $Ru_3(CO)_{12}$, $5Na_2Te_2$ and $6Ph_4PBr$. The methylation of the telluride ligands is a novel feature of these reactions. The process appears to be solvent driven in view of the observation of an analogous ethylation reaction occurring in ethanol to form $(Ph_4P)_2[Fe_4(Te_2)_2(Te)_2(TeEt)_2(CO)_8]$. The reaction of $Fe(CO)_5$ with Te^{2-} ligand gives a cluster $[Fe_3(CO)_9Te]^{2-}$ (**62**) whose Fe···Fe bond is quite reactive [87] as addition of protons may occur across it. Similarly electrophiles $(PPh_3)Au^+$ and $PhHg^+$ also add to a Fe···Fe edge and not the apical Te. The reaction of I_2 with **62** gives $[Fe_3Te_2(CO)_9I]^-$ which is isostructural with $Fe_3Te_2(CO)_{10}$. The aerial oxidation of mixture of $Fe(CO)_5$ and Te^{2-} gives $[(Fe(CO)_4)_2Te]^{2-}$. The alkylation of $[TeFe_3(CO)_9]^{2-}$ with methyl triflate or MeI occurs at the metal framework (Fe---Me=2.226(9) Å) [88], like its Se analog but contrary to the sulfur one, which is alkylated at S itself. The tetrahedral cluster $[TeFe_3(CO)_9]^{2-}$ on successive treatment with $Te_2Fe_3(CO)_9$ results in $[Te_4Fe_5(CO)_{14}]^{2-}$, $[Te_6Fe_8(CO)_{24}]^{2-}$ and $[Te_{10}Fe_8(CO)_{20}]^{2-}$ [89]. The last, on reaction with $[Cu_2(dppm)_2(MeCN)_4][BF_4]_2$, results in a cubic cluster $[Te_4Fe_4(CO)_{10}(dppm)]$ (**63**). The Fe---Te bond length is 2.605(2)–2.625(2) Å. The reaction [90] of $Co_2(CO)_8$, Na_2Te and Ph_4PCl in methanol gives clusters $(Ph_4P)_2[Co_{11}Te_7(CO)_{10}]$ (**64**) and $Co_4Te(CO)_{11}$ (**65**), when the reactants are in the molar ratio 3:2:3 and 4:2:1, respectively. The first cluster has a slightly elongated pentagonal prismatic array of cobalt atoms, which has one Co atom at the center. Each Co_4 square is capped by a μ_4-Te^{2-} ligand (mean Co---Te 2.510(3) Å). There are two μ_6-Te^{2-}, one at the center of each pentagonal Co face (mean Co---Te(3) 2.569 Å; Co(central)---Te 2.642(2)/2.647(2) Å). The formal core framework of this cluster is $[Co_{10}Co(\mu_6\text{-}Te)_2(\mu_4\text{-}Te)_5]$. The two μ_4-Te ligands make the core framework of $Co_4Te_2(CO)_{11}$. As mentioned earlier the $[Fe_2(CO)_6(\mu_2\text{-}Te_2)]$ has been used to obtain a variety of metal clusters. The room temperature reaction of $Fe_2(CO)_6(\mu_2\text{-}Te_2)$ with $Ru(CO)_4(C_2H_4)$ leads to an immediate and almost quantitative formation of $Fe_2Ru(CO)_9(\mu_3\text{-}Te)_2$ [92]. On treatment of the mixed-metal cluster with NaOMe and acidification of the solution, $Fe_2(CO)_6(\mu_2\text{-}Te)_2$ is obtained, suggesting that during the formation of $Fe_2Ru(CO)_9(\mu_3\text{-}Te)_2$ from $Fe_2(CO)_6(\mu_2\text{-}Te_2)$, the Fe-Fe bond is retained intact. Also, $Fe_2Ru(CO)_9(\mu_3\text{-}Te)_2$ reacts at room temperature with $Pt(PPh_3)_4$ to give $(CO)_6Fe_2(\mu_3\text{-}Te)_3Pt(PPh_3)_2$. $Ru_4(CO)_{10}(\mu\text{-}CO)(\mu_4\text{-}Te)_2$ reacts with PPh_3 to give $Ru_3(CO)_6(PPh_2)_3(\mu_3\text{-}Te)_2$ at room temperature. The interesting reactions of $Fe_2(CO)_6(\mu_2\text{-}Te)_2$ used to generate new cluster systems are summarized in Table 4. The mixed chalcogenide cluster $Fe_2(CO)_6(\mu_2\text{-}Te)(\mu_2\text{-}E)$ where E = S or Se is equally a good precursor for a variety of cluster containing tellurium [101-110] Its reaction when E = Se with $[Ru(CO)_4(C_2H_4)]$, results in mixed metal cluster $Fe_2Ru(CO)_9(\mu_2\text{-}Te)(\mu_2\text{-}Se)$ [101]. Other important reactions are also given in Table 4. A common feature noticeable in the reactions of Table 4 is that Fe-Fe bond of the starting cluster remains intact and reactions occur on Te-E bond. The $Fe_3Te_2(CO)_9$ has also been used to generate new carbonyl clusters containing Te. The $Fe_2Te_2M(CO)_6(PPh_3)_2$ (M = Ni, Pd, Pt) is formed on reacting it with $M(PPh_3)_4$ [110] At room temperature, $Fe_3(CO)_9(\mu\text{-}Te)_2$ reacts with CH_2N_2 to form $(CO)_6Fe_2(\mu\text{-}TeCH_2Te)$ and $(CO)_6Fe_2Te_2(CH_2)_3$ [111]. The excess of diazomethane results in $[(\mu\text{-}CH_3Te)Fe_2(CO)_6]_2[\mu\text{-}Te(CH_2)Te\text{-}\mu]$ [112]. Recently $[TeFe_3(CO)_9]^{2-}$ was treated with 1 equivalent of propargyl bromide in MeCN, the novel acyl complex $[Et_4N][(\mu_3\text{-}Te)Fe_3(CO)_9(\mu_3\text{-}\eta^1\text{:}\eta^1\text{:}\eta^3\text{-}C(O)C(H)CCH_2)]$ (Fig 5) was obtained which on reaction with $[Cu(MeCN)_4]BF_4$ gives $Te_2Fe_3(CO)_9$. When excess propargyl bromide in MeCN was taken to treat $[TeFe_3(CO)_9]^{2-}$ two butterfly clusters $Te_2Fe_2(CO)_6(CH{=}C{=}CH_2)_2$ and $Te_2Fe_2(CO)_6(CH_2C{\equiv}CH)(CH{=}C{=}CH_2)$ were formed [123]

Table 3. Some polynuclear metal carbonyls and clusters containing Te ligands reported up to 1985

Synthetic route	Compound	Ref.
$Fe_3(\mu_3\text{-}Te)_2(CO)_9$ + L	$Fe_3(\mu_3\text{-}Te)_2(CO)_nL_m$ (**39**) n = 9-7; m = 1-3; L = phosphine and arsine ligands	67
$Fe_3(\mu_3\text{-}Te)_2(CO)_9$ + L	$Fe_3(\mu_3\text{-}Te)_2(CO)_9L$ (**40**)(L = PPh_3, Me_3CNC, CO, amines)	68
$Fe_3(\mu_3\text{-}E)(\mu_3\text{-}Te)(CO)_9$ + L	$Fe_3(\mu_3\text{-}E)(\mu_3\text{-}Te)(CO)_nL_m$ (**41**) E = S / Se; n = 8, m = 1; n = 7, m = 2; L $AsPh_3$ $P(OPh_3)$	74
$Fe_3(\mu_3\text{-}Te)_2(CO)_9$ decomposition in polar solvents	$Fe_4(\mu_3\text{-}Te)_4(CO)_{12}$ (**42**)	75
$Co_2(CO)_8$ + $Fe_3(\mu_3\text{-}Te)_2(CO)_9$	$Co_2Fe(\mu_3\text{-}Te)_2(CO)_9$; $Co_4(\mu_3\text{-}Te)_2(CO)_{11}$ (**43, 44**)	75
$Fe_2(\mu\text{-}Te_2)(CO)_6$ + $Fe(CO)_5$ (in the presence of base)	$Fe_3(\mu_3\text{-}Te)_2(CO)_{10}$ (**45**)	76, 91
$Co_2(CO)_8$ + $Fe_3(CO)_{12}$ + Et_2Te	$FeCo_2(\mu_3\text{-}Te)(CO)_9$ (**46**)	77
$Ru_3(CO)_{12}$ + TeO_3 (in alkaline solution)	$H_2Ru_3(\mu_3\text{-}Te)((CO)_9$ (**47**)	78
$Os_3(CO)_{12}$ + Te (reflux in octane) $Ru_3(CO)_{12}$ + Te (reflux in octane) (1CO:1H_2 pressure of 35 atm.)	$[Os_3(\mu_3\text{-}Te)_2(CO)_9]$; $[H_2Os_4(\mu_3\text{-}Te)_2(CO)_{12}]$ (**48, 49**) $[H_2Ru_3(\mu_3\text{-}Te)_2(CO)_9]$; $[Ru_3(\mu_3\text{-}Te)_2(CO)_9]$ (**50, 51**)	79
$Co_2(CO)_8$ + $2TePh_2$	$Co_4(\mu_4\text{-}Te)_2(CO)_8(\mu_2\text{-}CO)_2$ (**52**)	20

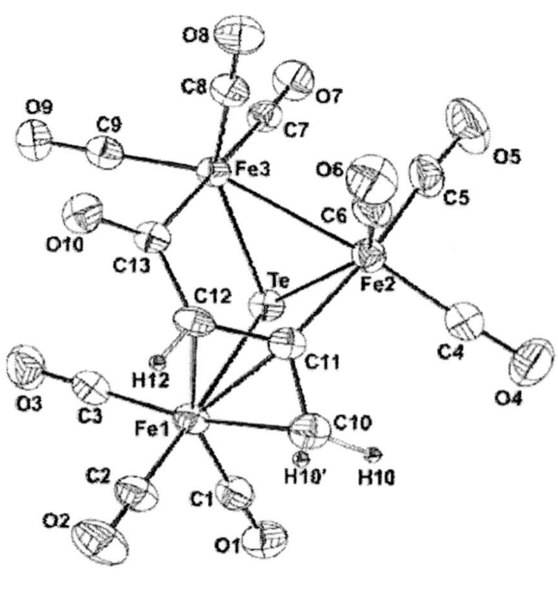

Figure 5. Structure of $[(\mu_3\text{-}Te)Fe_3(CO)_9(\mu_3\text{-}\eta^1{:}\eta^1{:}\eta^3\text{-}C(O)C(H)CCH_2)]^-$

Table 4. Reactions of $Fe_2(CO)_6(\mu_2\text{-}Te_2)$ and $Fe_2(CO)_6(\mu_2\text{-}Te)(\mu_2\text{-}E)$ (where E = S or Se) generating new clusters

Reagent	Product	Ref.
$Fe_2(CO)_6(\mu_2\text{-}Te_2)$		
CH_2N_2 CH_2N_2 (Excess)	$(CO)_6Fe_2(\mu\text{-}TeCH_2Te)$ (**53**) $[(\mu\text{-}CH_3Te)Fe_2(CO)_6]_2[\mu\text{-}Te(CH_2)Te\text{-}\mu]$ (**54**)	92, 93 99
$M(PPh_3)_4$ (M = Ni, Pd, Pt)	$Fe_2Te_2M(CO)_6(PPh_3)_2$ (**56**)	94, 95
$LiEt_3BH$ at –78 °C + $(PPh_3)_2MCl_2$ (M = Pt, Pd, Ni) Cp_2TiCl_2 R_2SnCl_2 (R = Me, Bu)	$[(\mu\text{-}Te)_2Fe_2(CO)_6]^{2-}$ (**57**) $(CO)_6Fe_2(\mu_3\text{-}Te_2)M(PPh_3)_2$ (**58**) $(CO)_6Fe_2(\mu_3\text{-}Te)_2TiCp_2$ (**59**) $(CO)_6Fe_2(\mu_3\text{-}Te)_2SnR_2$ (**60**)	98
$Ru_3(CO)_{12}$	$[\{Fe_2(CO)_6\}(\mu_4\text{-}Te)(\mu_3\text{-}Te)\{Ru_3(CO)_{11}\}$(**61**)	96
$Os_3(CO)_{11}(NCMe)$	$Fe_2Os_3(CO)_{17}(\mu_4\text{-}Te)(\mu_3\text{-}Te)$ (**62**) UV / Thermolysis → $Os_3(CO)_9(\mu_3\text{-}Te)_2$ (**63**)	97
$W(CO)_5(THF)$.	$Fe_2W(\mu_3\text{-}Te)_2(CO)_{10}$ (**64**)	100
$CH\equiv CH$	$(CO)_6Fe_2\{\mu\text{-}TeC(H){:}C(H)Te\}$(**65**)	106
$(CO)_5M{:}C(OEt)(C\equiv CPh)$ (M = Cr, W)	$[(CO)_6Fe_2Te_2\{\mu\text{-}PhC{:}CC(OEt)\}M(CO)_5]$ (**66**)	108
$Fe_2(CO)_6(\mu_2\text{-}Te)(\mu_2\text{-}E)$ E = Se or S		
CH_2N_2	$Fe_2(CO)_6(\mu\text{-}ECH_2Te)$ (**67**)	102
Phenylacetylene	$Fe_2(CO)_6\{\mu\text{-}SeC(H){:}C(Ph)Te\}$(**68**) $Fe_2(CO)_6\{\mu\text{-}SeC(Ph){:}C(H)Te\}$(**69**) $Fe_2(CO)_6(\mu\text{-}SC(H){:}C(Ph)Te)$(**70**) $Fe_2(CO)_6(\mu\text{-}SC(Ph){:}C(H)Te)$ (**71**)	103
$[M(CO)_5(THF)]$ (M = W, Mo)	$Fe_2M(CO)_{10}(\mu_3\text{-}S)(\mu_3\text{-}Te)$ (**72**) $[Fe_4Mo(CO)_{14}(\mu_3\text{-}E)_2(\mu_3\text{-}Te)_2]$ (**73**) $[Fe_3Mo(CO)_{11}(\mu_3\text{-}E)(\mu_3\text{-}Te)(\mu\text{-}Te\text{-}Te)]$ (**74**)	104 107
$CH\equiv CH_2OH$	$(CO)_6Fe_2\{\mu\text{-}EC(H){:}C(CH_2OH)Te\}$(**75**)	105
$CH\equiv CH$	$(CO)_6Fe_2\{\mu\text{-}EC(H){:}C(H)Te\}$ (**76**)	106
$(CO)_5M{:}C(OEt)(C\equiv CPh)$ (M = Cr, W)	$Fe_2(CO)_6\{\mu\text{-}EC(Ph){:}C(Te)[(OEt)C{:}M(CO)_5]\}$(**77**)	109

An excellent review describing many Te containing carbonyl clusters has been written by P. Mathur [12]. Many of the clusters described in Table 4 have been structurally characterized. The "hour-glass" structure of **73** is shown below.

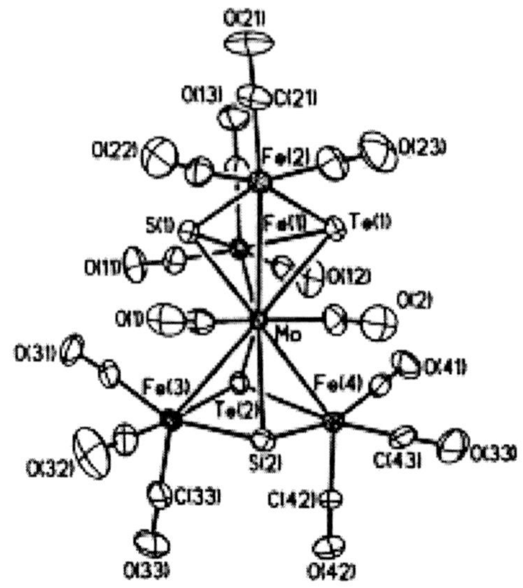

[$Fe_4Mo(CO)_{14}(\mu_3$-S/Se$)(\mu_3$-Te$)_2$]

The treatment of **77** with excess of Bu_3SnH in hexane at 0 °C gives enol ether derivative $(OC)_6Fe_2\{\mu$-EC(Ph)(H)C(Te):C(H)(OEt)$\}$[113]. Alkoxy groups in **66** are readily replaced by amino groups at room temperature as *syn-* and *anti*-isomers, [$(CO)_6Fe_2Te_2\{\mu$-PhC:CCN(H)R$\}M(CO)_5$] (R = Me, CH_2Ph, $CH_2CH:CH_2$) [108]. When a methanolic solution. of [$(CO)_6Fe_2ETe\{\mu$-PhC:CC(OEt)$\}M(CO)_5$] (**77**) was stirred at room temperature for 1-2 days, the ester and *ortho* ester derivatives $(CO)_6Fe_2\{\mu$-EC(Ph):C(CO)(OR)Te$\}$ and $(CO)_6Fe_2\{\mu$-EC(Ph):CC(OR)$(OMe)_2$Te$\}$ (R = Me, Et) were formed [108]. Thermolysis of the cluster-bound Fischer carbene complex, [$(CO)_6Fe_2ETe\{\mu$-C(Ph):CC(OEt):Cr$(CO)_5\}$] gives [$(CO)_6Fe_2ETe\{\mu$-(2-ethoxy-μ-naphthol)$\}$] [114]

Compounds Containing Cyclopentadienyl, Benzene and Related Ligands

Half sandwich compounds containing cyclopentadienyl group with $TeMe_2$ or MeTeTeMe as a co-ligand were synthesized in 1983 [115-116] but no structure was reported for the isolated compounds [$CpNi(TeMe_2)_2$]BF_4 or [CpNi(MeTeTeMe)]BF_4. The ditelluride ligand can be easily substituted with PBu_3. The oxidative addition of Ph_2Te_2 with $CpFe(\mu$-$CO)_2FeCp$ gives [$CpFe(CO)_2TePh$], which on refluxing in benzene and infrared irradiation gives $Cp(OC)Fe(\mu$-$TePh)_2Fe(CO)Cp$ [117]. The dimeric iron derivative exists as an isomeric mixture. Like this iron-cyclopentadienyl dimer containing bridging RTe^- group $Cp(NO)_2Cr(\mu$-$TeR)_2Cr(NO)_2Cp$ (R = Ph or *n*-Bu) has been reported [118]. The metal cyclopentadienyl derivatives containing RTe^- as ligand but not CO group viz. $Cp_2Nb(TePh)_2$, $Cp_2Mo(TeAr)_2$ and $Cp_2W(TeAr)_2$ (Ar = Ph, *p*-$MeOC_6H_4$) were also reported forty years ago[119] . The Cp group in all these compounds is presumably in η^5 bonding mode. The $CpNi(PR_3)(TeC_6H_4X)$ (R = *n*-Bu, X = H, *p*-Cl, *p*-MeO, *p*-Me, *m*-CF_3) is among the early compounds[120] having phosphines as co-ligand of RTe^-. In metal complexes having Cp, CO and Te^{2-} or polytelluride ions as building blocks the interest dates back few decades. Addition of Te to metal-metal double bonds of the dinuclear $[(\eta^5\text{-}C_5Me_2)M(\mu\text{-}CO)]_2$ (M = Co, Rh) resulted in $[\mu_2\text{-}Te\{C_5Me_2(OC)M]_2$ [121]. Treatment of $[(\eta^5\text{-}C_5Me_5)Mn(CO)_2(THF)]$ with Al_2Te_3 and aqueous HCl gave $[(\mu\text{-}Te)[(\eta^5\text{-}C_5Me_5)Mn(CO)_2]_2]$ which has a bent multiply bonded MnTeMn system as determined by X-ray analysis [122]. However, reaction of $[\eta^5\text{-}C_5H_5Mn(CO)_2(THF)]$ with Al_2Te_3 gave $[\mu_3\text{-}Te[Mn(CO)_2(\eta^5\text{-}C_5H_5)]_3]$ and $[Te_2[Mn(CO)_2(\eta^5\text{-}C_5H_5)]_3]$. The structure of first one was determined by X-ray analysis [124]. The square-pyramidal 50e cluster $Fe_3Te_2(CO)_9$ is also used to generate mixed metal clusters containing cyclopentadienyl goup and Te-ligand. Its reaction with $\eta^5\text{-}CpRh(CO)_2$ gave heterometallic 52e cluster $CpRhFe_2Te_2(CO)_7$ [125] which on Me_3NO-induced decarbonylation resulted in two isomers of $CpRhFe_2Te_2(CO)_6$ which on recarbonylated gave original compound in isomerically pure form. In contrast, addition of PPh_3 to $CpRhFe_2Te_2(CO)_6$ gave two isomers of $CpRhFe_2Te_2(CO)_6PPh_3$ that differ with regard to the placement of the PPh_3 ligand on Fe or Rh. On reacting $Fe_3Te_2(CO)_9$ with $C_5H_5Co(CO)_2$ and $Pt(C_2H_4)(PPh_3)_2$ good yields of $(C_5H_5Co)Fe_2Te_2(CO)_7$ and $Fe_2PtTe_2(CO)_6(PPh_3)_2$, respectively were obtained [126]. $(C_5H_5Co)Fe_2Te_2(CO)_7$ underwent reversible decarbonylation to give a mixture of two isomers of $(C_5H_5Co)Fe_2Te_2(CO)_6$ as established by ^{125}Te NMR spectroscopy.

The mononuclear $[(\eta^5\text{-}C_5H_5)Nb(CO)_3(L)]$ (L = tellurocyclopentane) and its reactions were reported by Basalo and Freeman [127] in 1991. A novel stannyl tellurolate R_2SnTe^- was generated by reacting $[Cp^*Ti]_2(\mu\text{-}Te)]$ or $Cp^*{}_2Ti(\eta^2\text{-}Te_2)$ with stoichiometric amounts of R_3SnH, which probably first results in an unstable compound $Cp^*{}_2Ti(H)TeSnR_3$ and finally $Cp^*{}_2TiTeSnR_3$ (R=Bu, Ph) which contains a three coordinate complex of Ti [128]. The reaction of elemental tellurium with $Cp^*{}_2TiH$ results in paramagnetic $[(C_5Me_5)Ti]_2(\mu\text{-}Te)$ [129]. The Ti---Te bond length is 2.697(3) -2.705(3) Å. The Te insertion in $Cp^*{}_2TiSnPh_3$ prepared from Ph_3SnH and $Cp^*{}_2TiH$ also results in the same product. The reaction of $(\eta^5\text{-}C_5Me_4R)_2Zr(CO)_2$ (R=Me/Et) with elemental Te in the presence of pyridine forms $[(\eta^5\text{-}C_5Me_4R)_2Zr(Te)NC_5H_5]$ [130a]. Its Hf analogue is also synthesized [130b] The $(\eta^5\text{-}C_5Me_5)ZrTe(NC_5H_5)$ on reaction with Na_2O gives an oxo derivative$(\eta^5\text{-}C_5Me_5)ZrO(NC_5H_5)$ and with $Ph(CO)CH_3$ forms $(\eta^5\text{-}C_5Me_4R)_2ZrTeH(\eta^1\text{-}OC(Ph)\text{=}CH_2)$. The Zr=Te bond length is 2.729(1) Å, as predicted on the basis of covalent radii. The double bond covalent radius of Zr has been concluded as 0.07 Å shorter than that of a single bond. The $[(\eta^5\text{-}C_5Me_5)_2Sc(TeCH_2C_6H_4)]$ [131a] and $\{(CH_3)_2Si[(t\text{-}C_4H_9)C_5H_3]_2Sc(PMe_3)\}_2(\mu\text{-}Te)C_6H_6$ [131b] are made by insertion of Te in metal-carbon bond They form TeR_2 on thermal decomposition, but the telluroether is expelled from the metal's coordination sphere in the process. The $[(\eta^5\text{-}C_5Me_5)_2Sc]_2(\mu\text{-}Te)$ has also been synthesized [131a]. The monomeric $(\eta^5C_5R_5)TiCl\{2,2'\text{-}Te(4\text{-}Me\text{-}6\text{-}t\text{-}Bu\text{-}C_6H_2O)_2\}$(R = H, Me) has four-legged piano-stool geometry (Fig 6) [132]. Upon addition of methylaluminoxane, these complexes catalyzed the polymerization of ethylene. The activities of the tellurium-bridged complexes were significantly higher than those of the corresponding methylene-bridged complex [133].

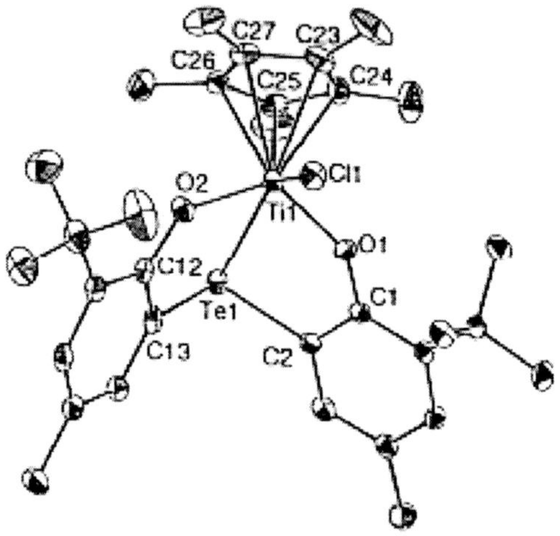

Figure 6. Molecular structure of ($\eta^5C_5Me_5$)TiCl{2,2'-Te(4-Me-6-t-Bu-$C_6H_2O)_2$}

In a recent study the X-ray crystal structure of [(C_5H_5)Fe{P(OPh)$_3$}$_2$($TeMe_2$)]BF_4 has been found to be on expected lines of half sandwich compounds [134]. Adducts of $TeMe_2$ and $TeEt_2$ with the metallocycle [($C_5H_5)_2Rh_2$(μ-CO)(CF_3CCCF_2)] where the added TeR_2 bonds to one Rh and the carbonyl group is now bonded η^1 to the other have been described [135]. Telluracyclopentane, $(CH_2)_4Te$, behaves similarly to $TeMe_2$ and the adduct has been characterized by an X-ray crystal structure. Rh(III) complexes [Rh(η^5-C_5Me_5){MeC(CH_2TeR)$_3$}][PF_6]$_2$ (R= Ph or Me) are formed by the reaction of [{(η^5-C_5Me_5)$RhCl_2$}$_2$], $TlPF_6$ and the ligands in refluxing methanol [136]. Comparison of their spectroscopic data with that of the [Rh(COD){MeC(CH_2TeR)$_3$}]PF_6 [136] shows that the tellurium ligands are superior σ -donors to Rh(I) than Se, but that to Rh(III) the donor ability is reversed, Se>Te. Similarly [Rh(η^5-C_5Me_5){Te{(CH_2)$_3$TePh}$_2$}][PF_6]$_2$ and [Rh(η^5-C_5Me_5){MeS(CH_2)$_3$Te(CH_2)$_3$SMe}][PF_6]$_2$ are prepared and characterized structurally by X-ray analysis [137] (See Fig. 7 for first). The Iridium analogues

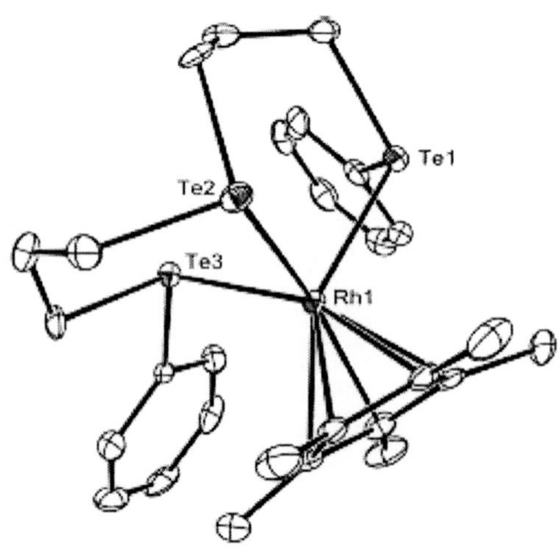

Figure 7. [Rh(η^5-C_5Me_5){Te{(CH_2)$_3$TePh}$_2$}]$^{2-}$

$[Ir(COD)\{MeC(CH_2TePh)_3\}]PF_6$ and $Ir(\eta^5\text{-}C_5Me_5)\{MeC(CH_2ER)_3\}][PF_6]_2$ were isolated and show parallel spectroscopic properties to the Rh(III) derivatives [136-137a]. Chiral cation $[CpFe(CO)(TeMe_2)L]^+$ where L = PPh_3, $AsPh_3$, $SbPh_3$, PMe_3, $P(OMe)_3$, $P(OPh)_3$ or $Ph_2P(CH_2)_nPPh_2$ (n =1, 2 or 3) is synthesized by direct reaction of L with $[CpFe(CO)(TeMe_2)]^+$ or $[CpFe(CO)_2(TeMe_2)]^+$ (after heating or irradiation) [138a]. In $[CpFe(CO)_2THF]BF_4$, THF is easily replaced by $TeMe_2$ resulting in $[CpFe(CO)_2TeMe_2]BF_4$ but for substitution of CO group photo-activation is essential. The $TeMe_2$ group can be substituted selectively by cyanide or iodide using KCN or LiI [138b]. The structure and dynamics of $[C_5H_5Fe(CO)_2(TeMe_2)]^+$ have been investigated by high-resolution solid-state ^{13}C NMR spectroscopy [139]. From the side bands in the ^{13}C CP-MAS NMR spectra, the principal values of the σ-tensors of the ring carbon atoms have been determined. They are axially symmetrical., pointing to mobility of the cyclopentadienyl ring. The static ^{13}C CP NMR spectra also reveal the dynamic nature of the C_5H_5 ring. From the spin lattice relaxation times of the ring carbon atoms it is concluded that the reorientation rate of the ring increases in the order S < Se < Te, probably due to the interaction of the lone pair electrons of the chalcogen with the electrons in the ring system. An extended H.ovrddot.uckel MO study [140] based on X-ray structure data of $[C_5H_5Fe(CO)_2(ER_2)]BF_4$ complexes (E = S, R = CH_3, C_6H_5; E = Se, Te, R = CH_3) has supported the rationalization of the inertness and stability of the Fe-E bond in these cations in the order E = Te > > Se > S > O. No important π-bonding effect is found. The reaction of $[Cp^{\#}{}_2NbTe_2H]$ ($Cp^{\#}$ = C_5Me_5, C_5Me_4Et} with $[Co_2(CO)_8]$ under different conditions gives Co carbonyl telluride clusters that contain different types of niobocene carbonyl fragments [141]. The $[Cp^{\#}{}_2NbTe_2Co_3(CO)_7]$ are two such examples $\{[Cp^{\#}{}_2Nb(CO)_2]\}_2[Co_9Te_6(CO)_8]$. With $[Fe_2(CO)_9]$ similar reaction in boiling toluene gives $[Fe_3(\mu_2\text{-}H)(\mu_3\text{-}Te)(CO)_9][Cp^{\#}{}_2Nb(CO)_2]$ [142]. $[Cp'_2Nb(Te_2)H]$ and $[Cp'_6Nb_4Te_4O]$ (Cp' = t-BuC_5H_4) formed by reaction of $[Cp'_2NbH_3]$ with Te powder in THF react with $[Cr(CO)_5THF]$ to give $[Cp'_2Nb(Te_2)H{\cdot}Cr(CO)_5]$ and $[Cp'_6Nb_4Te_4O{\cdot}2Cr(CO)_5]$ respectively [143]. The products are characterized by X-ray analysis. In the first a Te_2 unit and an H ligand are coordinated to a bent niobocene moiety; the $Cr(CO)_5$ group is attached to the lateral Te atom. Thermolysis of first leads to the formation of diamagnetic $[Cp'_4Nb_2Te_2]$ which contains a planar Nb_2Te_2 core. The reaction of $[Cp'_2Nb(Te_2)H]$ with two equivalents of $[Co_2(CO)_8]$ in boiling toluene gave the dark red neutral compound $[\{Cp'_2Nb(CO)\}_3Co_9(CO)_8Te_6]$ which on reacting with an excess of $[Cr(CO)_5(THF)]$ in THF gave the red brown and dark brown complexes $[\{Cp'_2Nb(CO)\}_3Co_9(CO)_8Te_6{\cdot}xCr(CO)_5]$ (x = 1or 3). The crystal structure of compound with x =3 is solved [144] The $[\{(C_5HR_4)NiTe\}_2]$ (R = $CHMe_2$) has been synthesized from $[\{(C_5HR_4)NiBr\}_2]$ and Na_2Te. Crystal structure analyses show four-membered Ni_2Te_2 rings with weak, but significant Te...Te interaction. Its electrochemical analysis has revealed nickel-centered oxidations, bridging ligand-centered reductions and the formation of mixed-valent Ni(II)/Ni(III) species [145]. The telluration of Cp^*WCl_4 by $Te(SiMe_2tBu)_2$ in $CHCl_3$ solution leads to dimeric products *anti*-$[Cp^*W(E)(\mu\text{-}E)]_2$ (E = Te). In the presence of air the two terminal telluride ligands are replaced quantitatively by oxido ligands (O) to give $[Cp^*W(O)(\mu\text{-}Te)]_2$ [146]. The $TeMe_2$ can bridge $[CpMn(CO)_2]$ when $[CpMn(CO)_2THF]$ is reacted with $TeMe_2$ in 2:1 ratio [35]. The $[CpMo(CO)_3TeMe_2]$ and $[CpMo(CO)_2(TeMe_2)_2]$ are obtained from $[CpMo(CO)_3I]$ (reacting with $AgBF_4$ and telluride) and $[CpMo(CO)_2]_2$ (oxidation by $[Cp_2Fe]BF_4$ in the presence of $TeMe_2$) [147]. The oxidation of $CpCo(CO)_2$ with $[Cp_2Fe]BF_4$ in the presence of SMe_2 gives $[CpCo(SMe_2)_3]^{2+}$ in which

SMe_2 is easily substituted by $TeMe_2$ giving $[CpCo(TeMe_2)_3]^{2+}$ [148]. Scantly known telluroaroyl ligand bridging is observed in $[WFe(\mu\text{-}TeCR^1)(CO)_5(\eta^5\text{-}C_5H_5)]$, $[WFe(\mu\text{-}TeCR^1)(\mu\text{-}dppm)(CO)_3(\eta^5\text{-}C_5H_5)]$, $[MoFe(\mu\text{-}TeCR^2)(CO)_5(\eta^5\text{-}C_5H_5)]$ and $[MFe_2(\mu\text{-}CR^2)(\mu_3\text{-}Te)(CO)_7(\eta^5\text{-}C_5H_5)]$ ($R^1 = C_6H_4Me\text{-}4$, $R^2 = C_6H_3Me_2\text{-}2,6$, M = Mo or W) [149]

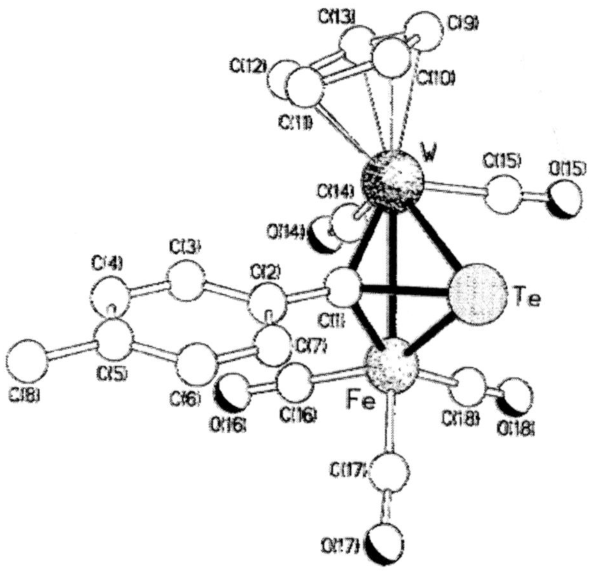

Figure 8. Molecular structure of $[WFe(\mu\text{-}TeCR^1)(CO)_5(\eta^5\text{-}C_5H_5)]$

The triply bridged dinuclear complex $[(\eta^5\text{-}MeC_5H_4)Mo_2(CO)_4(\mu\text{-}TePh)_3]$ is also formed by reaction of $[(\eta^5\text{-}MeC_5H_4)_2Mo_2(CO)_4]$ with Ph_2Te_2 along with $[(\eta^5\text{-}MeC_5H_4)_2Mo_2(CO)_4(\mu\text{-}TePh)_2]$ [150]. However, the doubly bridged dinuclear complex is formed when $[(\eta^5\text{-}RC_5H_4)_2M_2(CO)_4]$ (M= Mo or W) is treated with Ph_2Te_2 for other R groups.

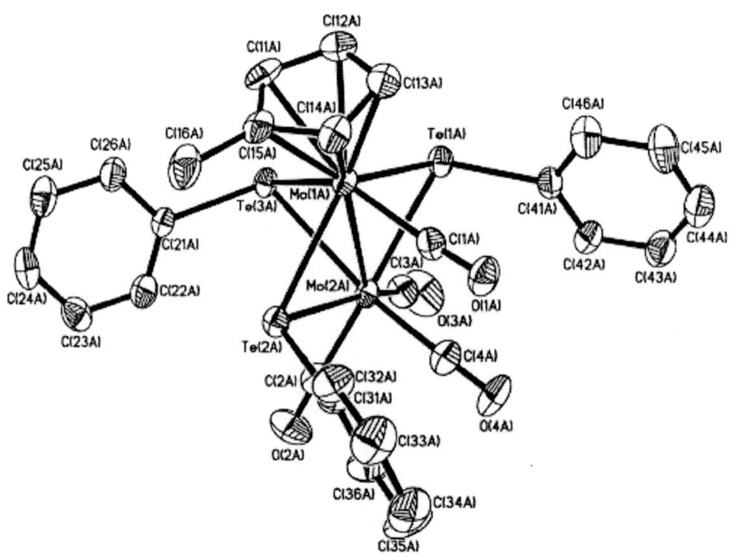

Figure 9. Molecular structure of $[(\eta^5\text{-}MeC_5H_4)Mo_2(CO)_4(\mu\text{-}TePh)_3]$

The reaction of triply bonded dimers [η^5-$RC_5H_4(CO)_2Mo]_2$ (R = MeCO, MeO_2C) with Ph_2Te_2 in xylene at 110 °C and subsequent treatment with Cp_2TiCl_2 at reflux gives novel tetrakis-bridged complexes (η^5-$RC_5H_4Mo)_2(\mu$-$PhTe)_4$ and (η^5-$RC_5H_4Mo)_2(\mu$-Cl)(μ-$PhTe)_3$ [151]. TheCp_2ZrCl_2 or Cp_2ZrBr_2 may be used in place of Cp_2TiCl_2 but the role of all the three is unclear. However, single crystal structure of only (η^5-$RC_5H_4Mo)_2(\mu$-Cl)(μ-$PhTe)_3$ was solved. Photolysis of the half sandwich tetracarbonylmetal complexes $CpV(CO)_4$, $Cp*V(CO)_4$ and $Cp*Ta(CO)_4$ in solution in the presence of Te_2R_2 (R = Me, Ph, Fc) leads to diamagnetic doubly organochalcogenolato-bridged compounds like, [$Cp*M(CO)_2(\mu$-ER)]$_2$ (E = Te)[152]. According to the X-ray structure determinations carried out for [$CpV(CO)_2(\mu$-TeMe)]$_2$ and [$Cp*V(CO)_2(\mu$-TePh)]$_2$ the molecular framework consists of a folded $M_2(\mu$-TeR$)_2$ ring with the cyclopentadienyl ligands in *cis*-configuration and the R in a *syn*-equatorial arrangement, thus forming a bowl-shaped molecule with the four terminal CO ligands protruding into the inner sphere. There is no V...V interaction.

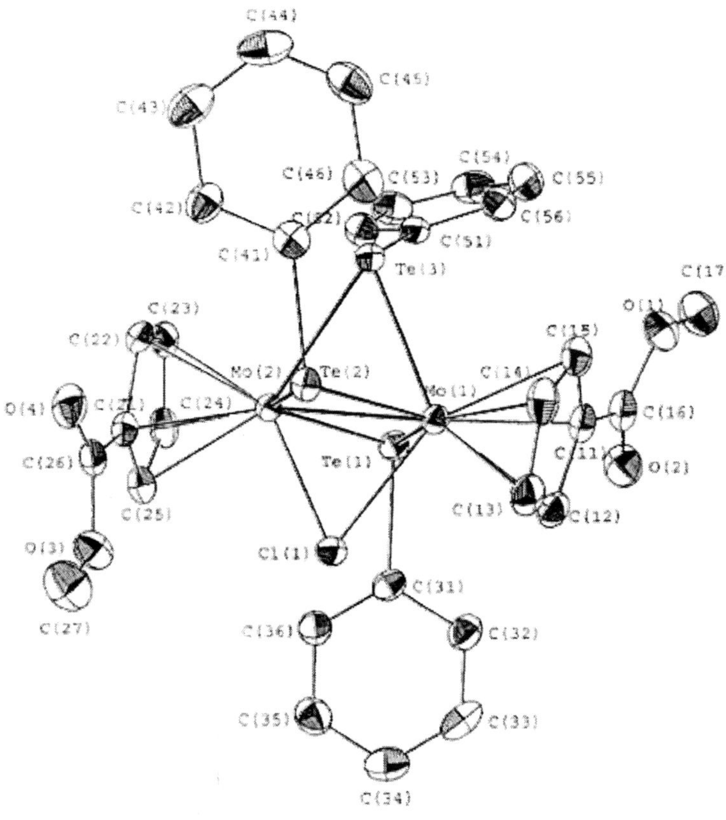

Figure 10. Molecular structure of (η^5-$RC_5H_4Mo)_2(\mu$-Cl)(μ-$PhTe)_3$

The [$CpOV(CO)_2(\mu$-ER)]$_2$ is completely decarbonylated in the presence of an excess of Te_2R_2 in boiling toluene, and the paramagnetic quadruply-bridged products, [$CpV(\mu$-ER$)_2]_2$, is isolated. In the reaction of [(η^5-$C_5Me_5)RuCl_2]_2$ with Me_3SiTeR (R=4-MeC_6H_4, C_6H_5), R_2Te_2 is generated at the metal center, resulting finally in (η^5-$C_5Me_5)Ru(\mu_2$-RTeTeR)(μ_2-TeR$)_2$Ru(η^5-C_5Me_5) [153] where the bridging by ditelluride ligand is very symmetrical (both Ru---Te=2.540(3) Å). The reaction [154] of [{Rh(η^5-$C_5Me_5)Cl_2\}_2$] with tellurophene

(C_4H_4Te) in the presence of $Ag(O_3SCF_3)$ gives $[Rh(\eta^5-C_5Me_5)(\eta^5-C_4H_4Te](O_3SCF_3)_2$ which, on reaction with $[Co(\eta^5-C_5H_5)_2]$, gives $[Rh(\eta^5-C_5Me_5)\{\gamma^5-C_5H_4Rh(\eta^5-C_5Me_5)\}]$, which reacts with $Fe_3(CO)_{12}$, to give another novel compound $[Rh(\eta^5-C_5Me_5)\{\eta^5-C_4H_4Fe(CO)_3\}]$. The reaction of $[\{Rh(\eta^5-C_5Me_5)Cl_2\}_2]$ with dibenzotellurophene, dbt, in the presence of $Ag(O_3SCF_3)$ gives $[Rh(\eta^5-C_5Me_5)(dbt)][O_3SCF_3]_2$ in which the mode of coordination of dbt in $CDCl_3$ solution appears to be η^1. It reacts with $[Co(\eta^5-Cp)_2]$ and $[Fe_3(CO)_{12}]$ to give a variety of products including $[Rh(\eta^5-C_5Me_5)(C_{12}H_8)(\eta^1-dbt)]$. Half sandwich compounds containing 2-telluraindane as ligand, $[(Cp^*Rh)(C_8H_8Te)][O_3SCF_3]_2$ and $[(Cp^*RhCl_2)(C_8H_8Te)]$ have been synthesized and their dissociation in solution investigated by ^{125}Te NMR [155]. Reaction of $Fe_3(CO)_9(\mu_3-Te)_2$ with $(\eta^5-C_5H_5)M(CO)_3(C\equiv CPh)$ (M = Mo, W), in the presence of trimethylamine N-oxide gives the novel mixed-metal clusters $(\eta^5-C_5H_5)MFe_2(\mu_3-Te)_2(CO)_7(\eta^1-C\equiv CPh)$ bearing η^1-acetylide groups [156]. The reaction of Mo containing cluster with dicobalt octacarbonyl at room temperature gives the mixed-metal clusters $(\eta^5-C_5H_5)MoFe_2Co_2(\mu_3-Te)_2(CO)_9(\mu-CCPh)$. Thermolysis of a toluene solution containing $[Fe_3(CO)_9(\mu_3-Te)_2]$ and $[W(\eta^5-C_5H_5)(CO)_3(C\equiv CPh)]$, resulted in coupling of coordinated acetylide ligands resulting in formation of mixed-metal clusters $[W_2Fe_3(\eta^5-C_5H_5)_2(CO)_6(\mu_3-Te)_2\{\mu-CC(Ph)C(Ph)C\}]$ [157].

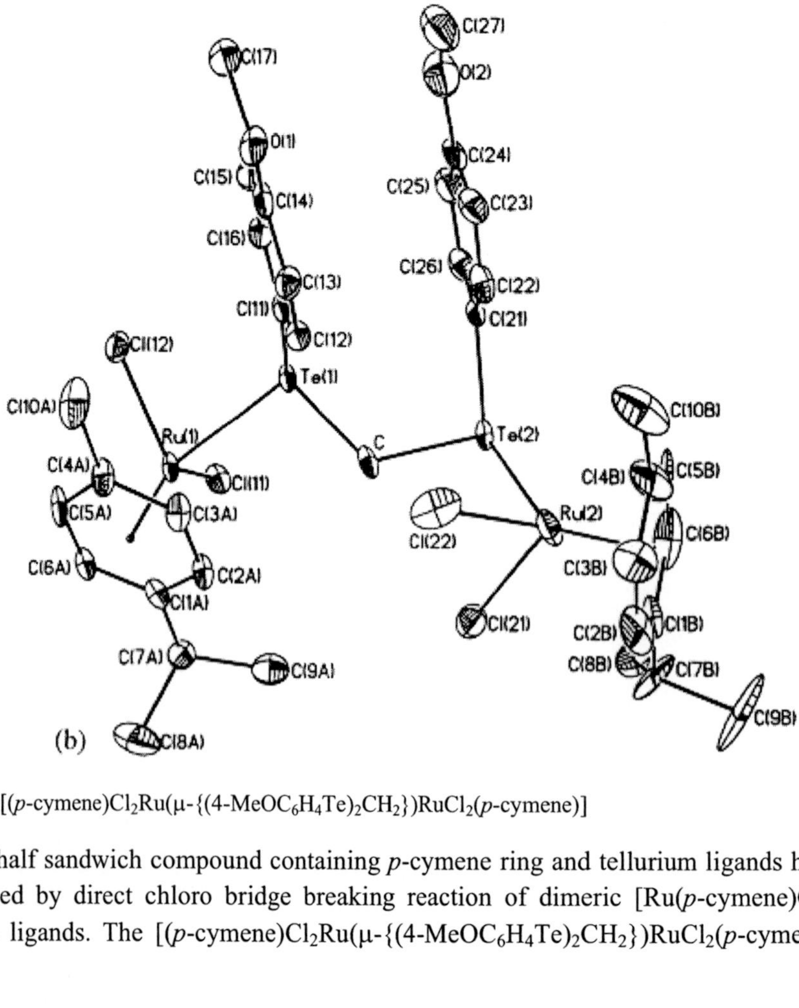

Figure 11. $[(p\text{-cymene})Cl_2Ru(\mu-\{(4-MeOC_6H_4Te)_2CH_2\})RuCl_2(p\text{-cymene})]$

The half sandwich compound containing *p*-cymene ring and tellurium ligands have been synthesized by direct chloro bridge breaking reaction of dimeric $[Ru(p\text{-cymene})Cl_2]_2$ and tellurium ligands. The $[(p\text{-cymene})Cl_2Ru(\mu-\{(4-MeOC_6H_4Te)_2CH_2\})RuCl_2(p\text{-cymene})]$ was first such compound [158] in which a ditelluroether ligand bridges two half sandwich ruthenium centers (Fig 11). The ligands *N*-[2-(4-methoxyphenyltelluro)ethyl]phthalimide [159], bis(2-{1,3-dioxan-2-yl}ethyl) telluride[160], 2-(4-ethoxyphenlytelluromethyl) tetrahydro-2*H*-pyran [161], 2-(2-{4-ethoxyphenyl}telluroethyl)-1,3-dioxane[161] and bis[2-(2-thienyl)ethyl] telluride [162] on reaction with $[Ru(p\text{-cymene})Cl_2]_2$ give compounds of the type $[Ru(p\text{-cymene})Cl_2{\cdot}L]$ which are characterized X-ray diffraction on their single crystals. In all such compounds the potentially polydentate Te-ligands coordinate through Te only. Structure of one such compound is shown in Fig. 12. However, the reaction of 4-$MeOC_6H_4TeCH_2CH_2NHCH(CH_3)C_6H_4$-2-OH with $[Ru(p\text{-cymene})Cl_2]_2$ has resulted in the substitution of *p*-cymene ring with the (Te, N,O) ligand [163]. The resulting molecule $[RuCl\{4\text{-}MeOC_6H_4TeCH_2CH_2NHCH\ (CH_3)C_6H_4\text{-}2\text{-}O^-\}]$ has interesting structural feature of Te····Cl (3.322(2) to 3.805(2) Å) secondary interaction (Fig. 13)

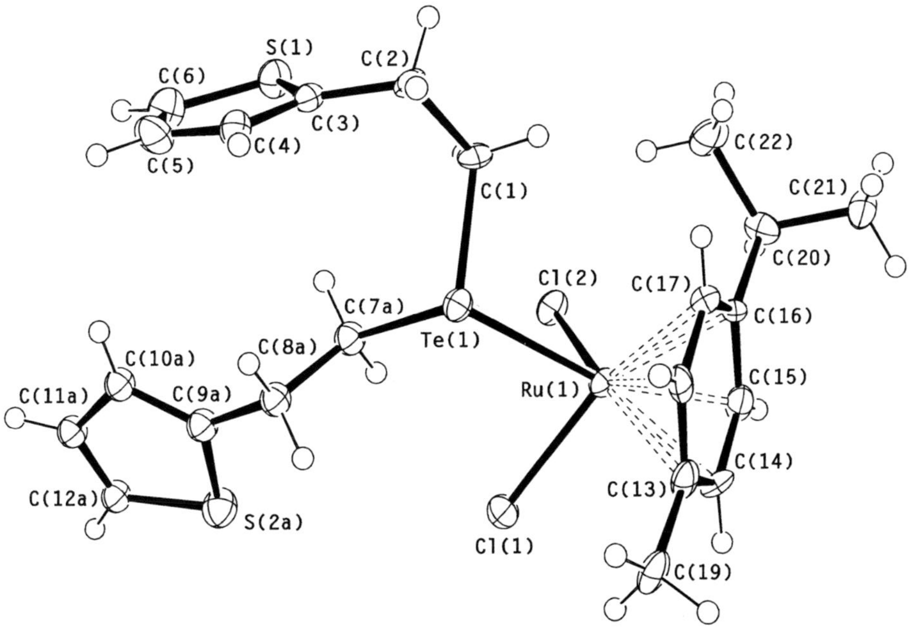

Figure 12. {Bis[2-(2-thienyl)ethyl] telluride}dichloro(*p*-cymene)ruthenium(II)

Miscellaneous Organometallic Compounds Containing Te-Ligands

The number of metal alkyls coordinated with tellurium ligands known so far is not large. $[PtIMe_3L]$ [L = $MeTe(CH_2)_3TeMe$, $PhTe(CH_2)_3TePh$ or $o\text{-}C_6H_4(TeMe)_2$] have been isolated as pure solids [164]. $^{195}Pt\{^1H\}$ and $^{125}Te\{^1H\}$ NMR studies show that ambient temperature solutions of the complexes with aliphatic ligand backbones consists predominantly of DL isomers. Pyramidal Te inversion inter-converts the isomeric species and precise rates of this process were measured at near-ambient temperatures by $^{195}Pt\{^1H\}$ and $^{125}Te\{^1H\}$ NMR 2-dimensional exchange spectroscopy or by ^1H 1-dimensional NMR band shape analysis at above-ambient temperatures Te pyramidal inversion-energy barriers were found 70-84 kJ mol^{-1}, reflecting the nature of the substituent at Te and the ligand backbone. The trend in inversion energies was Te > Se > S. Trimethylplatinum(IV) iodide reacts with R_2Te_2 (R = Me, p-$MeOC_6H_4$, p-$EtOC_6H_4$) or $RTeCH_2TeR$(R = p-$MeOC_6H_4$), to give bimetallic complexes [165] of the type [$(PtIMe_3)_2L$] Ditelluroether also forms [$(PtIMe_3)(RTeCH_2TeR)$] and [$(PtIMe_3)(RTeCH_2TeR)_2$] (R = C_6H_5). Trimethylplatinum(IV) complexes of the type [$PtMe_3X(Me_2NCH_2CH_2TePh)$] (X = Cl, Br, I) have been prepared [166] in 23-53% yield but characterized only by NMR (^1H and ^{13}C) spectral data.

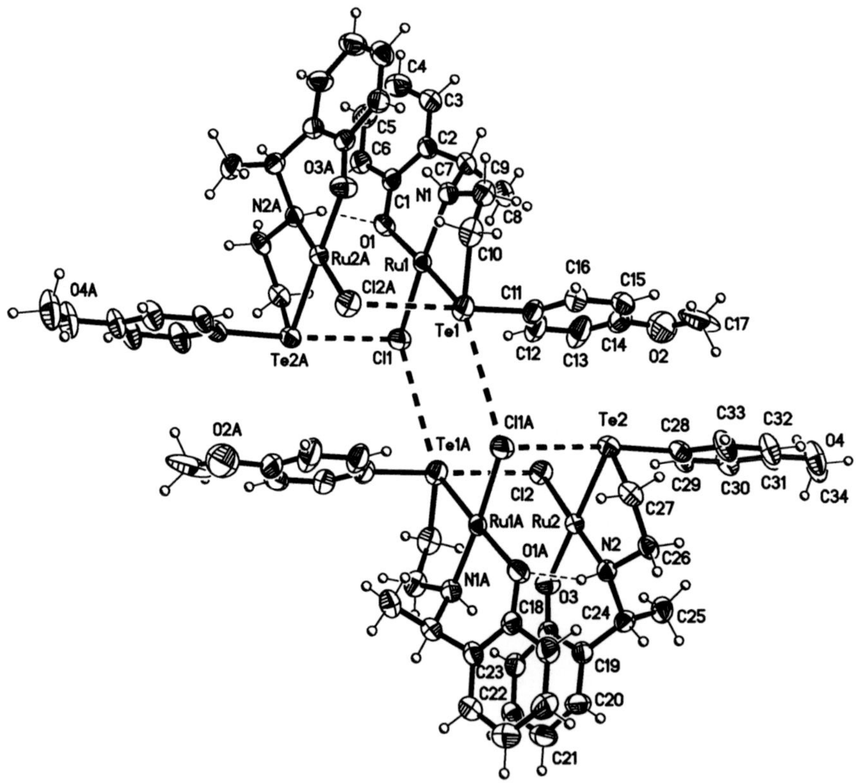

Figure 13. Molecular structure of [RuCl{4-$MeOC_6H_4TeCH_2CH_2NHCH(CH_3)C_6H_4$-2-$O^-$}]

Recently several complexes [$PdCl(TeCH_2CH_2NMe_2)(PR_3)$] (R = Et, *n*-Pr, *n*-Bu, Ph, tolyl or PR_3 = PMe_2Ph, $PMePh_2$), and [$PtCl(TeCH_2CH_2NMe_2)(PR_3)$] ($PR_3$ = PMe_2Ph, $PMePh_2$), are synthesized [167]. The structures of [$PdCl(TeCH_2CH_2NMe_2)$(*n*-Pr_3P)], [$PdCl(TeCH_2CH_2NMe_2)(PMePh_2)$], and [$PtCl(TeCH_2CH_2NMe_2)(PMePh_2)$] were determined crystallographically, revealing a typical pattern of *trans*-positioned neutral N and P donor atoms in an approximately square planar setting. The TD-DFT methodology is used to understand the origin of their weak long-wavelength bands (e.g. observed at about 460 nm for Pt/Te complexes and 600 nm for Pd/Te complexes). These transitions are identified as charge transfer transitions from the selenolato or tellurolato centers to unoccupied orbitals involving mainly the phosphine co-ligands for the Pt(II) compounds and more delocalized MOs for the Pd(II) analogues. The thermodynamic quantities for 1:1 complex of diethylzinc with diethyltellurium, has been studied between 13 and 306 K in a vacuum adiabatic calorimetric cryostat [168]. The $^{69,71}Ga$ and ^{115}In NQR spectra of complexes of alkyl derivatives of gallium and indium $R_3Ga/In{\cdot}Et_2Te$ were studied [169]. Not only electron transfer but also geometry of the complexes influences the behavior of the quadrupole coupling constants of gallium in the complexes. In a series of alkyl and haloalkyl gallium derivatives, the behavior of quadrupole coupling constants is determined to a great extent by chemical properties of the substituents at the central atom. The oxidative addition of Ph_2TeCl_2 to $[PtMe_2(bu_2bpy)]$ $[bu_2bpy$ = 4,4'-di-tert-butyl-2,2'-bipyridine] gives an organoplatinum(IV) complex which may be formulated as $[PtClMe_2(TePh_2)(bu_2bpy)]Cl$ or as a neutral complex $[PtClMe_2(TePh_2Cl)(bu_2bpy)]$ The complex contains an unusually long Te····Cl bond length of 3.43 Å. The weakly bound chloride ligand is easily removed by reaction with AgO_3SCF_3 to give the cationic complex $[PtClMe_2(TePh_2)(bu_2bpy)](CF_3SO_3)$ in which the triflate anion is not coordinated to tellurium. This complex reacts with a second one equivalent of AgO_3SCF_3 to give an aqua complex $[PtMe_2(OH_2)(TePh_2..O_3SCF_3)(bu_2bpy)]$ (CF_3SO_3), in which a triflate anion forms secondary bonds with both the aqua and $TePh_2$ ligands. In these platinum(IV) complexes, the magnitude of the coupling constant $^1J(Pt, Te)$ is strongly influenced by the presence of the weakly bonded Te·····X groups.

ACKNOWLEDGEMENT

Author thank Department of Science and Technology (India) and Council of Scientific and Industrial Research for financial support

REFERENCES

[1] Kanatzidis, M.G.; Huang, S.P. *Coord. Chem. Rev.* 1994, 130, 509 - 621.

[2] Dean, P.A.W.; Vittal, J. In *Metallothioneins*; Stillman, M.J.; Shaw III, F.C.; Suzuki, K.J.; Eds.; VCH, New York 1992; pp 346-386.

[3] Beck, J.. *Angew. Chem.* 1994, 106, 172-182.

[4] Ansari, M.; McConnachie, J.M.; Ibers, J.A.. *Acc. Chem. Res.* 1993, 26, 574- 578.

[5] Singh, A. K.; Srivastava, V. *J. Coord. Chem.,* 1992, 27, 237-253.

[6] Hope, E.G.; Levason, W. *Coord. Chem. Rev.* 1993, 122, 109-170.

[7] Levason, W.; Orchard, S. D.; Reid, G. *Coord. Chem. Rev.* 2002, 225, 159-199.

[8] Singh, A. K.; Sharma, S. *Coord. Chem. Rev.* 2000, 209, 49-98.

[9] Gysling, H. J. *Coord. Chem. Rev.* 1982, 42, 113-244.

[10] Gysling H.J. In *Chemistry of Organic Selenium and Tellurium Compounds;* Patai S.; Rappoport, Z. Eds, Wiley, New York 1986; Vol. 1, pp 680-855.

[11] Berry, F.J.. In *Comprehensive Coordination Chemistry,* Wilkinson, G.; Gillard R.D.; McCleverty, J.A. Eds, Pergamon, Oxford, 1987, pp. 661- 674.

[12] Mathur, P. *Adv. Organomet. Chem.* 1997, 41, pp 243-314.

[13] Sadekov, I. D.; Uraev, A. I.; Garnovskii, A. D. *Russian Chem. Rev.* 1999, 68, 415-433.

[14] Arnold, J. *Prog. Inorg. Chem.* 1995, 43, pp 353-417.

[15] Srivastava, V.; Batheja, R.; Singh, A. K. J. Organomet. Chem., 1994, 484, 93-96.

[16] Batheja, R.; Singh, A. K., *Polyhedron,* 1997, 16, 4337-4345.

[17] Batheja, R.; Singh, A. K., *Polyhedron*, 1997, 16, 2509-2516.
[18] Koepf, H.; Klapoetke, T. *J. Chem. Soc. Chem. Comm.* 1986, 1192-1193
[19] Meunier, P.; Gautheron, B.; Mazouz, A. *J. Organomet. Chem.*1987, 320, C39-C43.
[20] (a)G. Singh, S. Bali, A. K. Singh, J. E. Drake, C. L. B. Macdonald, M. B. Hursthouse, M. E. Little, *Inorg. Chim. Acta*, 2005, 358, 912-918 (b)Hieber, W.; Kruck, T. *Chem. Ber.* 1962, 95, 2027-2041.
[21] Hausmann, H.; Hoefler, M.; Kruck, T.; Zimmermann, H. W. *Chem. Ber.* 1981, 114, 975-981.
[22] Öfele, K.; Dotzauer, E. *J. Organomet Chem.* 1972, 42, C87-C90.
[23] Garnovskii, A. D.; Abakarov, G. M.; Sadekov, I. D.; Minkin, V. I.; Cherkasova, T. G.; Varshavskii, Yu. S. *Koord. Khim.* 1984, 10, 234; *Chem. Abstr.* 1984, 100, 203373t.
[24] Schumann, H.; Weis, R. *Angew. Chem. International Ed. Eng.*1970, 9, 246-247
[25] Lappert, M. F.; Martin, T. R.; McLaughlin, G. M. *J. Chem. Soc. Chem. Commun.* 1980, 635-637.
[26] Fischer, H.; Zeuner, S. *J. Organomet. Chem.* 1983, 252, C63-C65.
[27] Fischer, H.; Fischer, E. O.; Cai, R.; Himmelreich, D. *Chem. Ber.* 1983, 116, 1009-1016.
[28] Headford, C. E. L.; Roper, W. R. *J. Organometal. Chem.* 1983, 244, C53-C56.
[29] Calderazzo, F.; Vitali, D.; Poli, R.; Atwood, J. L.; Rogers, R. D.; Cummings, J. M.; Bernal, I. *J. Chem. Soc. Dalton Trans.* 1981, 1004-1009.
[30] Atwood, J. L.; Bernal, I.; Calderazzo, F.; Canada, L. G.; Poli, R.; Rogers, R. D.; Veracini, Carlo A.; Vitali, D. *Inorg. Chem.* 1983, 22, 1797-804.
[31] Schulze, W.; Ziegler, M. L. *Z. Anorg. Allgem. Chem* 1981, 481 78-94; (b) Schermer, E. D.; Baddley, W. H. J. Organomet. Chem. 1971, 30, 67-73.
[32] Küllmer, V.; Vahrenkamp, H. *Chem. Ber.* 1977, 110, 228-236.
[33] Al-Rubaie, Ali Z.; Al-Obaidi, Yassin N.; Yousif, Lina Z *Polyhedron* 1990, 9, 1141-1146.
[34] Shen, J. K.; Gao, Y.; Shi, Q.; Rheingold, A. L.; Basolo, F. *Inorg. Chem.* 1991, 30, 1868-1873.
[35] Bremer, G.; Boese, R.; Keddo, M.; Kruck, T. *Z. Naturforsch. Teil B* 1986, 41B, 981-986
[36] Fischer, H.; Gerbing, U. *J. Organomet. Chem.* 1986, 299, C7-C10.
[37] Jaitner, P.; Winder, W. *Inorg. Chim. Acta* 1987, 134, 201-202.
[38] Liaw, W.-F.; Chiang, M.-H.; Lai, C.-H.; Chiou, S.-J.; Lee, G.-H.; Peng, S.-M. *Inorg. Chem.* 1994, 33, 2493-2494.
[39] Liaw, W.-F.; Horng, Y.-C.; Ou, D.-S.; Chuang, C.-Y; Lee, C.-K. *J. Chin. Chem. Soc. (Taipei)* 1995, 42, 59-65.
[40] Gao, Y.; Yang, X.; Shi, Q. *Inorg. Chim. Acta* 1995, 240, 661-663.
[41] Coleman, A. P.; Ke, M.; Dickson, R. S.; Deacon, G. B.; West, B. O. *Polyhedron* 1994, 13, 2301-2310.
[42] Thaler, G.; Wurst, K.; Sladky, F. *Organometallics* 1996, 15, 4639-4641.
[43] Liaw, W.-F.; Ou, D.-S.; Li, Y.-S.; Lee, W.-Z.; Chuang, C.-Y.; Lee, Y.-P.; Lee, G.-H.; Peng, S.-M. *Inorg. Chem.* 1995, 34, 3747-3754.
[44] Liaw, W.-F.; Chiou, S.-J.; Lee, W.-Z.; Lee, G.-H.; Peng, S.-M. *J. Chin. Chem. Soc. (Taipei)* 1996, 43, 29-35.
[45] Liaw, W.-F.; Chuang, C.-Y.; Lee, W.-Z.; Lee, C.-K.; Lee, G.-H.; Peng, S.-M. *Inorg. Chem.* 1996, 35, 2530-2537.

[46] Levason, W.; Reid, G.; Tolhurst, V.-A. *J. Chem. Soc., Dalton Trans.* 1998, 3411-3416.
[47] Liaw, W. F.; Chiang, M. H.; Liu, C. J.; Harn, P. J.; Liu, L. K. *Inorg. Chem.* 1993, 32, 1536-1538.
[48] Liaw, W. F.; Chiou, S. J.; Lee, W. Z.; Lee, G. H.; Peng, S. M. *J. Chin. Chem. Soc. (Taipei) 1993*, 40, 361-365.
[49] Liaw, W. F.; Horng, Y. C.; Ou, D. S.; Lee, G. H.; Peng, S. M. *J. Chin. Chem. Soc. (Taipei)* 1993, 40, 367-371.
[50] Liaw, W. F.; Lai, C. H.; Chiang, M. H.; Hsieh, C. K.; Lee, G. H.; Peng, S. M *J. Chin. Chem. Soc. (Taipei)* 1993, 40, 437-444.
[51] Barton, A. J.; Levason, W.; Reid, G. *J. Organomet. Chem.* 1999, 579, 235-242.
[52] Levason, W.; Patel, B.; Reid, G.; Ward, A. J. *J. Organomet. Chem.* 2001, 619, 218-225.
[53] Barton, A. J.; Connolly, J.; Levason, W.; Mendia-Jalon, A.; Orchard, S. D.; Reid, G. *Polyhedron* 2000, 19, 1373-1379.
[54] Barton, A. J.; Levason, W.; Reid, G.; Ward, A. J. *Organometallics* 2001, 20, 3644-3649.
[55] Liu, X.; Gao, Y.-C.; Su, Z.-X.; Wang, Y.-Y.; Shi, Q.-Z.. *Transition Met. Chem.* 1999, 24, 666-668.
[56] Levason, W.; Orchard, S. D.; Reid, G.. *Organometallics* 1999, 18, 1275-1280.
[57] Connolly, J.; Genge, A. R. J.; Levason, W.; Orchard, S. D.; Pope, S. J. A.; Reid, G. *J. Chem. Soc. Dalton Trans.* 1999, 2343-2352.
[58] Levason, W.; Orchard, S. D.; Reid, G.. *J. Chem. Soc., Dalton Trans.* 1999, 823-824.
[59] Kaur, R.; Singh, H. B.; Butcher, R. J. *Organometallics* 1995, 14, 4755-4763.
[60] Oilunkaniemi, R.; Laitinen, R. S.; Ahlgren, M. *Inorg. Chem. Comm.* 2000, 3, 8-10.
[61] Singh, A. K.; Misra, M.; Kadarkaraisamy, M. *Phosphorus, Sulfur and Silicon* 1998, 143 101-114.
[62] Dickson, R. S.; Yin, P.; Ke, M.; Johnson, J.; Deacon, G. B. *Polyhedron* (1996), 15, 2237-2245.
[63] Schiemann, J.; Huebener, P.; Weiss, E. *Angew Chem.* 1983, 95, 1021.
[64] Lesch, D. A.; Rauchfuss, T. B. *J. Organomet. Chem.* 1980, 199, C6-C8.
[65] Day, V. W.; Lesch, D. A.; Rauchfuss, T. B. *J. Am. Chem. Soc.* 1982, 104, 1290-1295.
[66] Hieber, W.; Gruber, J. *Z. Anorg. Allgem. Chem.* 1958, 296, 91-103.
[67] Cetini, G.; Stanghellini, P. L.; Rossetti, R.; Gambino, O. *J. Organomet. Chem.* 1968, 15, 373-381.
[68] Lesch, D. A.; Rauchfuss, T. B. *Organometallics* 1982, 1, 499-506.
[69] Grobe, J.; Vetter, J.; Krebs, B.; Pascaly, M. *Z. Anorg. Allgem Chem* 2000, 626, 430-443.
[70] Carey, J.; Fettinger, J. C.; Poli, R.; Smith, K. M. *Inorg. Chim. Acta* 2000, 299, 118-122.
[71] Liaw, W.-F.; Hsieh, C.-K.; Lin, G.-Y.; Lee, G.-H. *Inorg. Chem.* 2001, 40, 3468-3475.
[72] Dyson, P. J.; Hill, A. F.; Hulkes, A. G.; White, A. J. P.; Williams, David, J. *Angew. Chem., International Edn* 1999, 38, 512-514.
[73] Rehder, D.; Ihmels, K.; Wenke, D.; Oltmanns, P. Inorg.Chim Acta 1985, 100, L13-L14.
[74] Rossetti, R.; Stanghellini, P. L.; Gambino, O.; Cetini, G. *Inorg. Chim. Acta* 1972, 6, 205-10.
[75] Bogan, L. E., Jr.; Lesch, D. A.; Rauchfuss, T B. *J. Organomet. Chem.* 1983, 250, 429-438.
[76] Lesch, D. A.; Rauchfuss, T. B. *Inorg. Chem.* 1981, 20, 3583-3585.

[77] Strouse, C. E.; Dahl, L. F. *J. Am. Chem. Soc.* 1971, 93, 6032-6041.

[78] Sappa, E.; Gambino, O.; Cetini, G. J. Organomet. Chem. 1972, 35, 375-380.

[79] Johnson, B. F. G.; Lewis, J.; Lodge, P. G.; Raithby, P. R.; Henrick, K.; McPartlin, M. J. Chem. Soc. Chem. Commun 1979, 719-20.

[80] Pergola, R.D.; Ceriotti, A.; Cinquantini, A.; deBiani, F.F.; Garlaschelli, L.; Manassero, M. M.; Piacentini, R.; Sansoni M.; Zanello, P. *Organometallics* 1998, 17, 802-806.

[81] Shieh, M.; Chen, H.-S.; Yang, H.-Y.; Ueng, C.-H. *Angew. Chem., Int. Ed* 1999, 38, 1252-1254.

[82] Braunstein, P.; Rose, J; Toussaint, D.; Jaaskelainen, S.; Ahlgren, M.; Pakkanen, T. A.; Pursiainen, J.; Toupet, L.; Grandjean, D. *Organometallics* 1994, 13, 2472-2479.

[83] Das, B. K.; Kanatzidis, M. G. *Inorg. Chem.* 1995, 34 , 5721-5725

[84] Das, B. K.; Kanatzidis, M. G. *Inorg. Chem.* 1995, 34, 1011-1012.

[85] Stauf, S.; Reisner, C.; Tremel, W. J. Chem. Soc. Chem. Commun. 1996, 1749-1750.

[86] Martin, T. M.; Wood, P. T.; Schimek, G. L.; Pennington, W. T.; Kolis, J. W. *Inorg. Chem.* 1995, 34 4385-4391.

[87] Roof, L.C.; Smith, D.M.; Drake, G.W.; Pennington, W.T.; Kolis, J.W.. *Inorg. Chem.* 1995, 34, 337-345.

[88] van Hal, J.W. ; Whitmire, K.H.. *Organometallics* 1998, 17, 5197 - 5201

[89] Huang, K.C.; Shieh, M.H.; Jang, R.J.; Peng, S.M.; Lee, G.H.; Shieh, M. *Organometallics* 1998, 17, 5202-5205.

[90] Seidel, R.; Kliss, R.; Weissgraebar, S.; Henkel, G. *J. Chem. Soc. Chem. Commun.* 1994, 2791- 2792.

[91] Gervasio, *G. J. Organomet. Chem.* 1992, 441, 271-276.

[92] Mathur, P.; Mavunkal, I. J.; Thimmappa, B. H. S.; Rugmini, V.; S., B. B. S. *Proc. Indian Acad. Sci., Chem. Sci.* 1990, 102, 395-402.

[93] Mathur, P.; Reddy, V. D. *J. Organomet. Chem.* 1990, 387, 193-196.

[94] Mathur, P.; Mavunkal, I. J. *J. Organomet. Chem.* 1988, 350, 251-256.

[95] Mathur, Pradeep; Mavunkal, I. J. *Inorg. Chim. Acta* 1987, 126, L9-L10.

[96] Mathur, P.; Mavunkal, I. J.; Rheingold, A L *J. Chem. Soc., Chem. Commun.* 1989, 382-384.

[97] Mathur, P.; Mavunkal, I. J.; Rugmini, V. *Inorg. Chem.* 1989, 28, 3616-3618.

[98] Mathur, P.; Reddy, V. D. *J. Organomet. Chem.* 1990, 385, 363-368.

[99] Mathur, P.; Reddy, V. D.; Das, K.; Sinha, U. C. *J. Organomet. Chem.* 1991, 409, 255-261

[100] Mathur, P.; Chakrabarty, D.; Hossain, M. M.; Rashid, R. S.; Rugmini, V.; Rheingold, A. L. *Inorg. Chem.* 1992, 31, 1106-1108.

[101] Chakrabarty, D.; Hossain, M. M.; Kumar, R. K.; Mathur, P. *J. Organomet. Chem.* 1991, 410, 143-148.

[102] Mathur, P.; Manimaran, B.; Hossain, M. M.; Shanbag, R. Murthy, J.; Saranathan, I. S.; Satyanarayana, C. V. V.; Bhadbhade, M. M. *J. Organomet. Chem.* 1995, 490, 173-178.

[103] Mathur, P.; Hossain, M. M.; Umbarkar, S.; Satyanarayana, C. V. V.; Tavale, S. S.; Puranik, V. G. *Organometallics* 1995, 14, 959-963.

[104] Mathur, P.; Trivedi, R.; Sekar, P.; Chakrabarty, D.; Hossain, M.; Puranik, V. G. *J. Cluster Sci.*, 1996, 7, 191-198.

[105] Mathur, P.; Hossain, M. M.; Umbarkar, S. B.; Satyanarayana, C. V.; Puranik, V. G. *J. Organomet. Chem.*, !996, 506, 313-319

[106] Mathur, P.; Dash, A. K.; Hossain, M. M.; Umbarkar, S. B.; Satyanarayana, C. V. V.; Chen, Yu-S.; Holt, E. M.; Rao, S. N.; Soriano, M. *Organometallics* 1996, 15, 1356-1361.

[107] Mathur, P.; Sekar, P.; Rheingold, A. L.; Liable-Sands, L. M. *J. Chem. Soc., Dalton Trans.*, 1997, 2949-2954.

[108] Mathur, P.; Ghosh, S.; Sarkar, A.; Satyanarayana, C. V. V.; Puranik, V. G. *Organometallics* 1997, 16, 4392-4398.

[109] Mathur, P.; Ghosh, S.; Sarkar, A.; Satyanarayana, C. V. V.; Rheingold, A. L.; Liable-Sands, L. M. *Organometallics* , 1997, 16, 3536-3540.

[110] Mathur, P.; Mavunkal, I. J. *J. Organomet. Chem.* 1988, 350, 251-256.

[111] Mathur, P.; Reddy, V.; Bohra, R.. *J. Organomet. Chem.* 1991, 401, 339-346.

[112] Mathur, P.; Reddy, V. D.; Das, K.; Sinha, U. C. *J. Organomet. Chem.* 1991, 409, 255-261.

[113] Mathur, P.; Ghosh, S.; Sarkar, A.; Satyanarayana, C. V. V.; Drake, J. E.; Yang, J. *Organometallics* 1997, 16, 6028-6031.

[114] Mathur, P.; Ghosh, S.; Sarkar, A.; Rheingold, A. L.; Guzei, I. A. *Tetrahedron* 2000, 56, 4995-5000.

[115] Kuhn, N.; Winter, M. *J. Organomet. Chem.* 1983, 249, C28-C30.

[116] Kuhn, N.; Winter, M. *J. Organomet. Chem.* 1983, 252, C86-C88.

[117] Schermer, E. D.; Baddley, W. H *J. Organomet. Chem.* 1971, 27, 83-88.

[118] Rott, J.; Guggolz, E.; Rettenmeier, A.; Ziegler, M. L. *Z. Naturforsch.* 1982, 37B, 13-23.

[119] Sato, M.; Yoshida, T. *J. Organomet. Chem.* 1975, 87, 217-222.

[120] Sato, M.; Yoshida, T. *J. Organomet. Chem.* 1973, 51 231-235.

[121] Herrmann, W A.; Bauer, C.; Weichmann, J. J. Organomet. Chem. 1983, 243, C21-C26.

[122] Herrmann, W. A.; Hecht, C.; Ziegler, M. L.; Balbach, B. *J. Chem. Soc., Chem. Commun* 1984, 686-687.

[123] Shieh, M.; Chen, H.-S.; Lai, Y.-W., *Organometallics,* 2004, 23, 4018– 4025.

[124] Herberhold, M.; Reiner, D.; Neugebauer, D. *Angew. Chem.* 1983, 95, 46-47.

[125] Lesch, D. A.; Rauchfuss, T. B. *Inorg. Chem.* 1983, 22 , 1854-1857.

[126] Bogan, L E., Jr.; Lesch, D. A.; Rauchfuss, T. B. *J. Organomet. Chem.* 1983, 250, 429-438.

[127] Freeman, J.W. ; Basolo, F.. *Organometallics* 1991, 10, 256-263

[128] Fisher, J.M.; Piers, W.E.; Batchilder, S.D.; Pearce, Z.; Zaworotko, M.J. *J. Am. Chem. Soc.* 1996, 118, 283-284.

[129] Fisher, J. M.; Piers, W.E.; MacGillivary, L.R.; Zaworotko, M.J. *Inorg. Chem.* 1995, 34, 2499- 2500.

[130] Howard, W.A.; Parkin, G.. *J. Am. Chem. Soc.* 1994, 116, 606-615; (b) Howard, W. A.; Parkin, G. *J. Organomet. Chem.* 1994, 472, C1-C4

[131] Piers, W. E.; Parks, D. J.; MacGillivray, L. R.; Zaworotko, M. J. *Organometallics* 1994, 13, 4547-4558; (b) Piers, W E.; Ferguson, G.; Gallagher, J. F. *Inorg. Chem.* 1994, 33, 3784-3787; (c) Piers W.E. *J. Chem. Soc. Chem. Commun.* 1994, 309-310

[132] Nakayama, Y.; Watanabe, K.; Ueyama, N.; Nakamura, A.; Harada, A.; Okuda, J. *Organometallics* 2000, 19, 2498-2503.

[133] Froese, R.D.J; Musaev, D.G.; Morokuma, K. *Organometallics* 1999, 18 373-379.

[134] Schumann, H.; Lam, E. J.W.; Smits, J.M.M.; Beurskens, P.T. *J. Cryst. Spectroscopic Res.* 1991 21, 667-672.

[135] Devery, M. P.; Dickson, R.S.; Skelton B.W.; White A.H. *Organometallics* 1999, 18, 5292-5298.

[136] Levason, W.; Orchard, S.D.; Reid, G. ; Street, J.M. *J. Chem. Soc. Dalton Trans.* 2000, 2537-2543.

[137] Barton, A.J.; Levason, W.; Reid, G.; Ward, A.J. *Organometallics* 2001, 20, 3644-3649; Hesford, M.; Levason, W.; Orchard, S.D.; Reid, G., *J. Organomet. Chem.*, 2002, 649, 214-218.

[138] Kuhn, N.; Schumann, H.; Zauder, E *J. Organomet. Chem.* 1987, 327, 17-29; (b) Kuhn, N.; Schumann, H, *J. Organomet. Chem.,* 1984, 276, 55-66.

[139] Kentgens, A. P. M.; Karrenbeld, H.; De Boer, E.; Schumann, H , *J. Organomet. Chem.* 1992, 429, 99-108.

[140] Schumann, H.; Arif, A. M.; Rheingold, A. L.; Janiak, C.; Hoffmann, R.; Kuhn, N. *Inorg. Chem.* 1991, 30, 1618-1625.

[141] Brunner, H.; Lucas, D.; Monzon, T.; Mugnier, Y.; Nuber, B.; Stubenhofer, B.; Stuckl, A. C.; Wachter, J.; Wanninger, R.; Zabel, M. *Chem. Eur. J.* 2000, 6 , 493-503.

[142] Brunner, H.; Wachter, J.; Wanninger, R; Zabel, M. *J.Organomet. Chem.* 2000, 603, 135-137.

[143] Brunner, H.; Cattey, H.; Evrard, D.; Kubicki, M. M.; Mugnier, Y.; Vigier, E.; Wachter, J.; Wanninger, R.; Zabel, M. *Eur. J. Inorg. Chem.* 2002, 1315-1325.

[144] Brunner, H.; Stuckl, A. C.; Wachter, J.; Wanninger, R.; Zabel, M. *Angew. Chem Internl. Edn.* 2001, 40, 2463-2465.

[145] Sitzmann, H.; Saurenz, D.; Wolmershaeuser, G.; Klein, A.; Boese, R. *Organometallics* 2001, 20, 700-705.

[146] Herberhold, M.; Jin, G.-X.; Milius, W. *Z. Anorg. Allg. Chem.* 2001, 627, 1759-1766.

[147] Kuhn, N. Schumann, H., Zauder, E., *J. Organomet. Chem.* 1988, 354, 161-167.

[148] Kuhn, N.; Brueggemann, H.; Winter, M.; De Bellis, V. M. *J. Organomet. Chem.* 1987 , 320, 391-400.

[149] Hulkes, A. J.; Hill, A. F.; Nasir, B. A.; White, A. J. P.; Williams, D. J., *Organometallics* 2004, 23, 679 –686.

[150] Song, L-C.; Shi, Y-C.; Hu, Q.-M.; Chen, Y.; Sun, J. *J. Organomet. Chem.* 2001, 626, 192-198.

[151] Song, L.-C.; Shi, Y.-C.; Zhu, W.-F.; Hu, Q.-M.; Huang, X.-Y.; Du, F.; Mao, X-A. *Organometallics* 2000, 19, 156-162.

[152] Herberhold, M.; Peukert, J.; Kruger, M.; Daschner, D.; Milius, W. *Z. Anorg. Allg. Chem.* 2000, 626, 1289-1295.

[153] Matsuzaka, H.; Ogino, T.; Nishio, M.; Hidai, M.; Nishibayashi, Y., Uemura, S. *J. Chem. Soc. Chem. Commun.* (1994) 223-224.

[154] Badyal, K.; McWhinnie, W. R.; Chen, H. L.; Hamor, T. A. *J. Chem. Soc., Dalton Trans.* 1997, 1579-1585.

[155] Badyal, K.; McWhinnie, W. R.; Homer, J.; Perry, M. C. *J. Organomet. Chem.* 1998, 555, 279-284.

[156] Mathur, P.; Bhunia, A. K.; Kumar, A.; Chatterjee, S.; Mobin, S. M. *Organometallics* 2002, 21, 2215-2218.

[157] Mathur, P.; Ahmed, Moawia O.; Dash, A. K.; Walawalkar, M. G.; Puranik, V. G. *Dalton* (2000), 2916-2922.

[158] Singh, A K.; Kadarkaraisamy, M.; Drake, J. E.; Butcher, R. J., *Inorg. Chim. Acta,* 2000, 304, 45-51.

[159] Singh, A. K.; Kadarkaraisamy, M.; Murthy, G. S.; Srinivas, J.; Varghese, B.; Butcher, R. J., *J. Organomet. Chem.* 2000, 605, 39-44.

[160] Singh, A. K.; Sooriyakumar, J.; Drake, J.E.; Hursthouse, M.B.; Light; M.E., *J. Organomet. Chem.* 2000, 617, 244-249.

[161] Singh, A. K.; Kadarkaraisamy, M.; Mishra, M.; Sooriyakumar, J.; Drake, J. E.; Hursthouse, M. B.; Light, M. E.; Jasinski, J. P., *Inorg. Chim. Acta* 2001, 320, 133-140.

[162] Bali, S.; Singh, A. K.; Sharma, P.; Toscano, R. A.; Drake, J. E.; Hursthouse M.B.; Light, M.E., *J. Organomet. Chem.* 2004, 689, 2346-2353.

[163] Raghavendra K. P.; Singh A. K.; Butcher, R.J.; Sharma, P.; Toscano, R. A., *Eur. J. Inorg. Chem.* 2004, 1107-1114.

[164] Abel, E.; Orrell, K. G.; Scanlan, S. P.; Stephenson, D.; Kemmitt, T.; Levason, W *J. Chem. Soc., Dalton Trans.* 1991, 591-595.

[165] Singh, H. B.; Regini V, A.; Jasinski, J. P.; Paight, E. S.; Butcher, R. J. *J. Organomet. Chem.* 1994, 466, 283-289.

[166] Jain, V. K.; Chaudhury, S.; Bohra, R. , *Polyhedron* 1993, 12, 2377-2385.

[167] Dey S.; Jain V. K; Knodler A.; Klein. ; Kaim W.; Zalis S., *Inorg. Chem.* 2002, 41, 2864-2870.

[168] Lebedev, B. V.; Kulagina, T. G., *J. Chem. Thermodyn.* (1991), 23, 1-9.

[169] Semin, G. K.; Bryukhova, E. V.; Golubinskaya, L. M.; Bregadze, V. I., *Metalloorg. Khim,* 1991, 4, 1397-1404.

[170] Janzen, M. C.; Jennings, M. C.; Puddephatt, R. J., *Inorg. Chem.* 2003, 42, 4553-4558.

In: Focus on Organometallic Chemistry Research
Editor: Marin A. Cato, pp. 111-132
ISBN 1-59454-495-6

Chapter 4

PREPARATION OF PLATINUM AND PALLADIUM COMPLEXES CONTAINING PHOSPHINE SULFIDE- AND THIOAMIDE-BASED SCS PINCER LIGANDS: MOLECULAR STRUCTURES AND PHOTOLUMINESCENT PROPERTIES OF THE PINCER COMPLEXES

Takaki Kanbara* and Takakazu Yamamoto
Chemical Resources Laboratory, Tokyo Institute of Technology, 4259 Nagatsuta, Midori-ku, Yokohama 226-8503, Japan

ABSTRACT

Cyclometalation of new pincer ligands constituted of centered toluene unit and two phosphine sulfides afforded platinum and palladium pincer complexes with η^3-S,C,S type coordination. Reaction of other pincer ligands, 1,3-benzenedicarbothioamides, with $PtCl_2(PhCN)_2$ also gave new η^3-S,C,S pincer platinum complexes. The molecular structures of the complexes were determined by X-ray crystallography. The pincer complexes exhibit strong emission in a glassy frozen state as well as in the solid state. Light-emitting diodes using the complexes as an active layer displayed red or reddish orange electroluminescence.

1. INTRODUCTION

Pincer ligands play an increasing role in coordination chemistry and catalysis [1]. One of their advantages comes from the possibility of controlling reactivity and stability of the metal

* Tel: +81-45-924-5222; FAX: +81-45-924-5976; e-mail: tkanbara@res.titech.ac.jp

center in a wide range by adjusting the electronic properties of the pincer ligands which have three coordination sites. For example, an NCN-pincer ligand (NCN=[2,6-$(CH_2NMe_2)_2C_5H_3]^-$) stabilizes unusual oxidation states of the metals such as Ni(III) intermediate in the Kharasch addition. On the other hand, cycloplatination of aromatic compounds attracted considerable attention in recent years, because cyclometalated platinum(II) complexes show interesting photoluminescent properties and can be utilized to electroluminescence devices [2]. CNN- and/or NCN-pincer platinum complexes, which are mainly obtained from 6-aryl-2,2'-bipyridine and 1,3-dipyridylbenzene analogues, have also been investigated for the luminophores [2e-l]. This situation have led to the continuous interest that has been devoted to the synthesis of new pincer ligands with various heteroatoms such as N, P, O, and S. We here report preparation of pincer platinum and palladium complexes containing new SCS pincer ligands with two phosphine sulfides and thioamides as ancillary ligands.

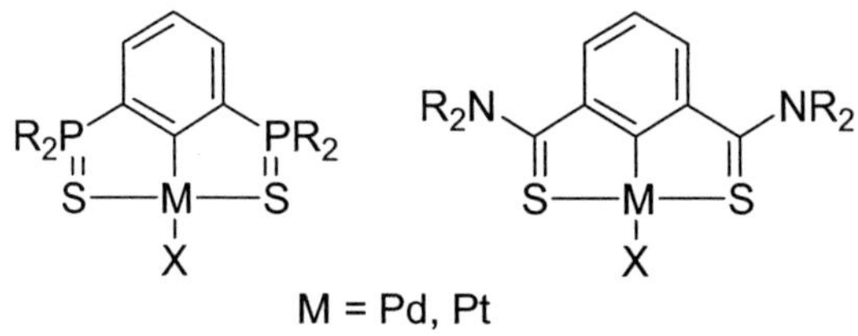

The above ligands consist of mixed donor sets (η^3-S,C,S coordination mode), and the phosphine sulfide and thioamide group could have an electronic contribution from exist as zwitterionic forms. Therefore, they could be prospective monoanionic pincer ligands comparable to the 1,3-bis[(thioether)methyl]benzene and its derivatives, which have been recently investigated for a number of catalytic reactions [1,3].

The coordination chemistry of phosphine sulfides [4] and thioamides [5] has well been investigated. However, only little attention has been paid to the SCS-pincer ligands constituted of the phosphine-sulfide and thioamide groups. Transition metal complexes containing phosphine sulfide-based tridentate ligands constituted of pyridine and phosphinene have recently been reported [6], whereas, to our knowledge, there has been no report on monoanionic pincer ligands featuring a centered *m*-phenylene unit and two phosphine sulfides as the ancillary ligands. On the other hand, Nonoyama and coworkers have investigated cyclopalladation of a series of thioamide-based aromatic ligands, and the thioamide-based pincer palladium complexes have also been reported [7].

In this chapter, preparation, molecular structures, photoluminescent properties of the pincer complexes, and their preliminary application in light-emitting diodes (LEDs) are described. Some of the results were reported in previous papers [8].

2. Preparation

2.1. Phosphine Sulfide-Based SCS-Pincer Complexes

3,5-Bisphosphinotoluenes (1a,b) was prepared by Pd-catalyzed aryl phosphination of 3,5-dibromotoluene with secondary phosphines [9]. Sulfurization of the *m*-arylenediphosphines with elemental sulfur afforded 3,5-bis(phosphinothioyl)toluenes (2a,b) [4,6,9c]. For reference, a related phosphine oxide analogue, 3,5-bis(diphenylphosphinoyl)toluene (3), was also prepared by oxidation of 1a with H_2O_2.

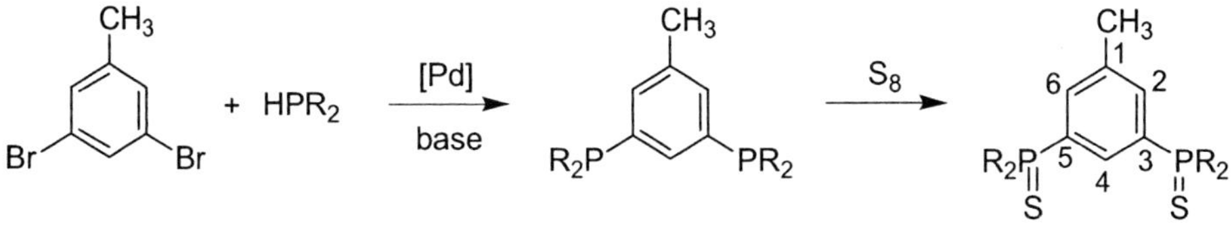

1a: R = Ph
1b: R = Cy

2a: R = Ph
2b: R = Cy

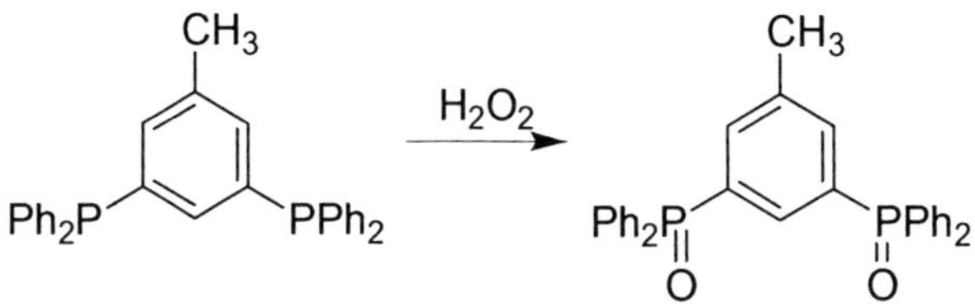

3

ortho,ortho-Cyclometalation of 2a,b with K_2MCl_4 (M = Pd, Pt) was carried out in acetic acid (AcOH) under reflux conditions. Orthopalladation is usually reversible in AcOH [10], and the corresponding SCS-pincer complexes are considered to be obtained if they were thermodynamically stable.

The reactions led to regioselective cyclometalation at the C-4 position of the central toluene ring with Pd(II) and Pt(II), and afforded the corresponding pincer complexes (4a,b: M = Pd; 5a,b: M = Pt) in good yields.

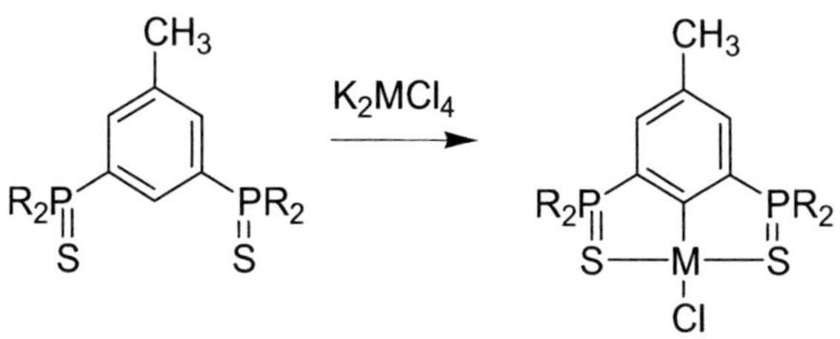

4a: M = Pd, R = Ph
4b: M = Pd, R = Cy
5a: M = Pt, R = Ph
5b: M = Pt, R = Cy

On the contrary, the related OCO-pincer ligand (3) did not afford the corresponding complex, indicating the coordination ability of phosphine sulfide groups toward Pd(II) and Pt(II) would be an important factor for the *ortho,ortho*-cyclometalation. The reaction of 2a with dichlorobis(benzonitrile)palladium ($PdCl_2(PhCN)_2$) in acetonitrile also gave 4a, whereas 5a was not obtained from the analogous reaction.

The obtained complexes are fairly stable to air and water. Solubility of the complexes is dependent on the substituent of the pincer ligands; i.e. the complexes 4a and 5a are soluble in DMSO and DMF, however, they are scarcely soluble in $CHCl_3$ and CH_2Cl_2. The complexes 4b and 5b show good solubility in DMSO, DMF, $CHCl_3$, and CH_2Cl_2. The structure of the complexes was confirmed by IR and NMR spectroscopy and elemental analysis. In the IR spectra of the complexes, lower frequency shifts of ν(P=S) was observed, indicating that the

S-coordination of the phosphine sulfide donor to Pd(II) or Pt(II) and decreases the double bond character of the P=S bond in the coordination form. In the ^{1}H NMR spectra of the complexes, the signal assigned to the proton at the 4-position of the centered toluene ring disappears and high-filed shifts of the signal assigned to the protons at the 2,6-positions of the centered toluene ring is observed. The ^{31}P NMR spectra of the complexes exhibit downfield shifts of the signal due to the *S*-coordination, and the two phosphine sulfide groups are magnetically equivalent in the complexes.

2.2. Thioamide-Based SCS-Pincer Complexes

Another class of new thioamide-based pincer ligands, 1,3-benzenedicarbothioamides (6a-c), was prepared by the Willgerodt-Kindler reaction by applying methods given in previous reports [7,11]. The η^3-SCS-pincer platinum complexes (7a-c) were obtained in moderate to good yields from the reactions of 6a-c with $PtCl_2(PhCN)_2$. On the contrary, a related OCO-pincer ligand (1,3-bis(1-piperidinocarbonyl)benzene) did not afford the corresponding complex, suggesting that the coordination ability of thioamide groups toward Pt(II) was a crucial factor for the *ortho,ortho*-cycloplatination.

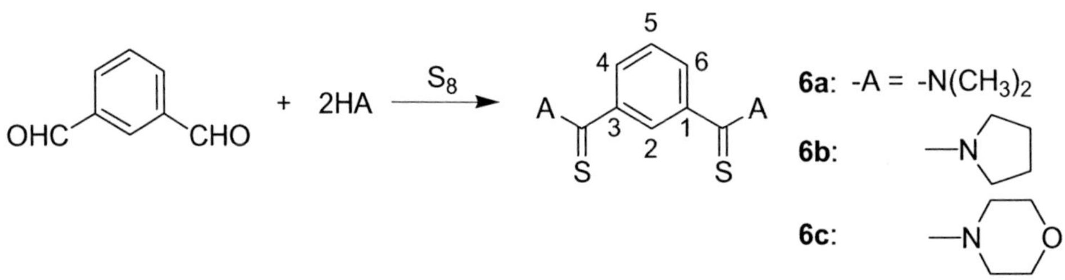

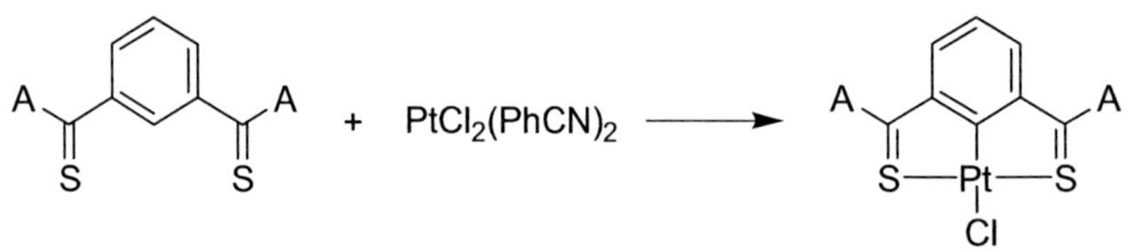

6a-c **7a-c**

The thioamide-based SCS-pincer complexes, 7a-c, are also fairly stable to air and water. The complexes 7a,b are soluble in DMSO and DMF, however, they are scarcely soluble in $CHCl_3$ and CH_2Cl_2. The complex 7c shows good solubility in DMSO, DMF, $CHCl_3$, and CH_2Cl_2. In the IR spectra of the complexes, shifts of ν(C-N) of the thioamide group to a higher frequency were observed, indicating that the *S*-coordination of the thioamide group increases double bond character of the C-N bond in the coordination form. In the ^{1}H NMR spectrum of 7b, the peak area of the aromatic proton signals were reduced by 1H, and downfield shifts of the signals assignable to the vicinal methylene protons of thioamide group were observed. In the ^{13}C NMR spectrum of 7b, the *ortho,ortho*-platinated carbon (C-2 carbon δ 128.3 in 6b) is deshielded by 37 ppm according to the σ-bonding to Pt and appeared at δ 165.3. The ^{195}Pt NMR spectrum of 7b shows a signal at –5281 ppm referred to H_2PtCl_6.

The position of the signal is located at a relatively low magnetic field compared with usual platinum complexes. A similar feature was observed with 7a,c.

We have also examined substitution reactions of 7b. Reaction of 7b with phenylacetylene in the presence of CuI and triethylamine easily led to a transmetalation to form a phenylacetylide complex (Pt(C≡CPh)(6b-H), 8b) in good yield. Treatment of 7b and 8b with an excess amount of CH_3I gave an iodo complex (PtI(6b-H), 9b), respectively. ^1H NMR spectra after the reaction with CH_3I showed the signals assigned to 9b and chloromethane (for the reaction of 7b), or 1-phenyl-1-propyne (for the reaction of 8b).

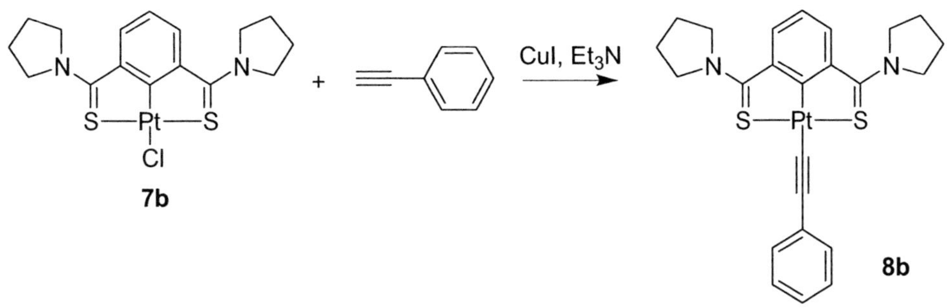

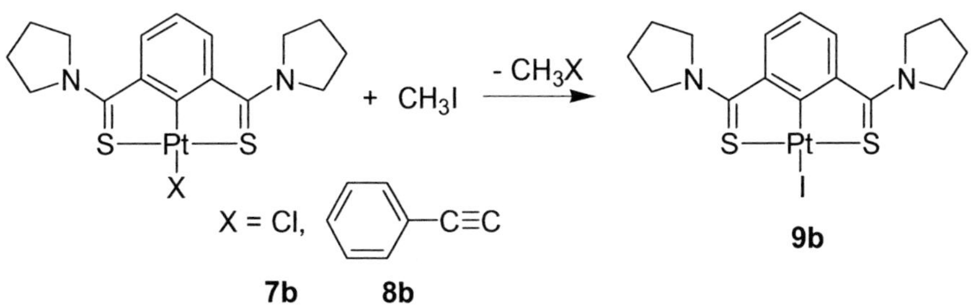

Several arenium-platinum complexes have been prepared and isolated from reactions of the pincer-type platinum complexes with CH_3I, which would be formed *via* S_N2-type nucleophilic substitution of the halide by the metal nucleophile followed by a 1,2-methyl migration between Pt and the *ortho,ortho*-platinated carbon of the aryl ligand [1,12]. The iodo complex 9b was also obtained from the reaction of 7b with $AgBF_4$ and following treatment with NaI.

3. Molecular Structure

3.1. Phosphine Sulfide-Based SCS-Pincer Complexes

Single crystals of 2b and 4b were obtained from a slow diffusion of ethanol into their chloroform solutions, whereas single crystals of 4a and 5a were obtained from a slow diffusion of ethanol into a DMSO solution of the complexes. A summary of the crystal structural refinements of the ligand 2b and the complexes is given in Table 1, and selected bond lengths, bond angles, and torsion angles for the ligand and the complexes are given in Table 2. The molecular structures of 2b, 4b, and 5a are shown in Figure 1.

Table 1. Crystal data and detail of the structure refinement of the ligand and complexes

Compound	**2b**	**4a**	**4b**	**5a**	**6b**	**7b**	**9b**
Formula	$C_{31}H_{50}P_2S_2$	$C_{31}H_{25}ClP_2PdS_2$	$C_{31}H_{49}ClP_2PtS_2$	$C_{31}H_{25}ClP_2PtS_2$	$C_{16}H_{20}N_2S_2$	$C_{16}H_{19}ClN_2PtS_2$	$C_{16}H_{19}IN_2PtS_2$
Mol wt	548.80	665.46	689.65	754.15	304.47	534.00	625.45
Cryst sust	monoclinic	orthorhombic	monoclinic	orthorhombic	orthorhombic	triclinic	orthorhombic
Space group	$P2_1/a$ (No.14)	$P2_12_12_1$ (No.19)	$P2_1/n$ (No.14)	$P2_12_12_1$ (No.19)	$P2_12_12_1$ (No.19)	*P*1bar (No. 2)	*Pbca* (No.61)
a / Å	15.172(7)	9.890(6)	9.876(2)	9.832(6)	12.012(9)	12.327(4)	22.589(5)
b / Å	9.865(5)	14.959(8)	19.832(4)	14.978(10)	17.707(9)	12.559(7)	12.862(2)
c / Å	20.581(10)	19.160(11)	15.749(3)	19.090(12)	7.492(7)	12.085(5)	12.211(4)
□ / deg						101.81(3)	
□ / deg	91.224(7)		92.091(3)			105.45(3)	
□ / deg						102.30(4)	
V / Å^3	3079.6(26)	2834.5(27)	3082.4(12)	2811.2(31)	1593.5(2)	1692.7(2)	3547(1)
Z	4	4	4	4	4	4	8
□ / cm^{-1}	2.95	10.30	9.49	53.49	3.26	86.57	98.52
F(000)	1192.00	1344.00	1440.00	1472.00	648.00	1024.00	2336.00
D_{calcd} / g cm^{-1}	1.184	1.559	1.486	1.782	1.269	2.095	2.342
No. unique reflns	7239	3582	7181	3222	2103	7771	4555
No. observations	3677	3079 (I>3□(I))	4067 (I>3□(I))	3022 (I>3□(I))	1306 (I>2□(I))	4983 (I>3□(I))	2207 (I>3□(I))
No. variables	316	334	334	334	181	397	199
R_1	0.085	0.070	0.031	0.089	0.037	0.030	0.040
R_w	0.123	0.118	0.042	0.139	0.048	0.042	0.050
GOF	1.033	0.940	0.783	0.980	0.850	0.930	0.802

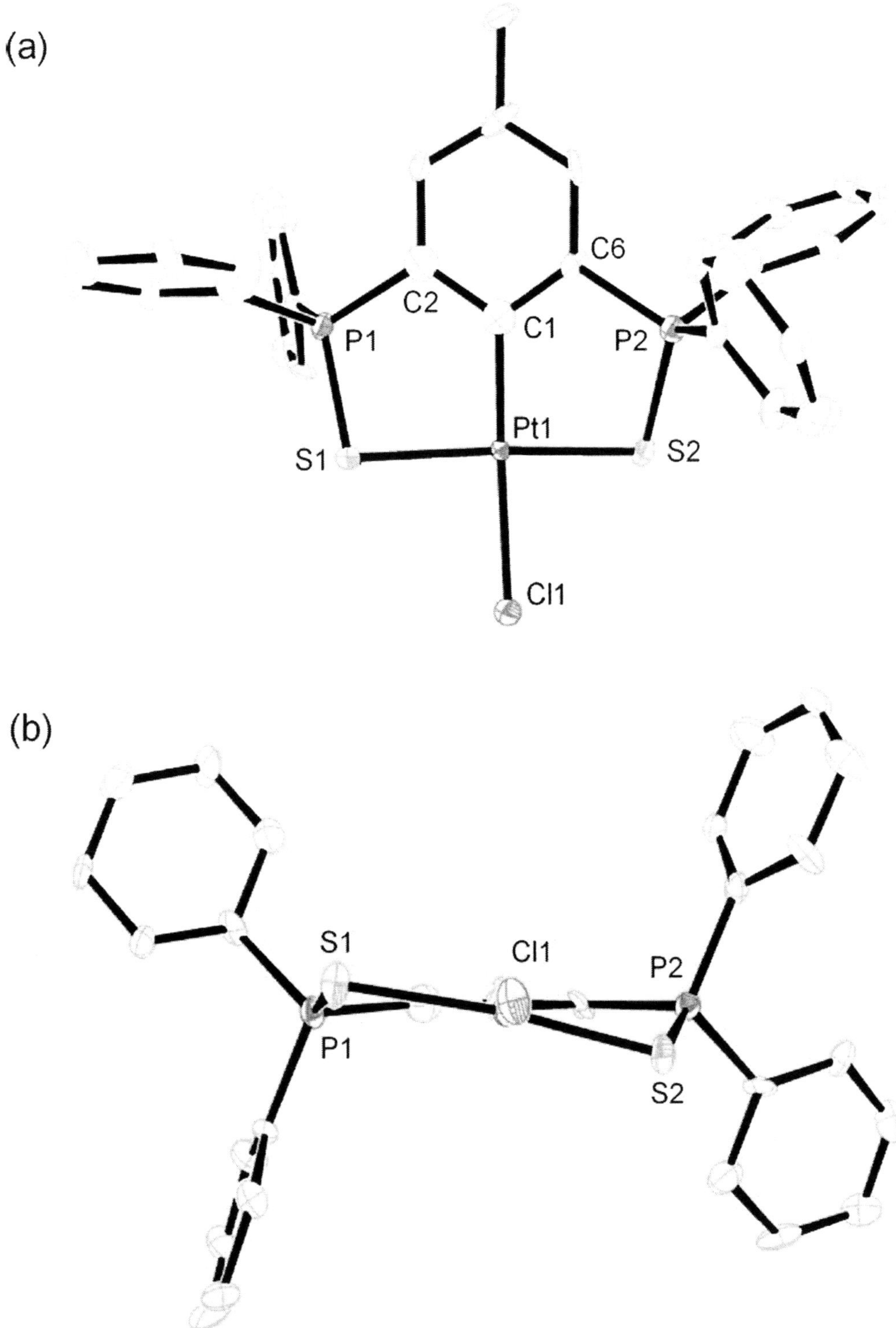

Figure 1. Continued on next page.

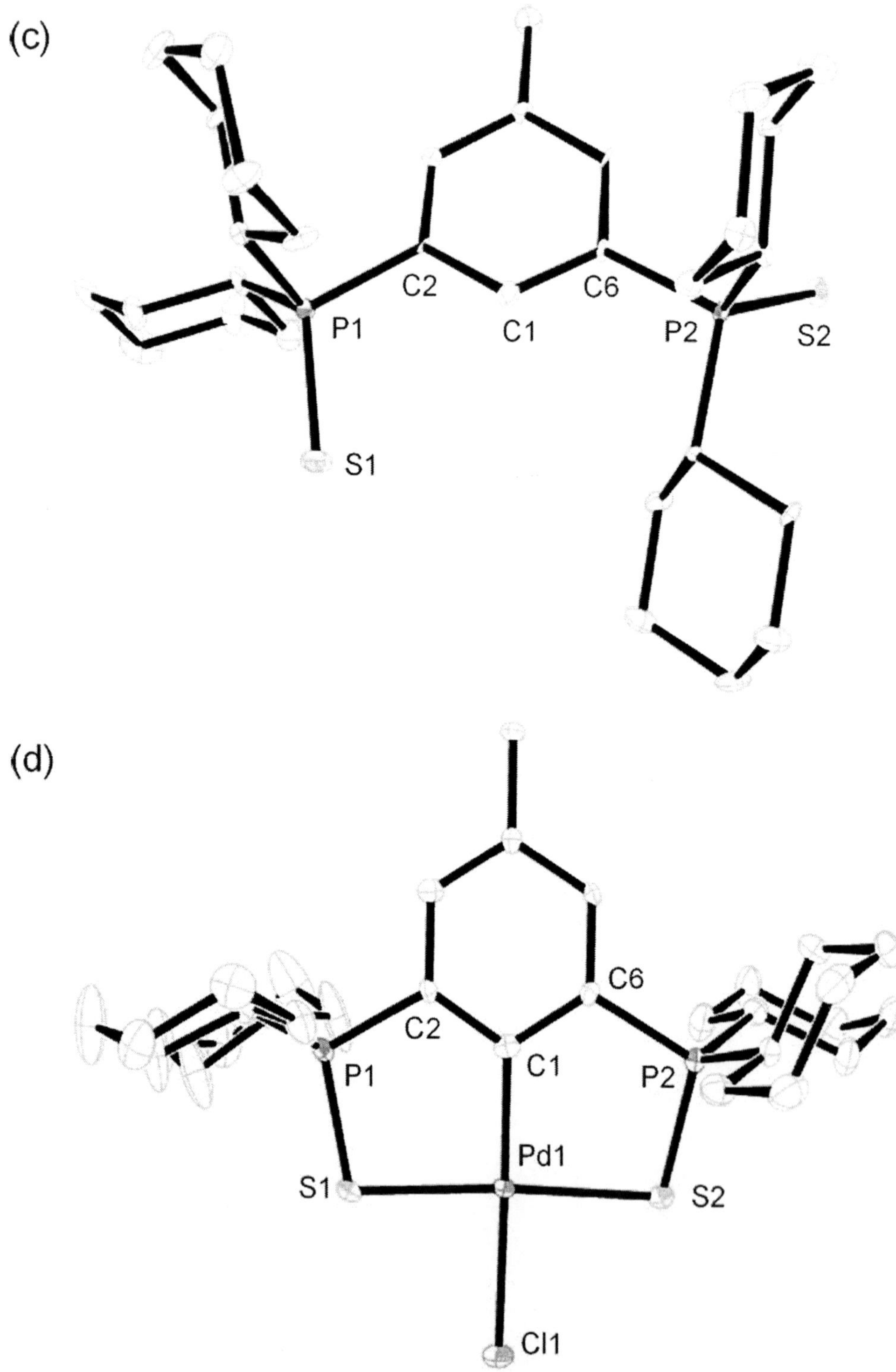

Figure 1. X-ray crystal structure of 5a (front (a) and side (b) view), 2b (c), and 4b (d) with thermal ellipsoids drawn at the 50% probability level. Hydrogen atoms are omitted for clarity.

Table 2. Selected bond lengths (Å), bond angles (°), and torsion angles (°) for complexes 2b, 4a, 4b, and 5a

2b: Bond lengths		Bond angles	
S(1)-P(1)	1.965(2)	C(2)-P(1)-S(1)	113.1(2)
S(2)-P(2)	1.962(2)	C(6)-P(2)-S(2)	111.7(2)
P(1)-C(2)	1.825(5)		
P(2)-C(6)	1.809(5)		
4a: Bond lengths		Bond angles	
Pd(1)-Cl(1)	2.381(4)	Cl(1)-Pd(1)-S(1)	86.88(13)
Pd(1)-S(1)	2.312(4)	Cl(1)-Pd(1)-S(2)	90.30(13)
Pd(1)-S(2)	2.321(4)	S(1)-Pd(1)-S(2)	173.3(1)
Pd(1)-C(1)	2.007(12)	Cl(1)-Pd(1)-C(1)	175.4(3)
S(1)-P(1)	1.995(5)	S(1)-Pd(1)-C(1)	91.5(4)
S(2)-P(2)	2.007(5)	S(2)-Pd(1)-C(1)	91.7(4)
P(1)-C(2)	1.786(13)	Pd(1)-S(1)-P(1)	97.9(2)
P(2)-C(6)	1.766(13)	Pd(1)-S(2)-P(2)	98.8(2)
		S(1)-P(1)-C(2)	107.4(5)
		S(2)-P(2)-C(6)	108.4(5)
Torsion angles			
S(1)-Pd(1)-C(1)-C(2)	14.7(10)		
S(2)-Pd(1)-C(1)-C(6)	4.9(10)		
S(1)-P(1)-C(2)-C(1)	-15.1(11)		
S(2)-P(2)-C(6)-C(1)	-15.8(11)		
4b: Bond lengths		Bond angles	
Pd(1)-Cl(1)	2.380(1)	Cl(1)-Pd(1)-S(1)	86.90(4)
Pd(1)-S(1)	2.317(1)	Cl(1)-Pd(1)-S(2)	88.85(4)
Pd(1)-S(2)	2.327(1)	S(1)-Pd(1)-S(2)	175.66(4)
Pd(1)-C(1)	2.000(4)	Cl(1)-Pd(1)-C(1)	178.8(1)
S(1)-P(1)	2.010(2)	S(1)-Pd(1)-C(1)	92.0(1)
S(2)-P(2)	2.016(2)	S(2)-Pd(1)-C(1)	92.3(1)
P(1)-C(2)	1.789(4)	P(1)-S(1)-Pd(1)	100.19(6)
P(2)-C(6)	1.794(4)	P(2)-S(2)-Pd(1)	99.64(6)
		C(2)-P(1)-S(1)	106.8(1)
		C(6)-P(2)-S(2)	107.7(2)
Torsion angles			
S(1)-Pd(1)-C(1)-C(2)	-4.3(3)		
S(2)-Pd(1)-C(1)-C(6)	-5.0(3)		
S(1)-P(1)-C(2)-C(1)	4.3(3)		
S(2)-P(2)-C(6)-C(1)	5.5(3)		
5a: Bond lengths		Bond angles	
Pt(1)-Cl(1)	2.370(5)	Cl(1)-Pt(1)-S(1)	86.4(2)
Pt(1)-S(1)	2.299(5)	Cl(1)-Pt(1)-S(2)	90.9(2)
Pt(1)-S(2)	2.307(5)	S(1)-Pt(1)-S(2)	174.4(2)
Pt(1)-C(1)	2.02(2)	Cl(1)-Pt(1)-C(1)	175.1(5)
S(1)-P(1)	2.002(7)	S(1)-Pt(1)-C(1)	91.4(5)
S(2)-P(2)	1.996(7)	S(2)-Pt(1)-C(1)	91.7(5)
P(1)-C(2)	1.77(2)	Pt(1)-S(1)-P(1)	97.5(3)

Table 2. Continued

P(2)-C(6)	1.73(3)	Pt(1)-S(2)-P(2)	99.3(3)
		S(1)-P(1)-C(2)	109.8(7)
		S(2)-P(2)-C(6)	105.7(7)
Torsion angles			
S(1)-Pt(1)-C(1)-C(2)	-16.8(15)		
S(2)-Pt(1)-C(1)-C(6)	-4.7(15)		
S(1)-P(1)-C(2)-C(1)	15.6(16)		
S(2)-P(2)-C(6)-C(1)	19.6(18)		

The two complexes 4a and 5a, which consist of same pincer ligand 2a, are isostructural and have the same crystal system and space group. The platinum complex 5a has a distorted square planar geometry similar to those of the reported pincer palladium and platinum complexes [1,2e-l,10b,13]. As shown in Fig. 1(b), the plane of coordination is somewhat twisted out of the plane of the centered toluene ring. The C-Pt-Cl and S-Pt-S axes are distorted as a consequence of strain of the coordinating plane. The Pt-C and Pt-Cl bond lengths lie in the range of lengths found in related pincer platinum complexes [1,10b,13]. A similar feature was observed with the palladium complex 4a and 4b. The P=S bond lengths of 4b (2.010(2)-2.016(2) Å) are somewhat longer than those of free ligand 2b (1.962(2)-1.965(2) Å), whereas the P-C connections of the phosphine sulfide and the centered toluene ring in 4b (P(1)-C(2) = 1.789(4), P(2)-C(6) = 1.794(4) Å) are shorter than those of free ligand 2b (1.809(5)-1.825(5) Å). These data indicate that delocalization takes place within the unsaturated part of the ligand (Scheme 1B) [6c].

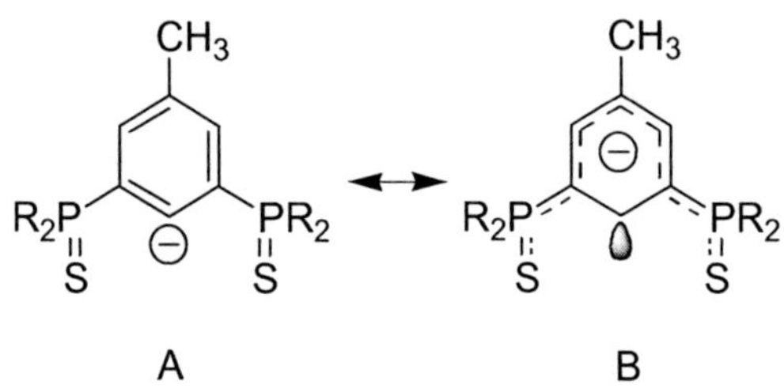

Scheme 1.

Solid state intermolecular d^8-d^8 and ligand-ligand interactions of the photoluminescent cyclometalated platinum(II) complexes have been widely investigated [2e-h,k,l]. In contrast, in all of the crystal lattices for the complexes, neither intercomponent π-stacking interactions of the centered toluene ring nor intermolecular d^8-d^8 interactions was observed with M···M separation distances of 9.17 (4a), 9.44 (4b), and 8.79 Å (5a), respectively.

3.2. Thioamide-Based SCS-Pincer Complexes

Single crystal of 6b was obtained from a slow diffusion of methanol into a chloroform solution, whereas single crystals of 7b and 9b were obtained from a slow diffusion of ethanol into a DMSO solution of the complexes. A summary of the crystal structural refinements of the ligand 6b and the complexes is also included in Table 1, and selected bond lengths, bond angles, and torsion angles for the ligand and the complexes are given in Table 3. The molecular structures of 6b and 7b are shown in Figure 2.

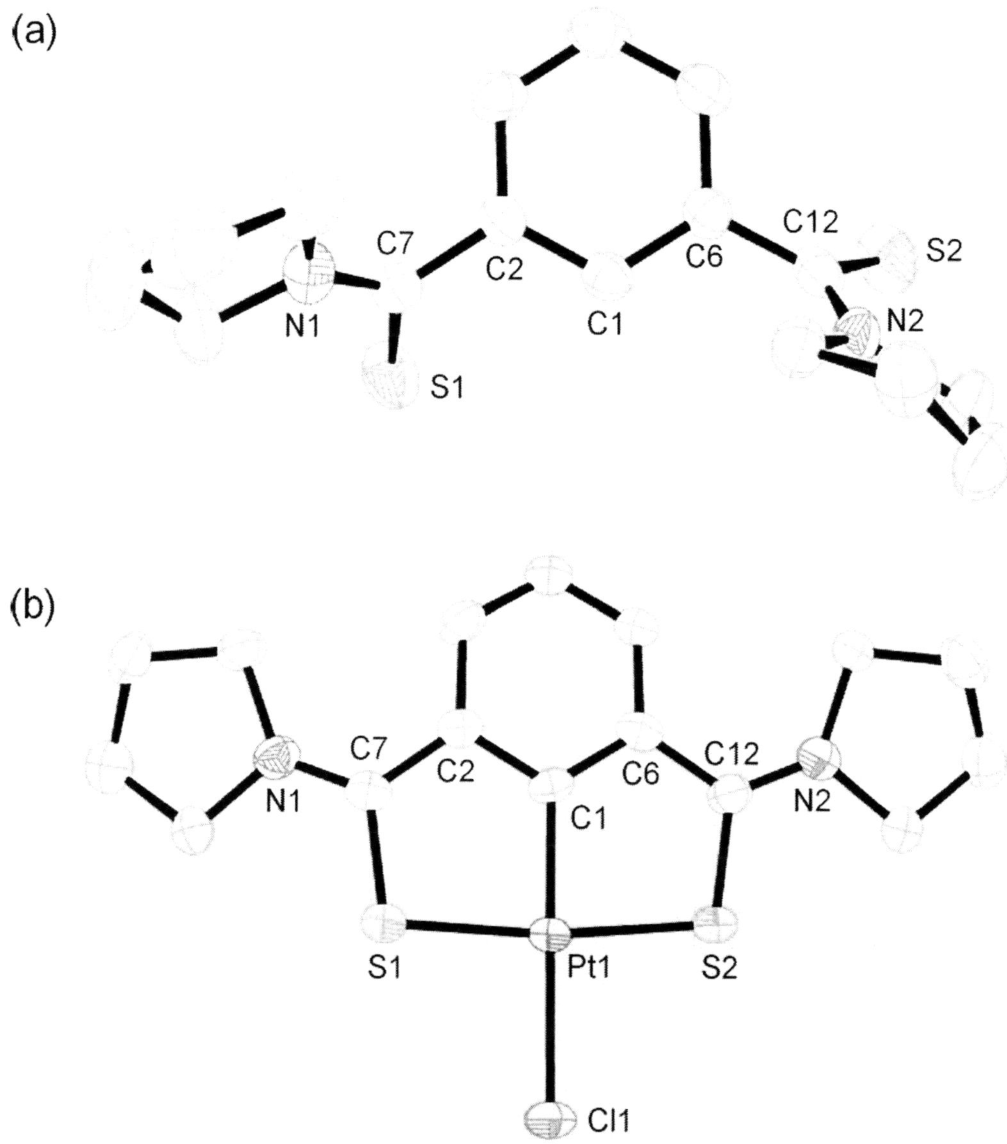

Figure 2. X-ray crystal structure of 6b (a) and 7b (b) with thermal ellipsoids drawn at the 50% probability level. One of the two crystallographically independent molecules of 7b is shown. Hydrogen atoms are omitted for clarity.

The molecular structures of the related palladium complexes (Pd(6a-H)Cl, and Pd(6b-H)Cl) have been reported by Nonoyama *et al.* [7b]. As for the related platinum complexes 7b and 9b, 7b is isostructural with Pd(6b-H)Cl and has the same crystal system and space group, whereas 9b has a different crystal system and space group as shown in Table 1.

Table 3. Selected bond lengths (Å), bond angles (°), and torsion angles (°) for complexes 6b, 7b, and 9b

6b: Bond lengths		Bond angles	
S(1)-C(7)	1.678(4)	S(1)-C(7)-C(2)	118.5(3)
S(2)-C(12)	1.676(5)	S(1)-C(7)-N(1)	122.4(3)
N(1)-C(7)	1.316(5)	S(2)-C(12)-C(6)	119.6(3)
N(2)-C(12)	1.319(5)	S(2)-C(12)-N(2)	123.0(3)
C(2)-C(7)	1.481(6)		
C(6)-C(12)	1.479(6)		
7b: Bond lengths		Bond angles	
Pt(1)-C(1)	1.945(7)	S(1)-Pt(1)-S(2)	172.56(7)
Pt(1)-Cl(1)	2.388(2)	Cl(1)-Pt(1)-C(1)	179.2(2)
Pt(1)-S(1)	2.265(3)	S(1)-C(7)-C(2)	116.5(6)
Pt(1)-S(2)	2.255(2)	S(1)-C(7)-N(1)	118.0(6)
S(1)-C(7)	1.716(8)	S(2)-C(12)-C(6)	116.9(6)
S(2)-C(12)	1.721(7)	S(2)-C(12)-N(2)	116.5(6)
N(1)-C(7)	1.31(1)	Pt(1)-S(1)-C(7)	100.3(3)
N(2)-C(12)	1.32(1)	Pt(1)-S(2)-C(12)	100.8(3)
C(2)-C(7)	1.48(1)		
C(6)-C(12)	1.49(1)		
Torsion angles			
C(1)-C(2)-C(7)-S(1)	9.3(1)		
C(1)-C(6)-C(12)-S(2)	1.3(1)		
9b: Bond lengths		Bond angles	
Pt(1)-C(1)	1.98(1)	S(1)-Pt(1)-S(2)	171.8(1)
Pt(1)-I(1)	2.682(1)	I(1)-Pt(1)-C(1)	175.3(3)
Pt(1)-S(1)	2.260(3)	S(1)-C(7)-C(2)	116.9(9)
Pt(1)-S(2)	2.259(4)	S(1)-C(7)-N(1)	117(1)
S(1)-C(7)	1.72(1)	S(2)-C(12)-C(6)	116.4(8)
S(2)-C(12)	1.72(1)	S(2)-C(12)-N(2)	116.9(8)
N(1)-C(7)	1.33(2)	C(7)-S(1)-Pt(1)	100.7(5)
N(2)-C(12)	1.33(1)	C(12)-S(2)-Pt(1)	101.1(4)
C(2)-C(7)	1.48(2)		
C(6)-C(12)	1.49(2)		
Torsion angles			
C(1)-C(2)-C(7)-S(1)	0(1)		
C(1)-C(6)-C(12)-S(2)	6(1)		

The complex 7b has a distorted square-planar geometry similar to those of the phosphine sulfide-based pincer complexes, and the Pt-C and Pt-Cl bond lengths lie in the range of lengths observed for related pincer platinum complexes [1,10b,13]. As shown in Table 3, the C=S bond lengths of 7b are longer than those of free ligand 6b, indicating that *S*-coordination to Pt decreases double bond character of the C=S bond, whereas the C-N bond lengths of 7b are not significantly altered upon coordination of the thioamide groups. In the crystal lattices of 7b and 9b, neither intercomponent π-stacking interactions of the centered benzene ring nor intermolecular d^8-d^8 interactions is observed with Pt···Pt separation distances of 6.59 (7b) and 7.08 Å (9b), respectively.

4. Photophysical Properties

The chloro and acetylide complexes are not light-emissive in solution at room temperature, whereas strong photoluminescence was observed in the glassy frozen state. The platinum complexes (5a,b, 7a-c, 8b) exhibit strong light-emission even in the solid state at room temperature. On the other hand, the iodo complex 9b was not light-emissive in solution and in the solid state, presumably due to a heavy atom effect of the iodo ligand. The absorption spectral data and emission data of the complexes are summarized in Table 4. The decay lifetimes of the emission from the complexes in the glassy frozen state were in the 10^{-5}-10^{-4} s time range for the phosphine sulfide-based SCS-pincer complexes and 10^{-6}-10^{-5} s for the thioamide-based pincer complexes. These data are indicative of phosphorescent emission. Figure 3 shows emission spectra of 2a, 4a,b, 5a,b, 7b, and 8b in the glassy frozen state and in the solid state. The phosphine sulfide-based ligand 2a also exhibits blue emission in the glassy frozen state. As for the platinum complexes, the small difference in energy between the solid sate at room temperature and the glassy frozen emission suggests the same electronic origins of the emitting states in both the states. These data suggest that the emission from the complexes is not attributed to excimeric species, because formation of the excimeric state in the frozen state is considered to be difficult. X-ray diffraction analysis of the complexes indicates that there is no intermolecular d^8-d^8 and ligand-ligand interactions of the complexes (see section 3). The emission of the complexes seems most likely to take place from a triplet metal-to-ligand charge transfer (^{3}MLCT) excited states and/or a mixed metal-perturbed triplet ligand-centered (^{3}LC) excited states [2]. The phosphorescent emission of the palladium complexes 4a,b is of interest, because there are not many reports on phosphorescence emission for the mononuclear palladium complexes [14]. However, the complexes exhibit only weak light-emission in the solid state at room temperature.

5. Electroluminescence

The strong emission from the triplet state of the metal complexes is attracting strong interest because of its high quantum efficiency in electroluminescence devices (or light-emitting diodes (LEDs)) [2,15]. We have thus preliminary examined usability of the complexes as light-emitting materials in LEDs. Figure 4 shows illustration of the LEDs fabricated in the present study, and electroluminescence spectra of the LEDs are shown in Fig. 5. Red (7b) and reddish orange (for 5a and 8b) electroluminescence was observed from the LEDs. The electroluminescence spectra of the LEDs essentially agree with the photoluminescence spectra of the complexes, revealing that the electroluminescence emission occurs from the complexes. The sub-peak at about 430 nm was attributable to the emission from poly(9-vinylcarbazole) matrix, indicating that the effective energy transfer from the matrix to the complex did not proceed. The present LEDs using 5a, 7b, and 8b have emission intensity of 190 cd m^{-2} (0.11% efficiency), 180 cd m^{-2} (0.20%), and 422 cd m^{-2} (0.20%) at 100 mA cm^{-2}, respectively. In order to make practicable LEDs, further optimization is required.

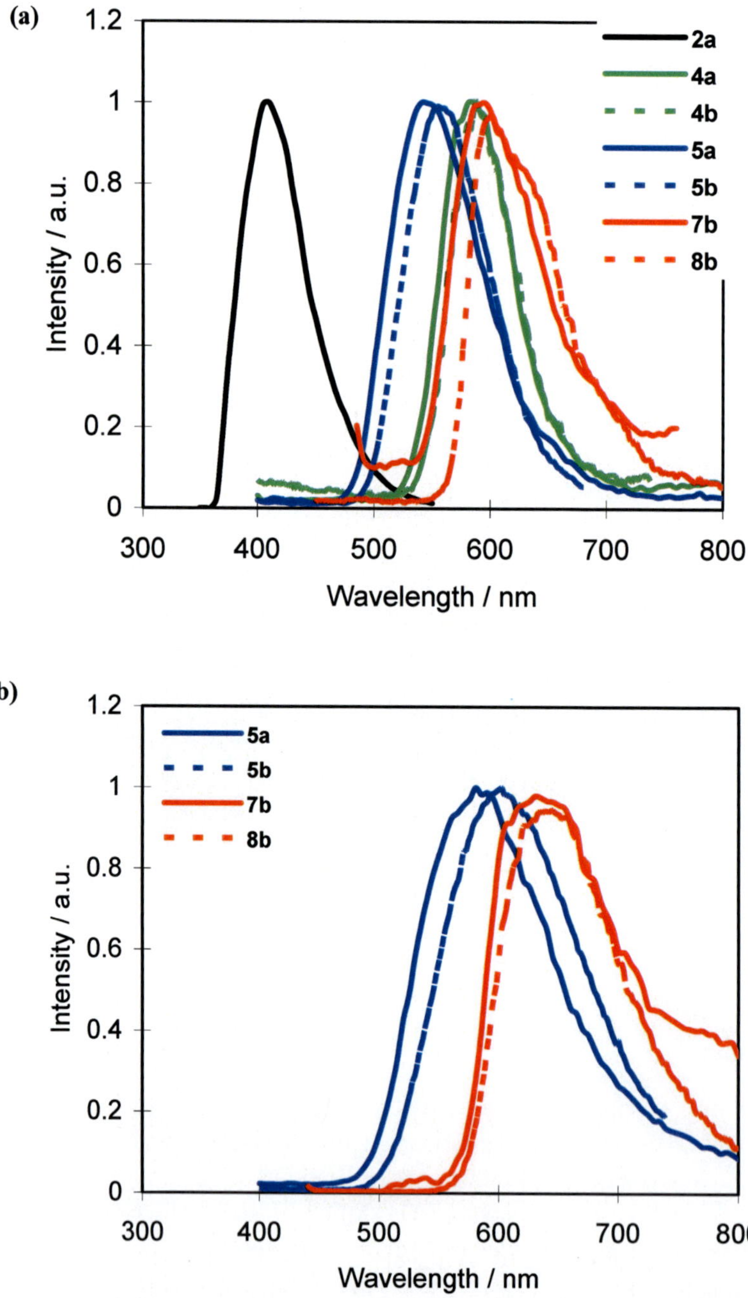

Figure 3. Emission spectra of 2a (black, solid), 4a (green, solid), 4b (green, dotted), 5a (blue, solid), 5b (blue, dotted), 7b (red, solid), and 8b (red, dotted): (a) in frozen CH_2Cl_2-MeOH-EtOH matrix (77 K) and (b) microcrystalline sample at room temperature.

Table 4. Photophysical properties of the pincer complexes

Complex	λ_{max}[a] / nm (ε[b], $M^{-1}cm^{-1}$)	λ_{em}[c] / nm	λ_{em}[d] / nm	ϕ_f[e]	τ[f] / μs
4a	265 (33100), 323 (5500)	—[g]	580	0.14	240
4b	256 (21700), 317 (3000)	—[g]	590	0.06	320
5a	304 (8900), 345 (3900)	580	540	0.24	110
5b	295 (7300), 342 (2300)	600	555	0.14	83
7a	287 (18500), 354 (4000), 447 (8100)	655	630	0.12	—[h]
7b	255 (33000), 345 (10400), 440 (4500)	630	595	0.11	11.0
7c	297 (15600), 360 (3500), 463 (6600)	710	680	0.05	—[h]
8b	247 (32000), 340 (10300), 445 (1400)	640	600	0.24	8.3

[a] Absorption maxima in CH_2Cl_2 at room temperature.

[b] Molecular absorption coefficient.

[c] Emission maximum of a microcrystalline sample.

[d] Emission maximum in a mixture of CH_2Cl_2, ethanol, and methanol matrix at 77 K.

[e] Fluorescence quantum yield compared with that of *fac*-tris(2-phenylpyridine)iridium (ϕ_f = 0.4 ±0.1) [18].

[f] Emission decay lifetime.

[g] Not determined.

[h] Not measured.

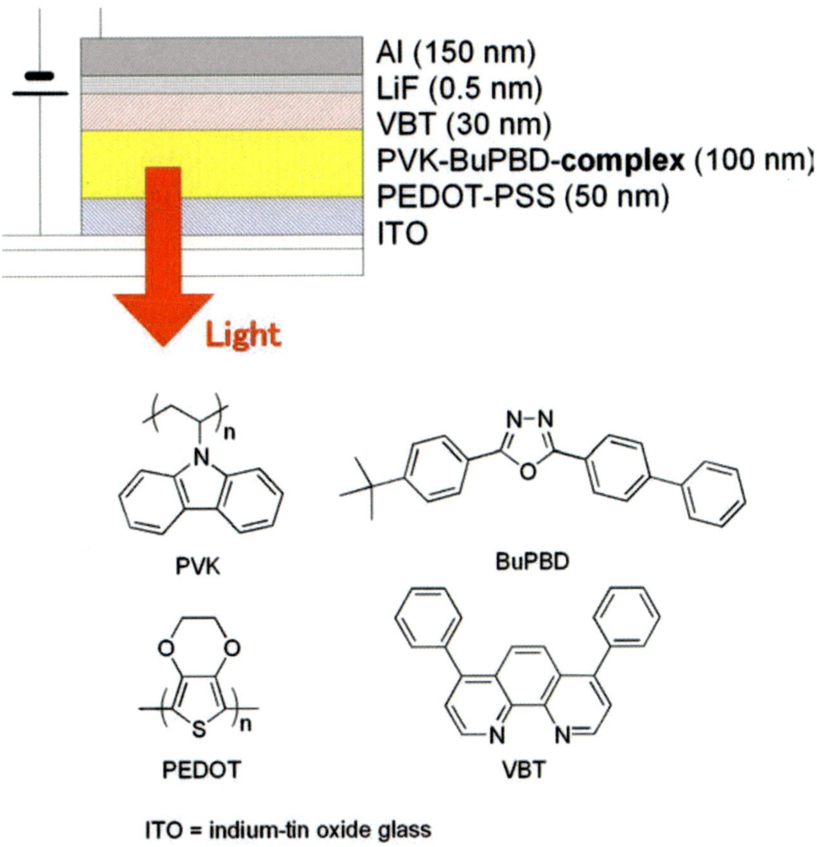

Figure 4. Schematic diagram of the LED.

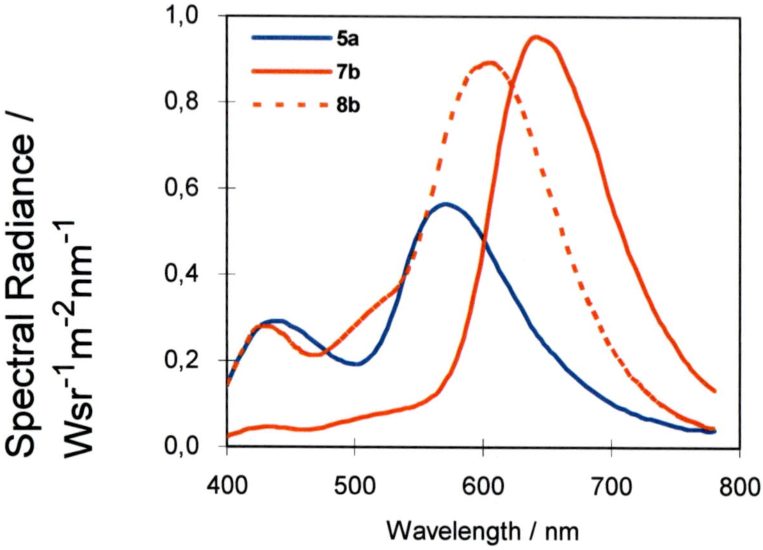

Figure 5. Electroluminescence spectra of the LEDs for 5a (blue), 7b (red, solid), and 8b (red, dotted).

6. Summary

As described above, new pincer platinum and palladium complexes containing phosphine sulfide- and thioamide-based SCS-pincer ligands have been prepared. The pincer complexes exhibit highly emissive properties in the glassy frozen state as well as in the solid state. The decay lifetimes of the emission from the complexes indicate phosphorescent emission. Substitution of the chloro ligand with various neutral and anionic ligands is possible. These features will make them good prospects for use in optoelectronic devices and sensors.

7. Experimental

7.1. General Procedures

All solvents were dried and distilled prior to use. $PdCl_2(PhCN)_2$, $PtCl_2(PhCN)_2$, and 1,3-bis(1-piperidinocarbonyl)benzene were prepared according to the previous reports [16,17]. The thioamide-based ligands (1,3-benzenedicarbothioamides, 6a-c) were prepared by the Willgerodt-Kindler reaction according to the previous reports [7,11].

IR and NMR spectra were recorded on a JASCO FTIR-460Plus spectrometer, and JEOL JNM-EX-400 NMR spectrometer, respectively. Elemental analyses were carried out with a Yanaco CHN Corder MT-5 and a Yanaco SX-Elements Micro Analyzer YS-10. UV-vis. absorption spectra and emission spectra were recorded on a Shimadzu UV-3100PC spectrophotometer and a Hitachi F-4500 fluorescence spectrophotometer, respectively. Fluorescence quantum yield of the complexes was estimated from comparison with that of *fac*-tris(2-phenylpyridine)iridium ($\phi_f = 0.4 \pm 0.1$) [18]. Emission lifetime measurements were performed with a Hamamatsu Photonics C4780 picosecond fluorescence lifetime measurement system.

7.2. Preparation of Phosphine Sulfide-Based Pincer Ligands (2a,B)

A typical preparation procedure is as follows. A mixture of 3,5-dibromotoluene (1.5 g, 6 mmol) and diphenylphosphine (2.24 g, 12 mmol) was dissolved in DMF (20 mL). Palladium chloride (53 mg, 0.3 mmol) and potassium acetate (1.77 g, 18 mmol) were added to the solution, and the reaction mixture was stirred at 130 °C for 24 h under nitrogen. After cooling to room temperature, sulfur (480 mg, 15 mmol as elemental sulfur) was added and the reaction mixture was stirred at 120 °C for 4 h under nitrogen. After cooling to room temperature, an aqueous solution of EDTA-2K (ca. 100 mL) and $CHCl_3$ (200 mL) was added, and the mixture was stirred for 30 min. The organic phase was separated and the solvent was evaporated. The crude product was purified by column chromatography and recrystallization from acetone to give 3,5-bis(diphenylphosphinothioyl)toluene (2a) as a white needle (2.21 g, 70% yield).

^{1}H NMR (400 MHz, DMSO-d_6): δ 7.75 (d, 2H, J = 13.7 Hz), 7.62-7.48 (m, 20H), 7.41 (t, 1H, J = 12.9 Hz), 2.38 (s, 3H). ^{31}P NMR (160 MHz, DMSO-d_6): δ 42.6. Anal. Found: C, 71.01; H, 5.17; S, 12.13%. Calc. for $C_{31}H_{26}P_2S_2$: C, 70.97; H, 5.00; S, 12.22%.

3,5-Bis(dicyclohexylphosphinothioyl)toluene (2b) was prepared analogously (2.49 g, 76% yield).

^{1}H NMR (400 MHz, $CDCl_3$): δ 7.87 (d, 2H, J = 12.2 Hz), 7.81 (t, 1H, J = 10.0 Hz), 2.48 (s, 3H), 2.25-1.00 (m, 44H). ^{31}P NMR (160 MHz, $CDCl_3$): δ 59.2. Anal. Found: C, 67.80; H, 9.05; S, 11.23%. Calc. for $C_{31}H_{50}P_2S_2$: C, 67.84; H, 9.18; S, 11.69%.

Preparation of 3,5-bis(diphenylphosphinoyl)toluene (3) was carried out in analogy with 2a using 1a and hydrogen peroxide (2.68 g, 91% yield).

^{1}H NMR (400 MHz, $CDCl_3$): δ 7.82 (d, 2H, J = 12.7 Hz), 7.67 (t, 1H, J = 14.0 Hz), 7.60-7.35 (m, 20H), 2.40 (s, 3H). ^{31}P NMR (160 MHz, $CDCl_3$): δ 28.8. Anal. Found: C, 75.28; H, 5.48%. Calc. for $C_{31}H_{26}P_2O_2$: C, 75.60; H, 5.32%.

7.3. Preparation of SCS-Pincer Complexes

7.3.1. Phosphine Sulfide-Based SCS-Pincer Complexes (4, 5)

A typical preparation procedure is as follows. A mixture of 2a (53 mg, 0.1 mmol), K_2PdCl_4 (33 mg, 0.1 mmol), and AcOH (3 mL) was heated at 115 °C for 24 h under nitrogen. The resulting yellow precipitate was filtered off and washed with methanol, water, and ether to give a yellow powder of 4a (48 mg, 72% yield).

^{1}H NMR (400 MHz, DMSO-d_6): δ 7.82-7.72 (m, 12H), 7.70-7.63 (m, 8H), 7.20 (d, 2H, J = 11.7 Hz), 2.16 (s, 3H). ^{31}P NMR (160 MHz, DMSO-d_6): δ 56.5. Anal. Found: C, 55.59; H, 3.96; Cl, 4.91; S, 9.51. Calc. for $C_{31}H_{25}ClP_2PdS_2$: C, 55.95; H, 3.79; Cl, 5.33; S, 9.64%.

Other phosphine sulfide-based η^3-SCS-pincer complexes (4b, and 5a,b) were prepared analogously.

4b: 81% yield. ^{1}H NMR (400 MHz, $CDCl_3$): δ 6.79 (d, 2H, J = 9.8 Hz), 2.30 (s, 3H), 2.30-1.15 (m, 44H). ^{31}P NMR (160 MHz, $CDCl_3$): δ 76.7. Anal. Found: C, 53.55; H, 6.99; Cl, 5.30; S, 8.97. Calc. for $C_{31}H_{49}ClP_2PdS_2$: C, 53.99; H, 7.16; Cl, 5.14; S, 9.30%.

5a: 89% yield. ^{1}H NMR (400 MHz, DMSO-d_6): δ 7.82-7.73 (m, 12H), 7.70-7.63 (m, 8H), 7.19 (d, 2H, J = 11.7 Hz), 2.09 (s, 3H). ^{31}P NMR (160 MHz, DMSO-d_6): δ 58.0. Anal. Found: C, 49.26; H, 3.50; Cl, 4.59; S, 8.42. Calc. for $C_{31}H_{25}ClP_2PtS_2$: C, 49.37; H, 3.34; Cl, 4.70; S, 8.50%.

5b: 88% yield. ^{1}H NMR (400 MHz, $CDCl_3$): δ 6.79 (d, 2H, J = 9.8 Hz), 2.35-1.10 (m, 47H). ^{31}P NMR (160 MHz, $CDCl_3$): δ 77.8. Anal. Found: C, 47.53; H, 6.23; Cl, 4.28; S, 8.24. Calc. for $C_{31}H_{49}ClP_2PtS_2$: C, 47.84; H, 6.35; Cl, 4.55; S, 8.24%.

4a was also prepared by another procedure. A mixture of 2a (53 mg, 0.1 mmol), $PdCl_2(PhCN)_2$ (38 mg, 0.1 mmol), and acetonitrile (10 mL) was heated at 65 °C for 24 h under nitrogen. The resulting yellow precipitate was filtered off and washed with acetonitrile, methanol, water, and ether to give a yellow powder of 4a (44 mg, 65% yield).

7.3.2. Thioamide-Based SCS-Pincer Platinum Complexes (7a-D)

A typical preparation procedure is as follows. An acetonitrile solution (25 mL) of 6b (30 mg, 0.1 mmol) was added to a chloroform solution (5 mL) of $PtCl_2(PhCN)_2$ (47 mg, 0.1 mmol), and the reaction mixture was stirred at 65 °C for 15 h under nitrogen. The resulting precipitate was filtered off, washed with acetonitrile, and dried under vacuum to give 7b as an orange solid (36 mg, 67% yield).

^{1}H NMR (400 MHz, DMSO-d_6): δ 7.76 (d, 2H, J = 8.3 Hz), 7.16 (t, 1H, J = 7.9 Hz), 4.29 (t-br, 4H), 3.97 (t-br, 4H), 2.08 (m-br, 8H). ^{13}C NMR (100 MHz, DMSO-d_6): δ 198.2, 165.3 (^{1}J(Pt) was not resolved), 143.0 (^{2}J(Pt) = 76.8 Hz), 129.8 (^{3}J(Pt) = 54.9 Hz), 119.1, 55.8, 54.8, 25.9, 22.9. ^{195}Pt NMR (107 MHz, DMSO-d_6, referred to H_2PtCl_6): δ -5281. Anal. Found C, 35.53; H, 3.83; N, 5.49; Cl, 6.74; S, 11.85. Calc. for $C_{16}H_{19}ClN_2PtS_2$: C, 35.99; H, 3.59; N, 5.25; Cl, 6.64; S, 12.01%.

Other thioamide-based η^3-SCS-pincer platinum complexes (7a,c) were prepared analogously.

7a: 79% yield. ^{1}H NMR (400MHz, DMSO-d_6): δ 7.63 (d, 2H, J = 7.6Hz), 7.12 (t, 1H, J = 7.6Hz), 3.80 (s, 6H), 3.61 (s, 6H). ^{13}C NMR (100 MHz, DMSO-d_6): δ 207.2, 202.6, 142.7, 131.3, 119.1, 47.0, 46.4. Anal. Found C, 30.02; H, 3.15; N, 5.88; Cl, 7.40; S, 13.44. Calc. for $C_{12}H_{15}ClN_2PtS_2$: C, 29.91; H, 3.14; N, 5.81; Cl, 7.36; S, 13.31%.

7c: 25% yield. ^{1}H NMR (400MHz, DMSO-d_6): δ 7.51 (d, 2H, J = 7.2Hz), 7.13 (t, 1H, J = 7.2 Hz), 4.25 (s, 8H), 3.85 (s, 8H). ^{13}C NMR (100 MHz, DMSO-d_6): δ 206.8, 197.3, 142.0, 131.0, 126.1, 65.8, 65.6, 52.4, 49.2. Anal. Found C, 33.72; H, 3.70; N, 4.68; Cl, 6.92; S, 11.01. Calc. for $C_{16}H_{19}ClN_2O_2PtS_2$: C, 33.95; H, 3.38; N, 4.95; Cl, 6.26; S, 11.33%.

7.3.3. Substitution of Chloro Ligand of 7b

A) Phenylacetylide Ligand

To a DMF solution (1 mL) of 7b (27 mg, 0.05 mmol) was added phenylacetylene (92 mg, 0.9 mmol) and CuI (4 mg, 0.02 mmol) and triethylamine (1 mL). The reaction mixture was stirred at room temperature for 5 days. The resulting precipitate was filtered off, washed with DMF, and dried under vacuum to give 8b as an orange solid (25 mg, 82% yield).

^{1}H NMR (400 MHz, DMSO-d_6): δ 7.86 (d, 2H, J = 7.3 Hz), 7.16 (m, 5H), 7.06 (t, 1H, J = 7.2 Hz), 4.32 (t-br, 4H), 4.03 (t-br, 4H), 2.09 (m-br, 8H). ^{195}Pt NMR (107 MHz, DMSO-d_6, referred to H_2PtCl_6): δ -5521. Anal. Found: C, 47.78; H, 4.63; N, 4.44; S, 9.83. Calc. for $C_{24}H_{24}N_2PtS_2$: C, 48.07; H, 4.03; N, 4.67; S, 10.69%.

B) Iodo Ligand

To a dichloromethane solution (15 mL) of 7b (16 mg, 0.03 mmol) was added an excess dido methane (0.1 mL). The solution mixture was stirred at room temperature for 8 days. The solvent was evaporated, and the crude product was washed with acetonitrile and dried under vacuum to give 9b as a red solid (12 mg, 65% yield).

^{1}H NMR (400 MHz, DMSO-d_6): δ 7.85 (d, 2H, J = 8.4 Hz), 7.24 (t, 1H, J = 8.2 Hz), 4.29 (t-br, 4H), 3.97 (t-br, 4H), 2.08 (m-br, 8H). Anal. Found: C, 31.00; H, 3.05; N, 4.91; I, 20.09; S, 9.34. Calcd. for $C_{16}H_{19}N_2IPtS_2$: C, 30.73; H, 3.06; N, 4.48; I, 20.29; S, 10.25%.

9b was also prepared by another procedure. 7b (16 mg, 0.03 mmol) was suspended in acetone (5 mL), and $AgBF_4$ (6 mg, 0.03 mmol) was added. After the dissolution of the starting complex, water (1 mL) was added. The precipitated silver chloride was filtered off. The addition of sodium iodide (15 mg, 0.1 mmol) and water (5 mL) to the filtrate gave a red precipitate. The precipitate was filtered off, washed with water and methanol, and dried under vacuum to give 9b as a red solid (7 mg, 39% yield).

7.4. X-Ray Crystallography

All measurements were made on Rigaku Saturn CCD and AFC5R diffractometers using graphite-monochromated Mo-Kα radiation (λ = 0.71071 Å) at room temperature (AFC5R) or 113 K (Saturn CCD), and processed using Crystal Clear (Rigaku). The data were collected for Lorentz and polarization effects, and an empirical absorption correction was applied. The structures of the compounds were solved by a direct methods (SIR 92) or heavy-atom Patterson methods, and expanded using Fourier techniques. These structures were refined by full-matrix least-square refinement on F. The non-hydrogen atoms were refined anisotropically. Hydrogen atoms were located by assuming the ideal geometry and included in the structure calculation without further refinement of the parameters. All calculations were performed using the Crystal Structure crystallographic software package.

7.5. Preparation and Measurement of LEDs

LEDs consisted of the indium-tin oxide (ITO) electrode, the hole injection layer of poly[3,4-(ethylenedioxy)thiophene]-poly(styrenesulfonic acid) (PEDOT-PSS), the emitting layer, the electron transporting layer of 4,7-diphenyl-1,10-phenanthroline (Bathophenanthroline, VPT), the electron injection layer of LiF, and the Al electrode. In all cases, the ITO anode was overcoated with PEDOT-PSS (about 50 nm). Thin films of the emitting materials (about 100 nm) were prepared by spin-casting from a mixture of chloroform-toluene (1:2) solution of a mixture of poly(9-vinylcarbazole) (PVK), 2-(4-biphenylyl)-5-(4-*tert*-butylphenyl)-1,3,4-oxadiazole (*tert*-BuPBD), and the platinum complex

(5a, 7b, or 8b). The thin layers of VPT (about 30 nm) and LiF (about 0.5 nm), and the Al cathode (about 150 nm) were prepared by vacuum deposition. Electroluminescence spectra and light-intensity characteristics of the LEDs were measured by using a monochromator with photodiode-array and luminance meter (light meter). All the measurements were done under vacuum (about 10^{-1} Pa) or argon atmosphere at room temperature.

ACKNOWLEDGEMENT

The authors are grateful to Dr. H. Ogawa and T. Inoue of TDK Corporation and K. Okada and M. Akaiwa of our laboratory for experimental support. This work has been partly supported by a Grant-in-Aid from the Ministry of Education, Culture, Sports, Science, and Technology of Japan and Ogasawara Foundation for the Promotion of Science and Engineering.

REFERENCES

[1] M. Albrecht, G. van Koten, *Angew. Chem. Int. Ed.* 40 (2001) 3750; (b) P. Steenwinkel, R. A. Gossage, G. van Koten, *Chem. Eur. J.* 4 (1998) 759; (c) M. H. P. Rietveld, D. M. Grove, G. van Koten, *New J. Chem.* 21 (1997) 751; (d) J. T. Singleton, *Tetrahedron* 59 (2003) 1837.

[2] J. Brooks, Y. Babayan, S. Lamansky, P. I. Djurovich, I. Tsyba, R. Bau, M. E. Thompson, *Inorg. Chem.* 41 (2002) 3055; (b) K. P. Bakashev, M. V. Puzyk, V. S. Kotlyar, M. V. Kulikova, *Coord. Chem. Rev.* 159 (1997) 109; (c) M. Maestri, *Coord. Chem. Rev.* 111 (1991) 117; (d) Q. Liu, L. Thorne, I. Kozin, D. Song, C. Seward, M. D'Iorio, Y. Tao, S. Wang, *J. Chem. Soc., Dalton Trans.* (2002) 3234; (e) T.-C. Cheung, K.-K. Cheung, S.-M. Peng, C.-M. Che, *J. Chem. Soc., Dalton Trans.* (1996) 1645; (f) W. Lu, B.-X. Mi, M. C.-W. Chan, Z. Hui, N. Zhu, S.-T. Lee, C.-M. Che, *Chem. Commun.* (2002) 206; (g) S.-W. Lai, M. C.-W. Chan, T.-C. Cheung, S.-M. Peng, C.-M. Che, *Inorg. Chem.* 38 (1999) 4046; (h) S.-W. Lai, M. C.-W. Chan, K.-K. Cheung, C.-M. Che, *Organometallics* 18 (1999) 3327; (i) V. W.-W. Yam, R. P.-L. Tang, K. M.-C. Wong, W. X.-X. Lu, K.-K. Cheung, N. Zhu, *Chem. Eur. J.* 8 (2002) 4066; (j) J. A. G. Williams, A. Beeby, E. S. Davies, J. A. Weinstein, C. Wilson, *Inorg. Chem.* 42 (2003) 8609; (k) W. Lu, B.-X. Mi, M. C. W. Chan, Z. Hui, C.-M. Che, N. Zhu, S.-T. Lee, *J. Am. Chem. Soc.* 126 (2004) 4958; (l) W. Lu, M. C. W. Chan, N. Zhu, C.-M. Che, C. Li, Z. Hui, *J. Am. Chem. Soc.* 126 (2004) 7639.

[3] J. Errington, W. S. McDonald, B. L. Shaw, *J. Chem. Soc., Dalton Trans.* (1980) 2312; (b) H. Nakai, S. Ogo, Y. Watanabe, *Organometallics* 21 (2002) 1674; (c) N. Lucena, J. Casabo, L. Escriche, G. Sanchez-Castello, F. Teixidor, R. Kivekäs, R. Sillanpää, *Polyhedron* 18 (1996) 3009; (d) D. R. Evans, M. Huang, W. M. Seganish, J. C. Fettinger, T. L. Williams, *Organometallics* 21 (2002) 893; (e) D. E. Bergbreiter, P. L. Osburn, A. Wilson, E. M. Sink, *J. Am. Chem. Soc.* 122 (2000) 9058; (f) D. E. Bergbreiter, P. L. Osburn, J. D. Frels, *J. Am. Chem. Soc.* 123 (2001) 11105.

[4] T. S. Lobana, in F. R. Hartley Ed., *The Chemistry of Organophosphorus Compounds*, John Wiley & Sons, Chichester, (1992) Vol. 2, Chapter 8, p. 409; (b) B. Walther, *Coord. Chem. Rev.* 60 (1984) 67; (c) L. Gonsalvi, H. Adams, G. J. Sunley, E. Ditzel, A.

Haynes, *J. Am. Chem. Soc.* 124 (2002) 13597; (d) H. Liang, S. Ito, M. Yoshifuji, *Org. Lett.* 6 (2004) 425.

[5] E. S. Raper, *Coord. Chem. Rev.* 129 (1994) 91; (b) M. A. Ali, S. E. Livingstone, *Coord. Chem. Rev.* 13 (1974) 101;

[6] P. J. Heard, A. E. Aliev, *Polyhedron* 17 (1998) 3981; (b) M. Doux, C. Bouet, N. Mézailles, L. Ricard, P. Le Floch, *Organometallics* 21 (2002) 2785; (c) M. Doux, L. Ricard, P. Le Floch, N. Mézailles, *Dalton Trans.* (2004) 2593.

[7] S. Takahashi, M. Nonoyama, M. Kita, Transition Met. Chem. 20 (1995) 528; (b) Y. Nojima, M. Nonoyama, K. Nakajima, *Polyhedron* 15 (1996) 3795.

[8] T. Kanbara, T. Yamamoto, J. Organomet. Chem. 688 (2003) 15; (b) T. Kanbara, K. Okada, T. Yamamoto, H. Ogawa, T. Inoue, *J. Organomet. Chem.* 689 (2004) 1860.

[9] O. Herd, A. Heßler, M. Hingst, M. Tepper, O. Stelzer, *J. Organomet. Chem.* 522 (1996) 69; (b) P. Machnitzki, T. Nickel, O. Stelzer, C. Landgrafe, Eur. J. *Inorg. Chem.* (1998) 1029; (c) T. Kanbara, S. Takase, R. Hayashi, S. Kagaya, K. Hasegawa, T. Yamamoto, *J. Polym. Sci.: Part A: Polym. Chem.* 40 (2002) 2637; (d) Z. Jin, B. L. Lucht, *J. Organomet. Chem.* 653 (2002) 167.

[10] A. D. Ryabov, *Inorg. Chem.* 26 (1987) 1252; (b) D. J. Cárdenas, A. M. Echavarren, *Organometallics* 18 (1999) 3337; (c) E. Díez-Barra, J. Guerra, I. López-Solera, S. Merino, J. Rodoríguez-López, P. Sánchez-Verdú, J. Tejeda, *Organometallics* 22 (2003) 541.

[11] E. V. Brown, Synthesis (1975) 358; (b) T. Kanbara, Y. Kawai, K. Hasegawa, H. Morita, T. Yamamoto, *J. Polym. Sci.: Part A: Polym. Chem.* 39 (2001) 3739.

[12] M. Albrecht, A. L. Spek, G. van Koten, *J. Am. Chem. Soc.* 123 (2001) 7233; (b) J. Terheijden, G. van Koten, I. C. Vinke, A. L. Spek, *J. Am. Chem. Soc.* 107 (1985) 2891. (c) D. M. Grove, G. van Koten, J. N. Louwen, J. G. Noltes, A. L. Spek, HH. J. C. Ubbels, *J. Am. Chem. Soc.* 104 (1982) 6609.

[13] B. F. M. Kimmich, W. J. Marshall, P. J. Fagan, E. Hauptman, R. M. Bullock, *Inorg. Chim. Acta* 330 (2002) 52; (b) D. Morales-Morales, C. Grause, K. Kasaoka, R. Redón, R. E. Cramer, C. M. Jensen, *Inorg. Chim. Acta* 300-302 (2000) 958; (c) C. M. Hartshorn, P. J. Steel, *Organometallics* 17 (1998) 3487.

[14] F. Neve, A. Crispini, C. D. Pietro, S. Campagna, *Organometallics* 21 (2002) 3511; (b) Y. Wakatsuki, H. Yamazaki, P. A. Grutsch, M. Santhaman, C. Kutal, *J. Am. Chem. Soc.* 107 (1985) 8153; (b) M. Ghedini, D. Pucci, G. Calogero, F. Barigelletti, *Chem. Phys. Lett.* 267 (1997) 341; (c) I. Aiello, D. Dattilo, M. Ghedini, A. Golemme, *J. Am. Chem. Soc.* 123 (2001) 5598; (d) R. Schwarz, G. Gliemann, Ph. Jolliet, A. von Zelewsky, *Inorg. Chem.* 28 (1989) 742: (e) M. Ghedini, I. Aiello, M. L. Deda, A. Grisolia, *Chem. Commun.* (2003) 2198; (f) D. Song, Q. Wu, A. Hook, I. Kozin, S. Wang, *Organometallics* 20 (2001) 4683; (g) Q. Wu, A. Hook, S. Wang, *Angew. Chem. Int. Ed.* 39 (2000) 3933; (h) F. Neve, A. Crispini, S. Campagna, *Inorg. Chem.* 36 (1997) 6150; (i) C. S. Consorti, G. Ebeling, F. Rodembusch, V. Stefani, P. R. Livotto, F. Rominger, F. H. Quina, C. Yihwa, J. Dupont, *Inorg. Chem.* 43 (2004) 530.

[15] M. A. Baldo, M. E. Thompson, S. R. Forrest, *Pure Appl. Chem.* 71 (1999) 2095; (b) B. W. D'Andrade, J. Brooks, V. Adamovich, M. E. Thompson, S. R. Forrest, *Adv. Mater.* 14 (2002) 1032; (c) Q. Zhang, Q. Zhou, Y. Cheng, L. Wang, D. Ma, X. Jing, *Adv. Mater.* 16 (2004) 432;

[16] J. R. Doyle, P. E. Slade, H. B. Jonassen, *Inorg. Synth.* 6 (1960) 218. C. H. Lindsay, L. S. Benner, A. L. Balch, *Inorg. Chem.* 19 (1980) 3503.
[17] S. Yu, H. Seino, M. Ueda, *J. Am. Chem. Soc*32 (1999) 1027.
[18] K. A. King, P. J. Spellane, R. J. Watts, *J. Am. Chem. Soc.* 107 (1985) 1431.

In: Focus on Organometallic Chemistry Research
Editor: Marin A. Cato, pp. 133-145
ISBN 1-59454-495-6

Chapter 5

CHARACTERISTICS AND APPLICATIONS OF METAL COMPLEXES WITH β-KETOIMINATE LIGANDS

Ik-Mo Lee*

Department of Chemistry, Inha University, 253 Yonghyundong, Namku, Incheon, 402-751, Korea

ABSTRACT

Up to date, β–diketonate complexes have been studied thoroughly and have been successfully used in various fields due to their high stability originated from chelate and resonance effects. β-Ketoiminate complexes have recently drawn interest because of their higher thermal stability than the β–diketonate analogues and their versatility, by changing the imine substituents, for tailoring their reactivity and volatility. In addition, polydentate β-ketoiminate derivatives have been adopted for the saturation or stabilization of coordination spheres to control the thermal and chemical stability. Easy control of charge on the ligand is another merit. Various designed complexes containing β-ketoiminate ligands have been successfully exploited as precursors for metal or metal oxide depositions and catalysts. In this monograph, preparation of ligands and complexes, characteristics, and applications of metal complexes as MOCVD (metalloorganic chemical vapor deposition) precursors and catalysts will be reviewed.

INTRODUCTION

Reactivity of transition metal and main group metal complexes can be controlled by oxidation state and nature of central metal and ligands. Stability of metal complexes has been improved by introducing stronger metal-ligand bonds or exploiting chelate effect. In this regard, various polydentate ligands have been adopted to increase the stability by anchoring the coordination sites and chelate effect. Among these polydentate ligands, β–diketonate

* Tel: 82-32-860-7682, FAX: 82-32-867-5604, imlee@inha.ac.kr

ligands have drawn much attention due to their resonance stability and unique structure and they have been investigated with virtually every metal and metalloid in the periodic table. These extensive works have been well reviewed in late 70s [1]. Recent interest on metalloorganic chemical vapor deposition (MOCVD) or atomic layer deposition (ALD) precursors (thermally stable, air- and moisture-stable, volatile, and easy to handle) incited by explosive development of information industry leads to reinvestigate the metal β–diketonates and complexes of special groups including group 2 were reviewed in 2000 [2].

Also the advent of single-site metallocene catalysts for olefin polymerization resulted in extensive investigation on the complexes with cyclopentadienes (Cp) and closely related analogues. Recent adoption of various ligands to replace the Cp-type ligands have witnessed that Cp ligands are not essential for the high activity towards olefin polymerization. Several groups have developed anionic nitrogen chelating ligands to complement Cp or phorphyrin ligands [3]. In most cases, nitrogen ligands have been arranged in a dianionic motif as analogues to bis(cyclopentadienyl) systems. Dianionic and trianionic ligand sets have been developed to meet the requirements demanding by the nature of metal.

With the goal of studying compounds supported by monoanionic ligands whose steric and electronic properties can be easily modified, β–diketiminate ligands, a subset of Schiff base ligands, have attracted much interest. In their deprotonated form, these uninegative, four-electron donors can be viewed as hemiporphyrinate ligands with lower coordination numbers than Cp and tris(pyrazolyl)borate ligands and these show some similarities with amidinate ligands. In addition to having a larger bite angle, β–diketiminate ligands have electrons in the π -manifold and can be viewed as heteroatom analogues of Cp ligands.

Applications of these ligands to organometallic chemistry have been limited despite several following attractive features; easy and efficient preparation from cheap and readily available building blocks, 2,4-pentanedione and primary amines, easy modification of steric and electronic requirements by changing the amines, and improved kinetic stability of metal complexes with several accessible oxidation states at the metal by changing the anionic charge of the ligand.

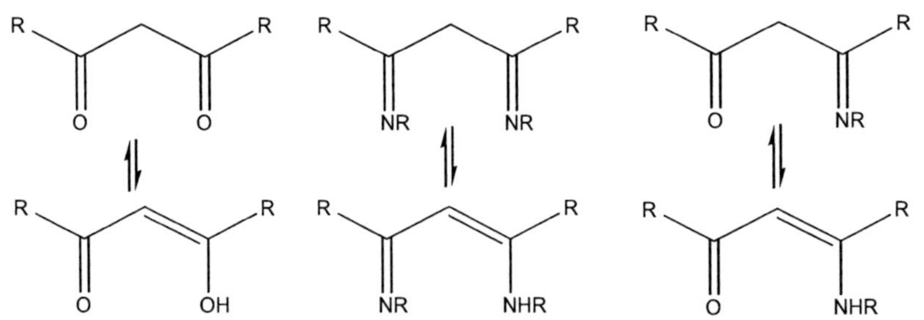

β-diketone β-diketimine β-ketoimine

Figure 1. Structures of β-diketone, β-diketimine and β-ketoimine

Asymmetric analogues, β–ketoiminate ligands are also interesting due to similar structures and features with β–diketonate and β–diketiminate ones (Figure 1) but exert different effects to trans ligands and experience ready introduction of chirality and further functionalization [4].

In this monograph, syntheses and characterization of β–ketoiminate ligands and complexes with these ligands and applications of these complexes are briefly reviewed.

PREPARATION OF B -KETOIMINE

β-Ketoimine and closely related β-diketimine have been generally prepared by the condensation reaction between β-diketone and primary amines with or without acid catalysts. Typical reaction pathways for β-diketimine are described as shown in the following reaction [5] and β–ketoimine can be prepared by controlling the amount of amines.

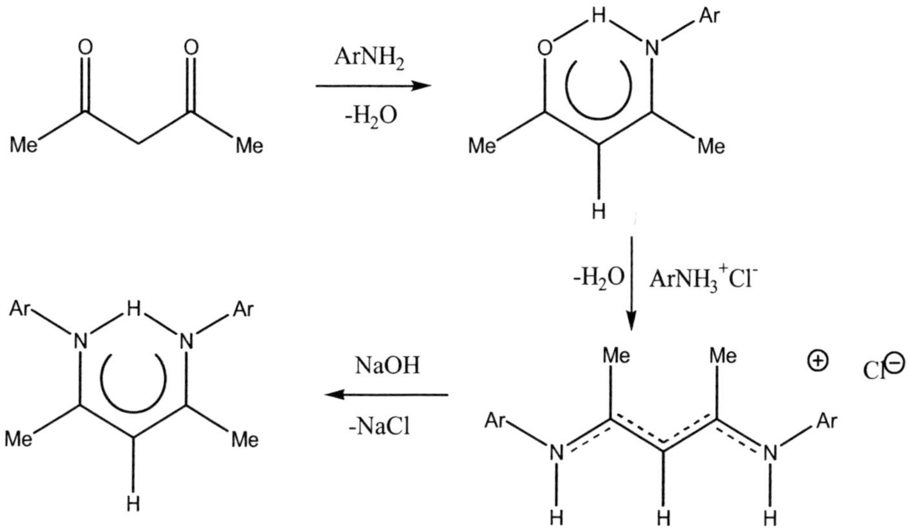

Reaction conditions and yields are in the wide range depending on the nature of β – diketone and primary amines, especially bulkiness of substituents in these compounds. Higher reaction temperature, longer reaction time and catalysts are required with increase of bulkiness of substituents and generally lower yields are resulted.

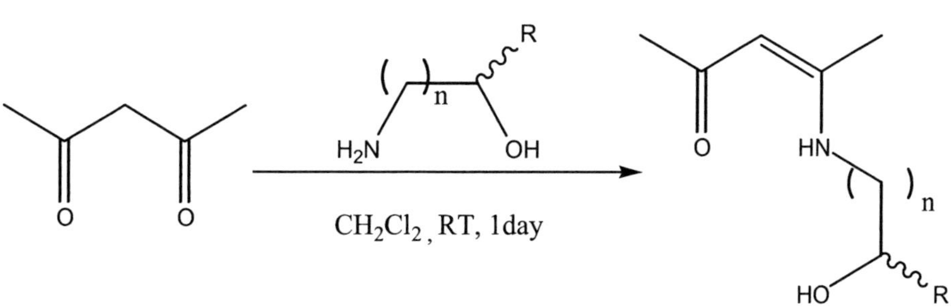

Ligand 1. n=1, R=H
2. n=1, R=Me
5. n=2, R=H

Scheme 1.

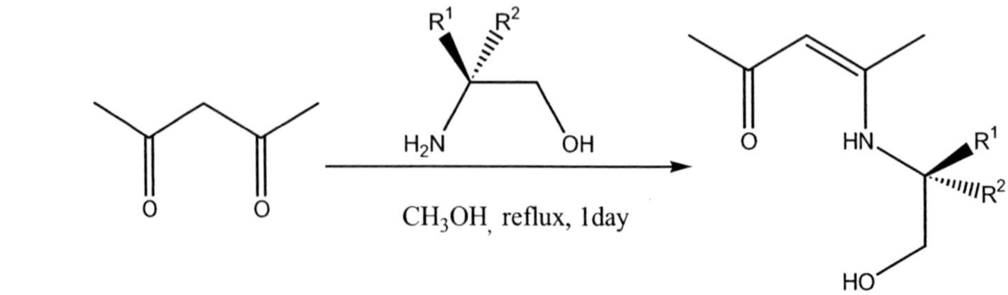

Ligand 3. R^1=Me, R^2=H
4. R^1=Me, R^2=Me

Scheme 2.

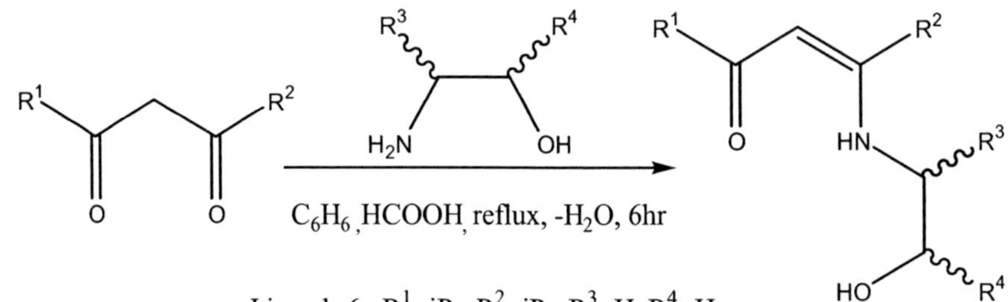

Ligand 6. R^1=iPr, R^2=iPr, R^3=H, R^4=H
7. R^1=iPr, R^2=iPr, R^3=H, R^4=Me
8. R^1=iPr, R^2=iPr, R^3=Me, R^4=H
9. R^1=t-Bu, R^2=Me, R^3=H, R^4=H

Scheme 3.

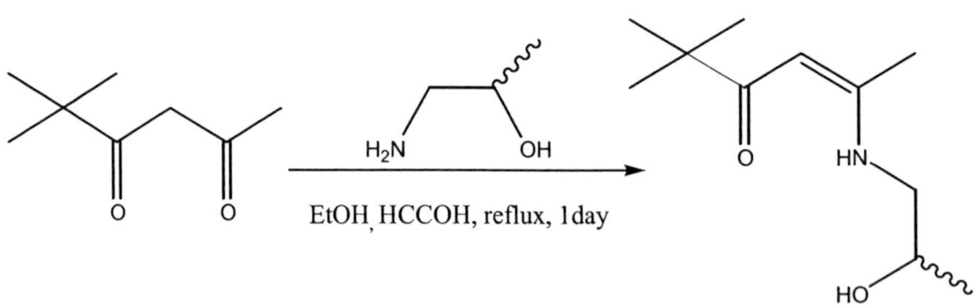

Ligand 10.

Scheme 4 .

Lithium β-diketiminate can be obtained by the reaction between $LiCHR_2$ and R'CN [6,7] or R'NC [8] and novel lithium 3-sila- and 3-germa- β −diketiminates are also prepared by modifying the starting materials [9].

Various derivatives such as tetradentate β -ketoimine [10], bis(β -ketoiminato) polyethers [11] and tripodal β -ketoimine [12] have been also prepared to introduce special properties to the metal complexes.

Systematic syntheses of various β -ketoimines shown below have been achieved to investigate the subtle change in the properties of metal complexes with these ligands [13-19].

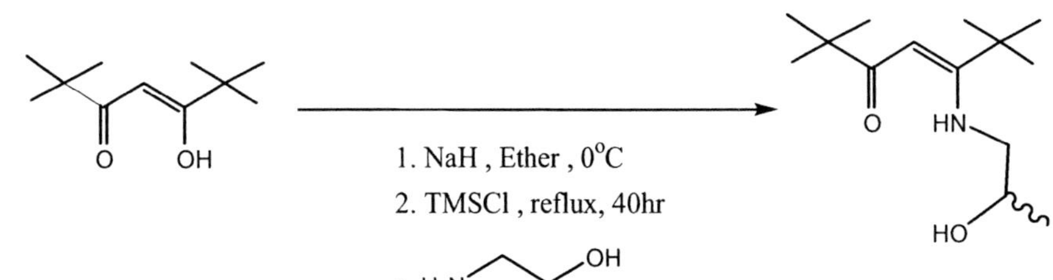

Ligand 11.

Scheme 5.

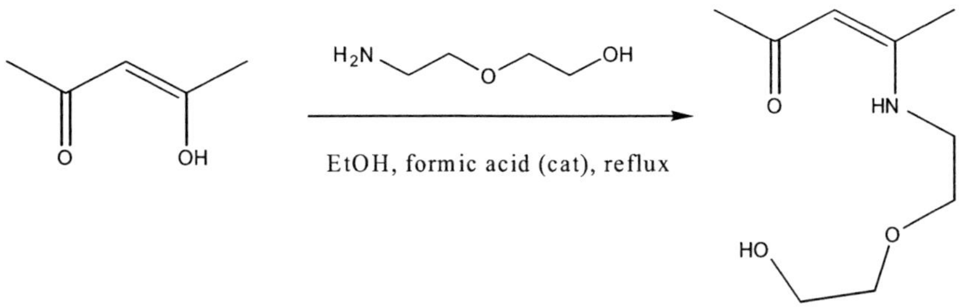

Ligand 12.

Scheme 6.

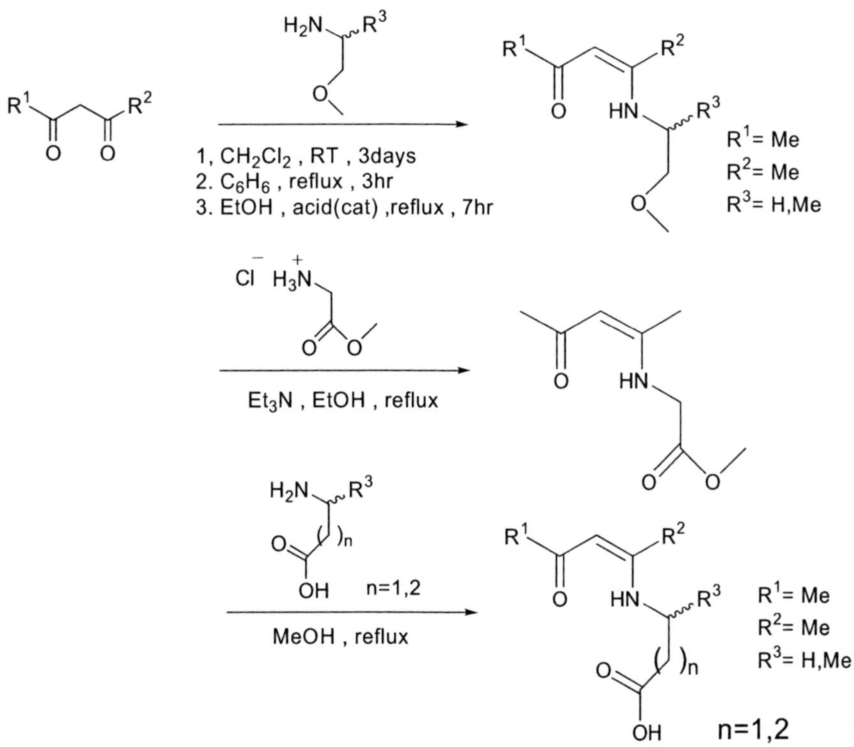

Ligand 13, 14, 15

Scheme 7.

SYNTHESIS OF β-KETOIMINATE COMPLEXES

Three synthons are useful for introducing β -ketoiminate ligands: the neutral ligand, iminium salts of the neutral ligand, and alkali metal ligand anions. For neutral and cationic ligands, β -ketiminate complexes can be prepared by amine elimination reactions from metal amides, while they are synthesized by the reaction between alkali ketiminates and metal halides [3].

For neutral ligands, additional techniques such as alkane [5], HX [4], alcohol [13,14] and water elimination reactions [20,21] from metal alkyls, halides, alkoxides and hydroxides have been also exploited.

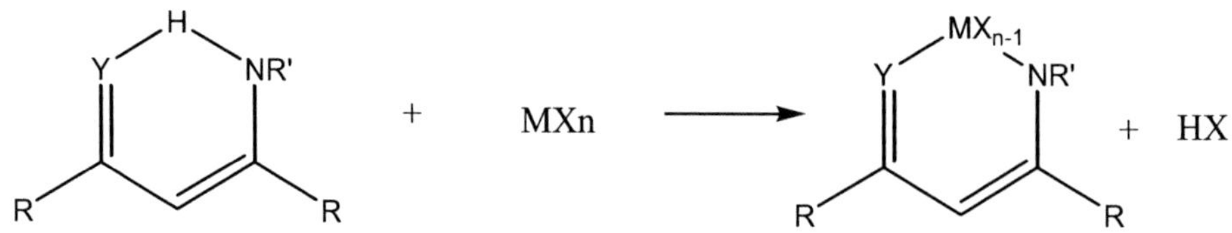

Y= O, NR' X=halide, R, OR, NR_2

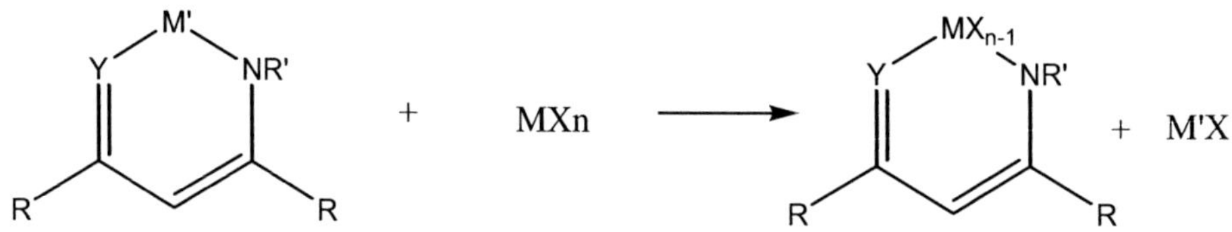

Y= O, NR'
M'=Li, Na, Tl

X=halide, R, OR, NR_2

CHARACTERIZATION OF β-KETOIMINE AND β-KETOIMINATE COMPLEXES

Characterization of β-ketoimine and β-ketoiminate complexes has been done by ^{1}H and ^{13}C NMR, IR, Mass spectroscopies, and X-ray crystallography. When the complexes are prepared for the application of CVD precursors, thermal analysis is critical in the characterization.

NMR spectroscopy is useful for the determination of solution structures of β-ketoimine and β-ketoiminate complexes. β-Diketones are well-known to show enol-keto tautomerization but β-ketoimines shown above (ligand 1-15) do not show this type of tautomerism and only enamino forms are observed. This was confirmed by the presence of the broad peaks at around 11 ppm in the ^{1}H NMR spectra. Also the proton attached to the nitrogen in the enamino form is further coupled with protons on the α-carbon to the nitrogen and increased multiplicity can be observed. However, assignments of methylene peaks in the backbone of the N-alkoxo portion and alkyl substituents in the β-ketoimine are not easy. Unambiguous assignments have been done through C-H correlation experiment, selective decoupling experiments and careful analysis of the spectrum of a dissymmetric ligand such as ligand 9 [22].

1) ^{1}H NMR Spectra

In the ligands, the methylene peak next to the nitrogen appears upfield than that next to oxygen in the backbone of the N-alkoxo portion as shown in the cases of ether and amine. Peaks for alkyl groups next to C=N locate downfield than those next to C=O in the β-ketoiminate portion. The methine peaks of β-ketoimines appear in the region of around 5 ppm but those of complexes generally go downfield by ~ 0.3ppm.

2) ^{13}C NMR Spectra

Methylene carbon peaks next to nitrogen appear upfield than those next to oxygen in the backbone of the N-alkoxo portion but location of carbon peaks in the β-ketoiminate depends on the distance from the C=O or C=N groups. For the α-carbon, peaks next to C=O group appear downfield than those next to C=N group but reverse trend is observed in the case of β-carbon. Methine carbon peaks in the complex appear at around 100 ppm, which are downfield than those in the ligand. Peaks for carbonyl group in the ligands locate downfield significantly than those in the complexes but peaks for imine group do not change their location on coordination.

3) IR Spectroscopy

IR spectroscopy has been utilized to find out the presence of carbonyl and imino groups but generally clear assignments have been not done. In the limited data, ν(C=N) and ν(C=O) are observed in the range of 1550-1600 cm^{-1} and ν(C=N) appear in the higher wavenumber than ν(C=O) [12,20].

4) Mass Spectroscopy

The electron impact mass spectroscopic data for the free ligands and their complexes are similar. The monomeric parent ion is observed for every compound. The major fragment usually corresponds to loss of a tert-butyl group when t-Bu is the substituent in the free ligands or complexes. Almost all the other fragments are derived from scission of C-C or C-O bonds in the polyether bridge of bis(β-ketoiminato) polyethers [11].

Structures of β-Ketoiminate Complexes

Crystal structures of many complexes with β -ketoiminate ligands have been determined and these structures show that backbones of the β-ketoiminate ligands are essentially planar. Imine-enolate form is confirmed in some β-ketoiminato complexes where bond lengths in the β-ketoiminate fraction [3,4,12,14,23,24] are reported, while significant delocalization is observed within the π -system in some β-diketiminato complexes[25,26] with some exceptions [7,26,27]. It is worth mentioning that β-diketiminate ligand functions as a six-electron donor in some cases [3] and the deviation from planarity for the ligand backbone implicates contributions from an allyldiimine resonance form (I), first suggested by Lappert [6].

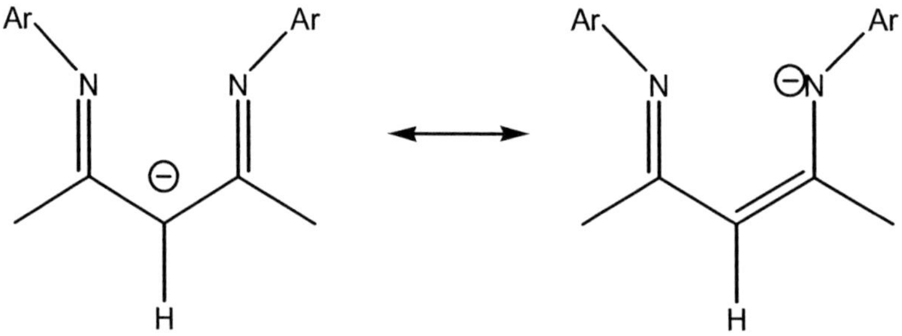

I II

Applications of β-Ketoiminate Complexes

Metal β-diketonates are known to exhibit higher thermal and chemical stabilities than metal alkyls and alkoxides and have shown moderate to high vapor pressure. Therefore, they, especially with fluorinated β-diketonate ligands, are widely employed as precursors for CVD. However, metal or metal oxide thin films produced by these fluorinated β-diketonates are inevitably contaminated by fluorinated impurities [10] and these have been used only in the limited cases.

Due to enhanced thermal stability of β-ketoiminato complexes with respect to the β-diketonates and potential advantages of modulating steric and electronic properties by varying the amine used in the ligand synthesis, β-ketoiminato complexes have attracted attention for the further application in CVD precursors.

A series of triphenylphosphine adducts of silver β-diketonates [AgL(PPh_3)] (L = β-diketonates) and β-ketoiminates have been synthesized and evaluated as precursors for the deposition of silver films using aerosol-assisted CVD methodology [28]. The results obtained suggest that the β-ketoiminates show greater potential, affording lower initial decomposition temperature and somewhat higher deposition rates. The films produced are smoother, of good thickness with relatively low impurity levels combined with acceptable reflectivities.

New class of linked β -ketoiminate-polyether-β -ketoiminate ligands have been designed and prepared to encapsulate alkaline-earth cations having low charges and large ionic radii and a limited success has been obtained in growing the thin films of CeO_2 and $BaTiO_3$ [11,29].

[M(N-alkoxo-β -ketoiminate)$(OEt)_3$] (M=Nb, Ta) obtained by partial replacement of ethoxide ligands from $M_2(OEt)_{10}$ showed enhanced thermal and chemical stability, especially toward moisture. These complexes demonstrate considerably higher deposition rates than $M_2(OEt)_{10}$, one of the most popular precursors for Ta_2O_5 and Nb_2O_5 films [13]. Even though some nitrogen and carbon impurities are found in the initial deposition, heat treatment at 700°C in oxygen removed all the impurities to the negligible level.

For the successful fabrication of thin films containing titanium oxides, suitable precursors with high volatility, enhanced thermal and chemical stability and easy decomposition are required. For (Ba,Sr)TiO_3 (BST) thin films fabrication by liquid source (LS) MOCVD method, enhanced compatibility of titanium precursors with barium and strontium complexes such as $Ba(thd)_2$ and $Sr(thd)_2$ (thd=2,2,6,6-tetramethyl-2,5-heptanedionate) is another important requirement [30,31,32]. New series of [Ti(N-alkoxo-β -ketoiminate)$_2$] are found to produce TiO_2 thin films without any sign of haziness and BST thin films with low carbon content and smooth surface morphology.

Novel ruthenium (III) complexes containng fluorinated β -ketoiminate ligands have been prepared and tested as potential CVD precursors for Ru and RuO_2 thin films [23]. These complexes show good volatility and thermal stability and CVD experiments using these complexes with either a mixture of 2% O_2 in argon or pure O_2 at 325-450°C produced Ru or RuO_2 successfully.

Several other β -ketoiminates and fluorinated β -ketoiminates of Cu, Ni, Ba, and Pd have been also successfully used as precursors for metal and metal oxide deposition [33-37].

There has been considerable interest in the preparation and chemistry of electron-deficient complexes of the group 4 transition metals in connection with their use in olefin polymerization and related processes. Although extensive research in the 80s and 90s deals with group 4 complexes with Cp analogues, recent attention has changed to isoelectronic alternatives to the Cp anions and late transition metal complexes. Various N-based chelate ligands have been tested but little attention has been paid to β -diketiminate or β -ketoiminate complexes. However, some recent promising results have been accumulated slowly but steadily.

New neutral nickel complexes with the formula of [Ni(β -ketoiminate)(Ph)(PPh_3)] or [Ni(β -diketiminate)(PPh_3)] have been prepared and it is found that these complexes show high catalytic activities of up to 3.16×10^7 g PNB/mol Ni h for the addition polymerization of norbornene in the presence of modified methylamluminoxane (MMAO) as cocatalyst [24].

Novel dialkylscandium complexes supported by β -diketiminato ligands are prepared and treatment of dibenzyl compound with $B(C_6F_5)_3$ produced the cationic scandium complex stabilized by a η^6-aryl interaction with the abstracted borate benzyl group [27]. Since this complex resembles the active species in the metallocene catalysts for olefin polymerization, it is very interesting to see the outcome of olefin polymerization but no results have been filed.

[MCl_2(β -diketiminates)]$(THF)_2$ (M=Ti, V, Cr) have been synthesized to witness the high activity toward olefin homo- and copolymerization in the presence of MAO [38].

Three coordinate, base-free aluminum alkyl cations with β -diketiminate ligands have been designed and prepared [39]. Contrary to expectation, these complexes are not active toward ethylene polymerization but only active for reversible ethylene cycloaddtion reaction [40].

Finally, new series of neutral Ni and Pd complexes containing functionalized β-ketoiminate ligands have shown to be very active toward ethylene and norbornene polymerization [18,19].

Conclusion

Applications of β -ketoiminates and β -diketiminates to organometallic chemistry have been limited despite several attractive features such as easy and efficient preparation from cheap and readily available building blocks, easy modification of steric and electronic requirements, and improved kinetic stability of metal complexes with several accessible oxidation states at the metal by changing the anionic charge of the ligand.

Fortunately, chemistry information of metal β -ketoiminates or β -diketiminates has been recently accumulated and increased examples of application in the fields of MOCVD precursors and catalysts for olefin polymerization have been reported with anticipation of further success.

Moreover, asymmetric character of β–ketoiminate ligands is surely to introduce some interesting features such as different effects to the trans ligands and ready introduction of chirality and further functionalization and expansion of their applications in more fields will be witnessed in near future.

References

[1] Mehrotra, R.C.; Bohra, R.; Gaur, D.P. Metal βDiketonates and Allied Derivatives, London: Academic Press; 1978.

[2] Otway, D.J.; Rees, Jr. W.S. (2000) Group II Element βDiketonate Complexes: synthetic and structural investigations, *Coord. Chem. Rev., 210*, 279-328.

[3] Kakaliou, L.; Scanlon, IV, W.J.; Qian, B.; Baek, S.W.; Smith, III, M.R.; Motry, D.H. (1999) Five- and Six-Coordinate Group 4 Compounds Stabilized by βKetiminate and Diketiminate Ligands: Syntheses and Comparisons between Solid-State and Solution Structures, *Inorg. Chem., 38*, 5964-5977 and references therein.

[4] Doherty, S.; Errington, R.J.; Housley, N.; Ridland, J.; Clegg, W.; Elsegood, M.R.J. (1999) N-Alkoxy-βKetoiminate Complexes of Group 4 and 5: Synthesis and Characterization of the Complexes $[(\eta^5\text{-}C_5H_4R)M\{CH_3C(O)CHC(NCH_2CHR'O)CH_3\}Cl_n]$ (M = Ti, n=1, M=Nb, n=2; R=H, Me; R'=H, Me), $[Ti[CH_3C(O)CHC(NCH_2CHR'O)CH_3]Cl_2(thf)]$, and $[Ti[CH_3C(O)CHC(NCH_2CHR'O)CH_3]_2]$, *Organometallics, 18*, 1018-1029.

[5] Qian, B.; Ward, D.L.; Smith, III, M.R. (1998) Synthesis, Structure, and Reactivity of βDiketiminato Aluminium Complexes, *Organometallics, 17*, 3070-3076.

[6] Hitchcock, P.B.; Lappert, M.F.; Liu, D.S. (1994) Transformation of the Bis(trimethylsilyl)methyl into Aza-allyl and β-Diketiminato Ligands, *J. Chem. Soc. Chem. Commun.*, 2637-2638.

[7] Hitchcock, P.B.; Lappert, M.F.; Liu, D.S. (1994) Transformation of the Bis(trimethylsilyl)methyl into a β-Diketiminato Ligand, *J. Chem. Soc. Chem. Commun.*, 1699-1700.

[8] Hitchcock, P.B.; Lappert, M.F.; Layh, M. (1998) Lithium Enamides, β-Diketiminates and 1,3-diazaallyls from the 1:1 insertion of an Isonitrile into the Li-C Bond of LiCH(SiMe$_3$)$_2$, *Chem. Commun.*, 201-202.

[9] Hitchcock, P.B.; Lappert, M.F.; Layh, M. (1998) Novel Lithium 3-Sila- and 3-Germa-β-diketiminates, *Chem. Commun.*, 2179-2180.

[10] Bastianini, A.; Battiston, G.A.; Benetollo, F.; Berbasi, R.; Porchia, M. (1997) Synthesis and Crystal Structure of Zirconium Complexes with Fluorinated tetradentate -□Ketoiminate Ligands, *Polyhedron, 16(7)*, 1105-1110.

[11] Studebaker, D.B.; Neumayer, D.A.; Hinds, B.J.; Stern, C.L.; Marks, T.J. (2000) Encapsulating Bis(β-Ketoiminato) Polyethers. Volatile, Fluorine-Free Barium Precursors for Metal-Organic Chemical Vapor Deposition, *Inorg. Chem. 39*, 3148-3157.

[12] Sato, T.; Takeda, H.; Sakai, K.; Tsubomura, T. (1996) Two Novel Types of Metal Complexes Derived from Tripod Amines and Acetylacetone, *Inorg. Chim. Acta, 246*, 413-421.

[13] Lim, S.K.; Lee, J.C.; Sohn, D.S.; Lee, W.I.; Lee, I.M. (2002) A Study on the Development of Chemical Vapor Deposition Precursors. 4. Syntheses and Characterization of New N-Alkoxy-□Ketoiminate Complexes of Niobium and Tantalum, *Chem. Mater., 14*, 1548-1554

[14] Lim, S.K.; Choi, B.H.; Min, Y.S.; Kim, D.S.; Yoon, I.; Lee, S.S.; Lee, I.M. (2004) A Study on the Development of Chemical Vapor Deposition Precursors V - Syntheses and Characterization of New N-Alkoxy-□Ketoiminate Complexes of Titanium, *J. Organomet. Chem., 689*, 224-237.

[15] Lee, I.M.; Lee, W.I.; Lim, S.K. (2002) Korea Patent 10-1999-0045872

[16] Min, Y.S.; Cho, Y.J.; Kim, D.S.; Choi, B.H.; Lee, I.M.; Lee, W.I.; Lim, S.K. (2002) Korea Patent 10-0367346

[17] Min, Y.S.; Cho, Y.J.; Kim, D.S.; Choi, B.H.; Lee, I.M.; Lee, W.I.; Lim, S.K. (2004) US Patent 6689427 B2

[18] Lee, I.M..; Whang, Y.J.; Seo, Y.K.; Lee, D.W. (2002) Korea Patent (in publication)

[19] Lee, I.M.; Lee, D.W. (2003) Korea Patent (in publication)

[20] Santana, M.D.; Rufete, A.; Garcia, G.; Lopez, G.; Casabo, J.; Cabrero, A.; Molins, E.; Miravitlles, C. (1997) Five-coordinate β-ketoiminate derivatives of the (N$_3$-macrocyle)nickel(II) fragment (N$_3$-macrocycle = 2,4,4-trimethyl-1,5,9-triazacyclodocec-1-ene and its 9-methyl derivative), *Polyhedron, 16(21)*, 3713-3721 and references therein.

[21] Sanchez, G.; Serrano, J.L.; Perez, J.; Lopez, G. (1997) *J. Fluorine Chem., 86*, 143-147.

[22] Lim, Sun Kwon (2001) Master Thesis, Inha University.

[23] Chou, T.Y.; Lai, Y.H.; Chen, Y.L.; Chi, Y.; Prasad, K.R.; Carty, A.J.; Peng, S.M.; Lee, G.H. (2004) Synthesis and Characterization of Tris(β-ketoiminato)ruthenium(III) Complexes: Potential Precursors for CVD of Ru and RuO$_2$ Thin Films, *Chem. Vap. Deposition, 10(3)*, 149-158.

[24] Zhang, D.; Jin, G.X.; Weng, L.H.; Wang, F. (2004) Synthesis, Molecular Structures, and Norbornene Addition Polymerization Activity of the Neutral Nickel Catalysts

Supported by β-Diketiminato [N,N], Ketiminato [N,O], and Shiff-Base [N,O] Ligands, *Organometallics, 23*, 3270-3275.

[25] Rahim, M.; Taylor, N.J.; Xin, S.; Collins, S. (1998) Synthesis and Structure of Acyclic Bis(ketenimine) Complexes of Zirconium, *Organometallics, 17*, 1315-1323.

[26] Qian, B.; Baek, S.W.; Smith III, M.R. (1999), Synthesis, Structure, and Reactivity of β-Diketiminate Boron(III) Complexes, *Polyhedron, 18*, 2405-2414

[27] Lee, L.W.M.; Piers, W.E.; Elsegood, M.R.J.; Clegg, W.; Parvez, M. (1999) Synthesis of Dialkylscandium Complexes Supported by β-Diketiminato Ligands and Activation with Tris(pentafluorophenyl)borane, *Organometallics, 18*, 2947-2949.

[28] Edwards, D.A.; Harker, R.M.; Mahon, M.F.; Molloy, K.C. (1999) Aerosol-assisted Chemical Vapor Deposition (AACVD) of Silver Films from Triphenylphosphine Adducts of Silver β-Diketonates and β-diketoiminates, including the Structure of [Ag(hfac)(PPh_3)], *J. Mater. Chem., 9*, 1771-1780.

[29] Edleman, N.L.; Wang, A.; Belot, J.A.; Metz, A.W.; Babcock, J.R.; Kawaoka, A.M.; Ni, J.; Metz, M.V.; Flaschenriem, C.J.; Stern, C.L.; Liable-Sands, L.M.; Rheingold, A.L.; Markworth, P.R.; Chang, R.P.H.; Chudzik, M.P.; Kannewurf, C.R.; Marks, T.J. (2002), Synthesis and Characterization of Volatile, Fluorine-Free β-Ketoiminate Lanthanide MOCVD Precursors and Their Implementation in Low-Temperature Growth of Epitaxial CeO_2 Buffer Layers for Superconducting Electronics, *Inorg. Chem., 41*, 5005-5023.

[30] Min, Y.S.; Cho, Y.J.; Kim, D.S.; Lee, J.H.; Kim, B.M.; Lim, S.K.; Lee, I.M.; Lee, W.I. (2001), MOCVD of $Ba_xSr_{1-x}TiO_3$ (BST) Thin Films Using a N-alkoxy-β-ketoiminato Titanium Complex, *Chem. Vap. Deposition 7(4),* 146-149.

[31] Min, Y.S.; Cho, Y.J.; Kim, D.S.; Lee, J.H.; Kim, B.M.; Lim, S.K.; Lee, I.M.; Lee, W.I. (2001) Influence of thermal decomposition behavior of titanium precursors on (Ba,Sr)TiO_3 thin films, *J. Phys., IV Fr., 11*, 675-682.

[32] Hong, S.H.; Lim, S.K.; Lee, I.M.; Min, Y.S.; Kim, D.S.; Lee, W.I. (2002) Decomposition Behaviors of Bis(N-alkoxy-β-ketoiminate)titanium Complexes in the Deposition of Titanium Oxide and Barium Strontium Titanate Thin Films, *Thin Solid Films, 409*, 82-87.

[33] Zhang, Y.; Yuan, Z.; Puddephatt, R.J. (1998), Allyl(β-diketonato)palladium(II) Complexes, including Liquid Precursors for Chemical Vapor Deposition of Palladium, *Chem. Mater., 10*, 2293-2300.

[34] Tung, Y.L.; Tseng, W.C.; Lee, C.Y.; Hsu, P.F.; Chi, Y.; Peng, S.M.; Lee, G.H. (1999), Synthesis and Characterization of Allyl(β-ketoiminato)palladium(II) Complexes: New Precursors for Chemical Vapor Deposition of Palladium Thin Films, *Organometallics, 18*, 864-869.

[35] Shin, H.K.; Hampden-Smith, M.J.; Kodas, T.T.; Rheingold, A.L. (1992) *J. Chem. Soc. Chem. Commun.*, 217.

[36] Gerfin, T.; Bechi, M.; Dahmen, K.H. (1993) Growth-kinetics of Copper Thin-films in Different MOCVD Systems, *J. Phys., Colloq.*, C5, 465.

[37] Schulz, D.L.; Hinds, B.J.; Stern, C.L.; Marks, T.J. (1993), New Precursors for Barium MOCVD. .beta.-Ketoiminate Complexes containing Appended Polyether "Lariats", *Inorg. Chem.*, 32, 249.

[38] Kim, W.K.; Fevola, M.J.; Liable-Sands, L.M.; Rheingold, A.L.; Theopold, K.H. (1998) $[(Ph)_2nacnac]MCl_2(THF)_2$ (M=Ti, V, Cr): A New Class of Homogeneous Olefin Polymerization Catalysts Featuring β-Diiminate Ligands, *Organometallics, 17*, 4541-4543.

[39] Radzewich, C.E.; Guzei, I.A.; Jordan, R.F. (1999) Three-Coordinate Cationic Aluminum Alkyl Complexes Incorporating β-Diketiminate Ligands, *J. Am. Chem. Soc., 121*, 8673-8674

[40] Radzewich, C.E.; Coles, M.P.; Jordan, R.F. (1998) Reversible Ethylene Cycloaddition Reactions of Cationic Aluminum β-Diketiminate Complexes, *J. Am. Chem. Soc., 120*, 9384-9385

In: Focus on Organometallic Chemistry Research
Editor: Marin A. Cato, pp. 147-166
ISBN 1-59454-495-6

Chapter 6

ORGANIC CHEMISTRY FOR INORGANIC CHEMISTS. UNSYMMETRICAL SCHIFF-BASES DERIVED FROM 3-FORMYL-SALYCILIC ACID AS VERSATILE LIGANDS IN COORDINATION CHEMISTRY

Marius Andruh[1] and Floriana Tuna[2]
[1]University of Bucharest, Faculty of Chemistry, Inorganic Chemistry Laboratory, Str. Dumbrava Rosie nr. 23, Bucharest, Romania
[2]Institute of Physical Chemistry of the Romanian Academy, Coordination Chemistry Laboratory, Splaiul Independentei nr. 202, Bucharest, Romania

Dedicated to the memory of Professor Olivier Kahn, who inspired this chemistry

ABSTRACT

The spectacular development of modern coordination chemistry relies upon programmed ligands, which are organic molecules designed to generate complexes with pre-established properties. This paper reviews our work on a family of tetradentate unsymmetrical Schiff-base ligands, which have been employed to synthesize homo- and heteropolynuclear complexes with various nuclearities and topologies of the metallic centers. Such ligands are obtained by reacting the 3-formyl-salycilic acid with organic molecules bearing primary and tertiary amino groups [*e. g.* 2-(aminomethyl)-pyridine; 2-(2-aminoethyl)-pyridine; N,N-dimethyl-ethylenediamine]. The resulted complexes have relevance in molecular magnetism, as they have been used as models to study the factors governing the nature (ferro- or antiferro), and magnitude of the exchange interactions.

1. INTRODUCTION

The design and synthesis of polynuclear metal complexes with predictable magnetic properties is a challenge for inorganic chemists interested in the field of molecular magnetism [1, 2]. One of the crucial points in this area is the understanding of the structural and chemical factors that determine the exchange coupling between paramagnetic metal centers. Along this line, magnetically condensed transition metal complexes, particularly heterometallic systems, play an extremely important role as they provide a unique opportunity to study the magneto-structural correlations in a systematic way.

Apart from their fascinating structures and magnetic properties, exchange-coupled transition metal complexes exhibit interesting electronic and catalytic properties [3]. They are also active centers in a number of metalloproteins and enzymes, such as iron–sulphur proteins [4], iron–oxo proteins [5], cytochrome *c* oxidase [6], particulate methane monooxygenase [7], nitrous oxide reductase [8], and in the oxygen evoluting complex (OEC) of Photosystem II [9].

In this context, the development of new synthetic strategies leading to discrete polynuclear complexes is of great importance, particularly when the syntheses have to be performed in a controlled manner with the aim at obtaining compounds with specific properties. These strategies must address two important problems: *i*) control of the nuclearity, which is the number of the metallic ions within the molecular entity; *ii*) control of the topology of the metallic centers, which are usually paramagnetic ions. An efficient means to control both nuclearity and spin topology is to employ specifically tailored organic molecules that can act as receptors with highly specific recognition of metal ions. In the language of supramolecular chemistry, these species are described as programmed ligands, whose information will be read by the metallic ions, according with their coordination algorithm [10]. Among them, we mention here the compartmental ligands, which are organic molecules able to hold together two or more metal ions in close proximity (Scheme 1). The Schiff bases derived from 2,6-diformyl-4-methyl-phenol (Robson-type ligands; **I**) and from 3-formyl-salicylic acid (**II**) are among the most popular ligands belonging to this family [11]. These ligands are designed to generate either homo- or heterobinuclear complexes [12], but they also can serve as useful building blocks for the construction, in a controlled manner, of discrete high-nuclearity species [13] or extended supramolecular structures [14].

Within the binuclear entities the two metal ions are held together by two phenoxide-oxygen bridges, and can interact with each other through the bridging atoms in a ferromagnetic or antiferromagnetic way. The type and magnitude of the exchange interaction depend on a number of structural parameters, such as the distance between the paramagnetic centers, the angles at the bridging atoms, the metal-bridge oxygen bond lengths or the stereochemistry around the metal ion, and can be monitored by inducing flexibility into the ligand sidearms. Thus these compounds constitute an excellent basis for magneto-structural studies. For example, intensive magneto-structural investigations on bis(phenoxide)-bridged macrocyclic binuclear copper(II) complexes have concluded that the leading factor in determining the exchange interaction is the Cu-O(Ph)-Cu angle, α, and that the singlet-triplet energy gap, J, is linearly dependent on α [15]. Furthermore, one of the most important tools that operate in the molecular magnetism field, namely: the principle of the strict orthogonality of the magnetic orbitals, originated from magnetic investigations on homo- and heterobinuclear 3d metal complexes containing compartmental, acyclic side-off ligands, of type II [16]. Such a situation of imposing a ferromagnetic interaction between nearest neighbour magnetic centers owing to the strict orthogonality of the magnetic orbitals may be observed in heterometallic Cu^{II}-VO^{IV}, Cu^{II}-Cr^{III}, Ni^{II}-Cr^{III}, or low-spin-Fe^{III}-Cu^{II} systems.

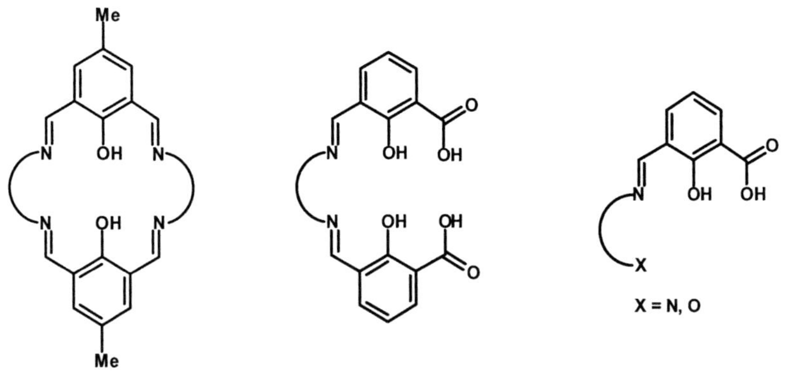

Scheme 1.

In the search for rationally designed polynuclear complexes with higher nuclearities and predetermined spin topologies, we have championed the use of a new family of ligands, which are unsymmetrical tetradentate Schiff bases derived from 3-formyl-salicylic acid (**III**). The main advantage of using such ligands is that their denticity cannot saturate the coordination sphere of the metal ions. In this case the vacant coordination sites will be occupied either by anions or through self-assembly to polynuclear arrays. The new strategy was successfully used to generate bi-, tri-, tetra-, and hexanuclear clusters.

In this chapter we show how unsymmetrical Schiff-bases derived from 3-formyl-salicylic acid can be employed as versatile ligands in designing multimetallic complexes. The structures of the complexes are discussed on the basis of various structure-directing factors, including the metal coordination preferences, bridging groups, steric effects, and conformational flexibility of the ligands.

2. Homonuclear Complexes

Using 3-formyl-salicylic acid as a starting material, we have prepared a wide variety of unsymmetrical tetradentate Schiff bases (Scheme 2), which can act as potentially bi- or trianionic ligands [17,18]. Being incapable of tetradentate ligation to a single metal center, these ligands rather self-assemble in a 1:1 or 2:3 ligand:metal ratio to give homobinuclear or homotrinuclear complexes, respectively. In particular circumstances, mononuclear complexes may be also assembled from these ligands, and further used as versatile building blocks for constructing molecular species with higher nuclearities. The preferred coordination environment of the metal ions, the flexibility of the ligand backbone, the metal:ligand molar ratio, the level of deprotonation of the ligand and the ability of the functional groups to bridge multiple metal ions, all play key role in directing the self-assembly process and the solid-state structure of the resulting compound.

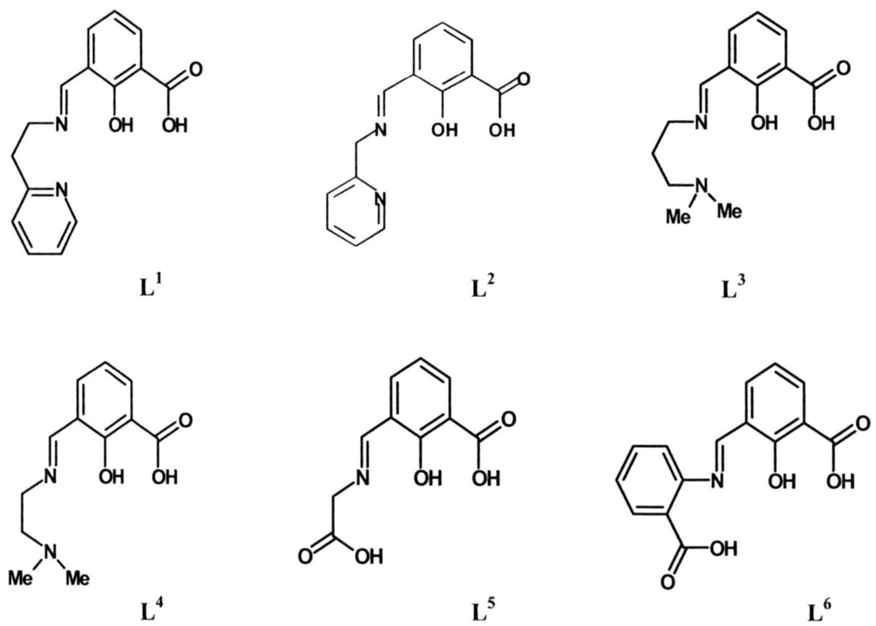

Scheme 2.

Our discussion will be mainly focused on the chemistry of the bianionic ligands. In a series of papers we have shown that unsymmetrical tetradentate Schiff-bases derived from 3-formyl-salicylic acid (H_2fsa) and various diamines containing one primary amine and one tertiary amino group ($L^1 - L^4$, Scheme 2) are versatile ligands for the synthesis of homo- and heterometallic complexes. For example, in the case of the ligand 3-[N-2-(pyridylethyl)formimidoyl]salicylic acid ($H_2fsaaep$; L^1), which has been obtained by reacting one equivalent of 3-formylsalicylic acid with one equivalent of 2-(2-aminoethyl)pyridine, we have obtained a series of bi- and trinuclear copper(II) complexes [18, 19], and demonstrated that 3d - 4f hexanuclear clusters [20] may also be synthesized by using binuclear species as precursors. The metal-to-ligand molar ratio as well as the presence of a base (Na_2CO_3 or LiOH) or, alternatively, of an acid (HCl) clearly influenced the nature of the polynuclear complexes which were formed. In a basic medium, with metal-to-ligand molar rations of 1:1 and 3:2, respectively, the dianionic ligand can generate either binuclear neutral, **2**, or trinuclear, **3**, species (Scheme 3). Conversely, in HCl solution, a chloro-bridged dimer **1** is obtained. It is interesting that the formation of two chelate rings with the copper(II) ion favours most probably the deprotonation of the phenolic group instead of the carboxyl one. The chloro-bridged dimer **1** can be converted into complex **2** by treating it with a Na_2CO_3 or LiOH aqueous solution. The reaction of **2** with copper(II) perchlorate yields the trinuclear complex **3**.

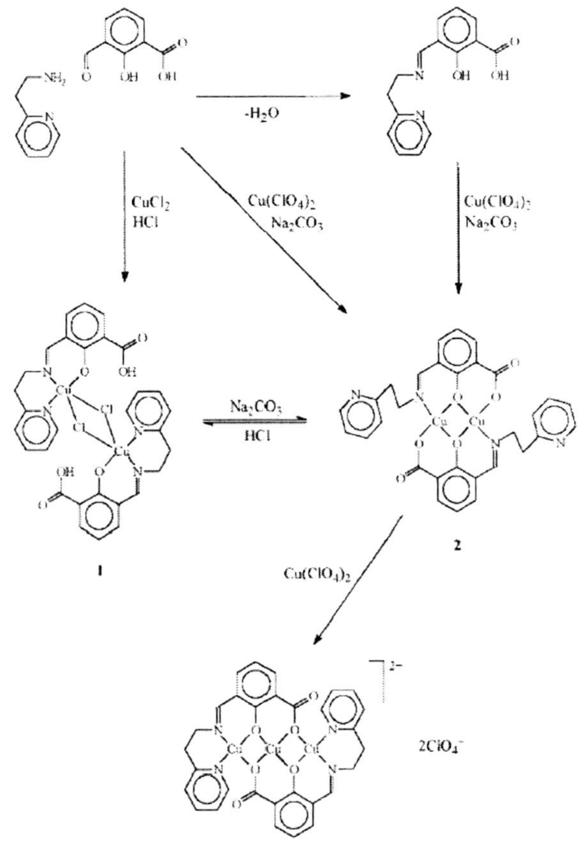

Scheme 3.

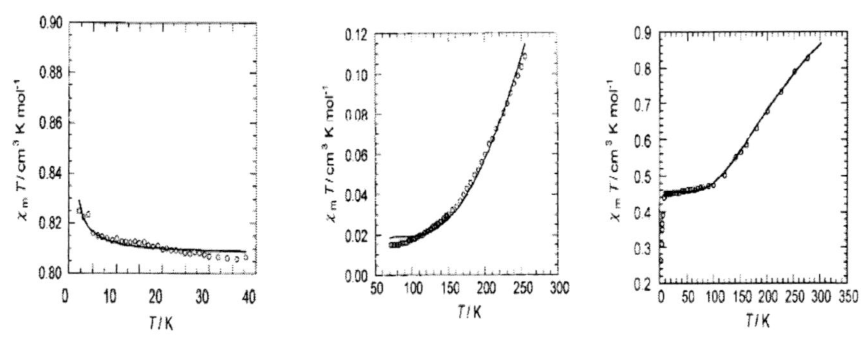

Figure 1. $\chi_M T$ *vs* T graphs of compounds **1** (*a*), **2** (*b*) and **3** (*c*)

Variable-temperature magnetic susceptibility measurements (Figure 1) indicated a very weak ferromagnetic coupling of the copper(II) ions in compound **1** (J = + 0.15 cm^{-1}) and an antiferromagnetic coupling in compounds **2** (J = -617 cm^{-1}) and **3** (J = -228 cm^{-1}).

The dramatic diference in magnetic behaviours of the two dimers is due to their different molecular structures. The X-ray crystallographic analysis of compounds **1** and **2** (Figure 2) revealed that:

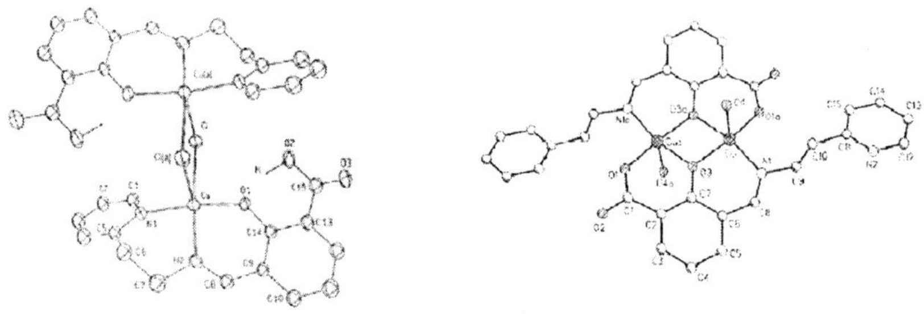

a *b*

Figure 2. The crystal structures of the dimers [{Cu(Hfsaaep)Cl}$_2$] **1** (*a*), and [{Cu(fsaaep)(H_2O)}$_2$] **2** (*b*)

For compound **1**: The copper(II) ions are doubly bridged by chloride anions. Each of the two copper centers adopts a distorted square pyramidal geometry, being surrounded by two chloride atoms, two nitrogen atoms and one oxygen atom. The Cu···Cu distance in the dimeric entity is 3.825(1) Å. The Cu-Cl-Cu bridging angles are 95.27(4)°. Hatfield *et al.* have shown that the singlet-triplet gap in such compounds varies in a regular way with the □/R_0 quotient, where □ is the angle at the bridging ligand, Cu(II) – Cl – Cu(II), and R_0 is the long, out-of-plane Cu(II) – Cl bond distance [21]. It was found that for values of this ratio that are lower than 32.6 or higher than 34.8 deg/Å, the exchange interaction is antiferromagnetic. For values falling between these limits, the exchange interaction was found to be ferromagnetic. In the case of the chloro-bridged dimer, [Cu(Hfsaaep)Cl]$_2$ **1**, the value of the quotient □/R_0 (33.6 deg/ Å) suggests a ferromagnetic interaction between the two copper(II) ions.

For compound **2**: The two copper(II) ions are bridged by phenolic oxygen atoms. The pyridyl groups are not co-ordinated. The co-ordination geometry around the copper atoms is square pyramidal. The basal positions are occupied by three oxygen atoms, two bridging phenolic oxygens and from the carboxylato group, while the apical position is occupied by the aqua ligand. The Cu···Cu distance within the binuclear unit is 3.028 Å. The Cu-O(Ph)-Cu bridging angles are 101.81(5)°. Magneto-structural correlations in bis(μ_2-phenoxide)-bridged macrocyclic binuclear copper(II) complexes have shown that the leading factor in determining the exchange interaction is the Cu – O(Ph) – Cu angle (α): $-J = (31.95\alpha - 2462)$ cm^{-1} [15]. It has been shown that, for Cu – O(Ph) – Cu angles ranging from 98.8 to 104.7°, the values of the coupling constant, $-J$, encompass the domain 689 - 902 cm^{-1}. The Cu – O(Ph) – Cu angle in compound **2** is 101.8°. The predicted value of J is –790 cm^{-1}, higher than the experimental one (-617 cm^{-1}). This difference is most probably an effect of the electron-withdrawing substituents bound to the phenolic groups.

As observed, there is an intimate relationship between spin coupling and molecular structure. We were further interested in investigating the structural impact of changing the length of the alkylidene spacer between imino and pyridine nitrogen atoms. For this proposal, we introduced 3-(N-(2-pyridylmethyl)formimidoyl)salicylic acid (H_2fsaamp, L^2) as a suitable tetradentate ligand [17c]. The reaction of this ligand with Cu(II) and Ni(II) perchlorates in a 2:3 metal/ligand molar ratio afforded trinuclear complexes of formula $[M_3(\text{faaamp})_2(H_2O)_x](ClO_4)_2$ (x = 2 or 4), with the ligand units fully deprotonated. The magnetic susceptibility data supported the presence of three paramagnetic centers within the homotrinuclear entities. Identical species have been obtained from template procedures, which consist in the direct assembling of two ligand units with 3 metal ions, in the presence of 4 equivalents of LiOH.

On the other hand, Erxleben has shown that the reaction of this ligand with $Zn(NO_3)_2$, in the presence of NaOH (*p*H = 9), led to a rather unexpected structure, namely a tetranuclear compound $[Zn_4(\text{fsaamp})_4].4.5H_2O$ **4** (Figure 3) with a cubane-like Zn_4O_4 core [22].

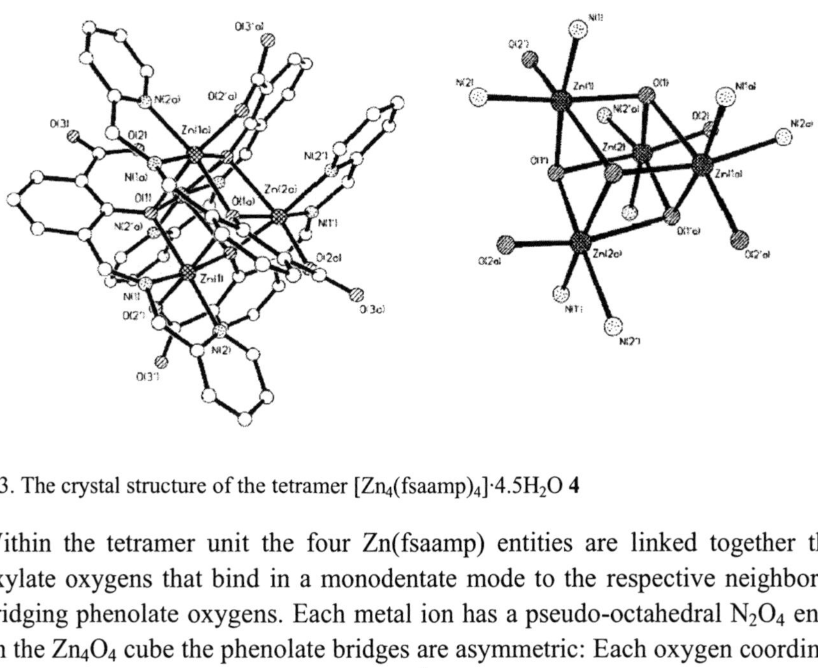

Figure 3. The crystal structure of the tetramer $[Zn_4(\text{fsaamp})_4]\cdot 4.5H_2O$ **4**

Within the tetramer unit the four Zn(fsaamp) entities are linked together through the carboxylate oxygens that bind in a monodentate mode to the respective neighboring Zn ion and bridging phenolate oxygens. Each metal ion has a pseudo-octahedral N_2O_4 environment. Within the Zn_4O_4 cube the phenolate bridges are asymmetric: Each oxygen coordinates to one Zn at a "normal" distance (2.022-2.056 Å), while the distance to the second Zn is significantly longer (2.158-2.172 Å) and the interaction with the third Zn is even weaker (2.453-2.496 Å). The angles between the six planes of the Zn_4O_4 cube are close to 90°. The Zn···Zn separations are in the range 3.202-3.594 Å.

To investigate the structural impact of substituting the pyridine ring with a doubly-substituted amino group, we synthesized two new ligands, namely: 3-(N-(N',N'-dimethylamino-propyl)formimidoyl)salicylic acid (H_2fsadmpn; L^3) and 3-(N-(N',N'-dimethylaminoethyl)-formimidoyl)salicylic acid (H_2fsadme; L^4). We expected that the reactions between these ligands and copper(II) perchlorate, in presence of a base, and keeping a 1:1 metal-to-ligand molar ratio, should always lead to neutral binuclear complexes. This was indeed the case of the H_2fsaaep ligand, when $[Cu(\text{fsaaep})(H_2O)]_2$ **2** was obtained. The formation of this neutral binuclear complex was suggested by the absence of the characteristic bands of the perchlorate anions, and has been further confirmed by crystallographic studies. Surprisingly, by reacting $Cu(ClO_4)_2$ with H_2fsadmpn, under the same conditions, we obtained a compound whose infrared spectrum clearly showed the characteristic bands of the perchlorate anion (1120, 1087, and 625 cm^{-1}). At first glance, we suspected the formation of a cationic trinuclear complex, $[Cu_3(fsadmpn)_2](ClO_4)_2 \cdot nH_2O$, but the results of the chemical analysis did not support such a formulation. The X-ray crystallographic investigation revealed the presence of cationic binuclear phenoxo-bridged $[\{Cu(fsaamdpn)\}_2]^{2+}$ **5** species (Figure 4) and perchlorate anions [19a]. The carboxylic groups of the two ligands are fully deprotonated as in the case of $[\{Cu(fsaaep)(H_2O)\}_2]$ **2**, but the uncoordinated tertiary amino groups are protonated. Phenoxide oxygen atoms form a double bridge between the metal centers. The coordination geometry around the copper atoms is square pyramidal. The basal positions are occupied by one nitrogen atom arising from the azomethine group and by three oxygen atoms (two bridging phenolic oxygen and one from the carboxylato group). The apical position at each copper atom is occupied by the aqua ligand. The Cu···Cu distance within the binuclear unit is 3.008(2) Å. The Cu-O(Ph)-Cu bridging angle is 101.23(15)°, very close to the one found in $[Cu(fsaaep)(H_2O)]_2$ [101.81(5)°]. The packing diagram of compound **5** shows interesting interactions at the supramolecular level. The aqua ligands are involved in strong hydrogen bond interactions with the carboxylato groups from neighboring binuclear entities (O-H···O = 2.684 Å), resulting in a hydrogen bonded polymer with a binuclear repeating unit (Figure 5).

The cryomagnetic investigation of compound **5** (Figure 6) revealed strong antiferromagnetic interactions between the metal ions (J = -646 cm^{-1}).

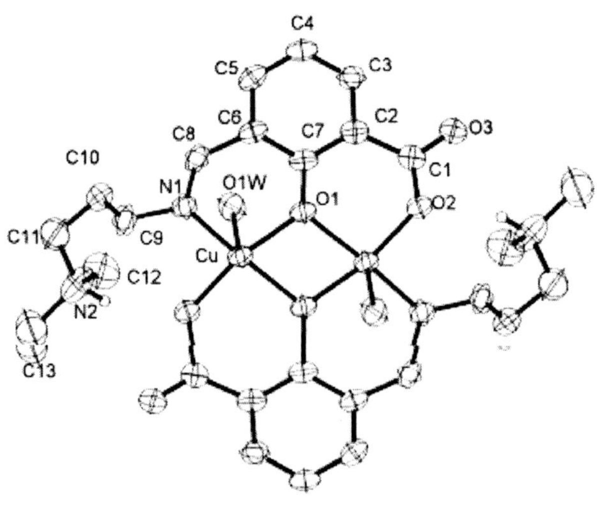

Figure 4. Crystal structure of dimer **5**

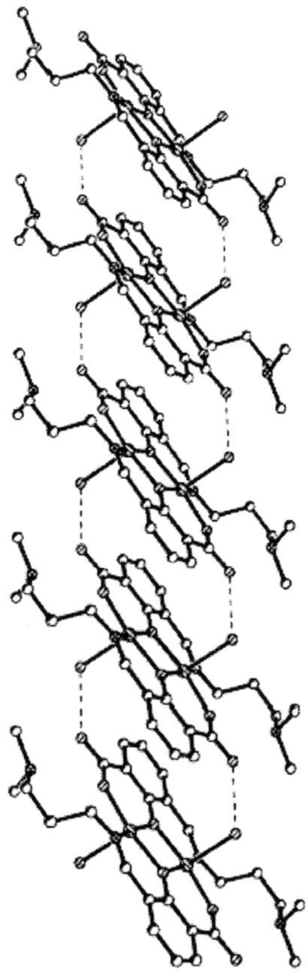

Figure 5. Packing diagram for compound **5** showing the formation of the hydrogen bonded polymer

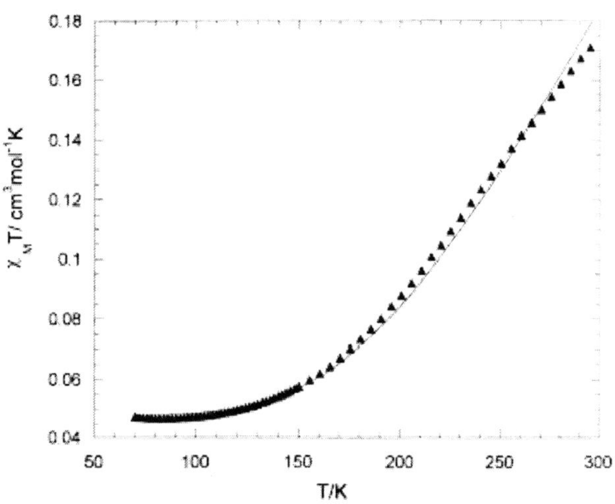

Figure 6. The graph $\chi_M T$ *vs* T for **5**

Similar zinc complexes were described by Erxleben [22]. Treating 3-formylsalicylic acid with *N,N*-dimethylethylenediamine and $Zn(CH_3CO_2)_2$ or $Zn(NO_3)_2$ in a 1:1 molar ratio affords the dinuclear zinc complexes: $[\{Zn(fsadmp)(CH_3COO)\}_2]$ **6a** and $[\{Zn(fsadmp)(H_2O)\}_2](NO_3)_2$ **6b** (Figure 7). The centrosymmetric cation is built up of two Schiff base ligands that bridge a pair of Zn atoms at a distance of 3.126(1) Å through the deprotonated phenolic oxygens. Further coordination sites of the ligands are the imine nitrogens and carboxylate oxygens, while the amino groups of the ligand sidearms are protonated. The donor atoms of the ligands span the basal plane of a square pyramid, the apical position of which is occupied by a water molecule. The basic structure of **6b** is identical to that of **6a**. The metal centers are five-coordinate (square-pyramidal geometry), with the water ligands being replaced by acetate groups. From the alkaline solution of **6b** (pH = 9.9, adjusted with 1N NaOH) yellow cubes crystallized. The crystals contained neutral molecules of composition $[Zn_4(fsadme)_4]$ **7**, whose structure is displayed in Figure 8. The four Zn ions are located at the corners of a nearly planar rectangle. Each ligand binds one Zn through the phenolate oxygen, the imine nitrogen, and the amine nitrogen. Neighboring Zn ions are connected through phenolate bridges. The carboxylate group of the ligand bound to Zn(2) forms an asymmetric, single-atom bridge between Zn(3) and Zn(4). Likewise, Zn(1) and Zn(2) are linked through a carboxylate oxygen of the ligand of Zn(4) in addition to a phenolate bridge. This type of monodentate bridging by carboxylate groups is relatively rare in metal carboxylate complexes, and formation of the monatomic bridges may be supported by the overall geometry of the tetranuclear complex. By contrast, the carboxylate groups of the remaining ligands coordinate only terminally to Zn(2) and Zn(4). Owing to the different coordination modes of the carboxylate groups, distorted octahedral N_2O_4 coordination spheres are observed for Zn(2) and Zn(4), with the ligands being *mer*, while Zn(1) and Zn(3) are five-coordinate. The N_2O_3 coordination polyhedra of Zn(1) and Zn(3) are best described as distorted square pyramids with bridging carboxylate oxygens occupying the apical positions. The Zn ions and phenolate oxygens form an eight-membered ring in which the phenolate oxygens lie alternating below and above the plane defined by the four Zn ions. The mean distance of 3.6 Å between opposite rings suggests some π-interactions and it was speculated that intramolecular stacking is the driving force for the formation of the tetranuclear complex in the solid state.

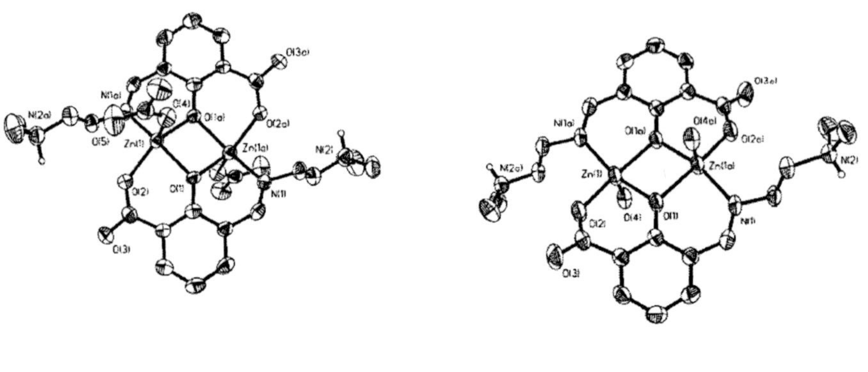

Figure 7. Crystal structures of dimers **6a** and **6b**

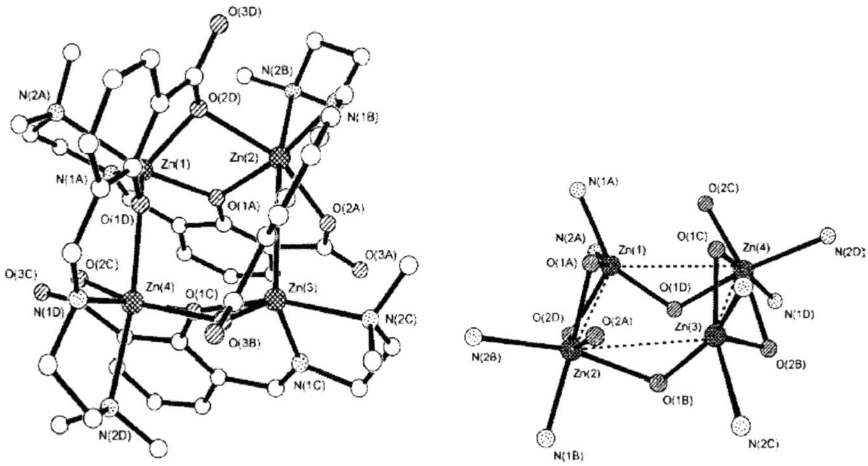

Figure 8. Crystal structure and tetranuclear core of compound **7**

3. HOMODINUCLEAR SPECIES WITH OTHER BRIDGING LIGANDS

By changing the two chloro-bridges in **1** with other bridging ligands (*e.g.* oxalato, azido, dicyanamido, or the nicotinato ion) it is possible to obtain new polynuclear complexes in which the Schiff-base acts only as a chelating ligand [19, 25].

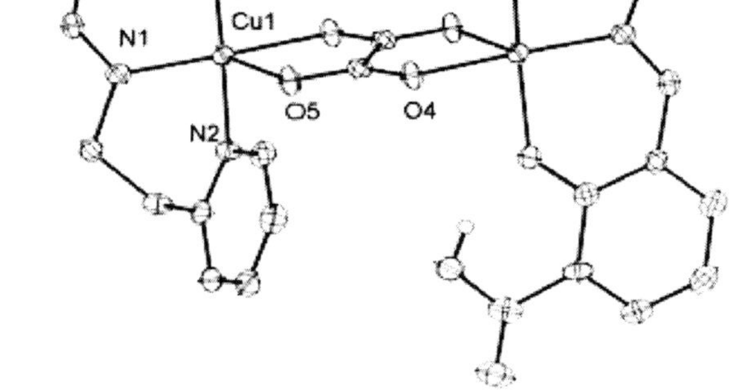

Figure 9. Crystal structure of dimer **8**

The reaction between the Schiff-base ligand, H_2fsaaep, oxalic acid and copper(II) perchlorate led to the dimer [Cu_2(Hfsaaep)$_2$(C_2O_4)] **8**, in which the two copper(II) ions are connected by an oxalato group (classical bisbidentate-bridging mode) [19a]. The crystal structure of this dimer is presented in Figure 9. The Cu···Cu distance within the binuclear entity is 5.482(1) Å. Each copper atom is coordinated by two oxygen atoms arising from the oxalato bridge, one phenolic oxygen and two nitrogen atoms arising from the organic ligand (one nitrogen atom from the azomethine group, and the other one from the pyridyl ring). The coordination polyhedron of the copper atom may be described as a distorted square-planar pyramid. Cryomagnetic investigation in the temperature range 1.9-300 K (Figure 10) revealed an antiferromagnetic interaction between the copper ions (J = -9 cm^{-1}).

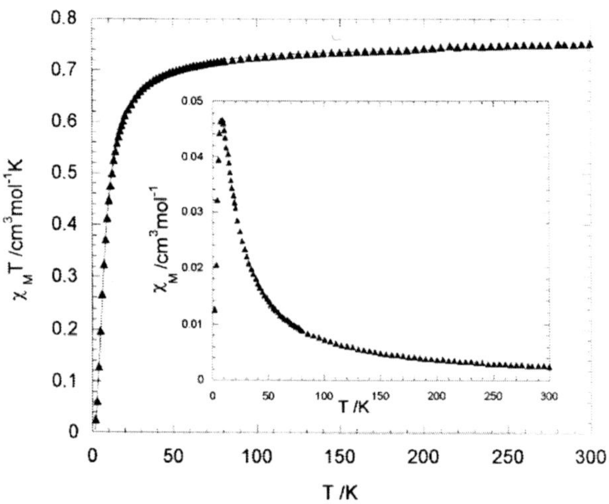

Figure 10. The graph $\chi_M T$ *vs* T for **8** (inset: temperature dependence of the magnetic susceptibility)

Pseudohalides are versatile and efficient ligands for making bridges. Among them, azido is the most efficient ligand with regard to superexchange pathways between paramagnetic metal centers. The versatility and efficiency of the azido moiety lies in its functionality as a terminal monodentate or bridging bi- or tridentate ligand [23]. The azide group is used to study the interaction between small molecules and the metal centers of metalloenzymes, and it has been observed that the azido group can bridge the metal centers in multicopper enzymes [24].

As expected, the reaction of compound 1 with NaN_3 leads to the dimer, **9**, in which the chloride bridges are replaced by $\mu_{1,1}$-N_3^- bridges [25]. Similarly, the reaction of 1 with $Na[N(CN)_2]$ affords a binuclear complex, **10**, with $\mu_{1,5}$-dicyanamido bridges (Scheme 4). A rectangular tetranuclear complex, **11**, resulted when compound **1** was treated with sodium salt of the nicotinic acid (Scheme 5).

4. HETEROMETALLIC COMPLEXES

In the last two decades or so, a large number of heterometallic compounds have been described. Their synthesis relies on the clever use of multifunctional organic molecules able to interact selectively with different metal ions. The archetype could be considered the compartmental Schiff-base ligand, **II**, with dissymmetric coordination sites (Scheme 1), which can generate, through stepwise synthesis, heterobinuclear complexes. The studies of these compounds have often been performed in relation to the modeling of some metalloenzymes containing several kinds of metal ions. A particular emphasis has been brought to the magnetic properties. The main idea emerging from those studies is that the interaction between non-equivalent magnetic centers may lead to situations that cannot be encountered with species containing a unique kind of spin carrier. Recent investigation of the magnetic properties of heteropolymetallic compounds has represented quite an important contribution to the development of molecular ferromagnetism as a whole, and has allowed the introduction of several important new concepts [1]. Moreover, recently it has been shown that high nuclearity clusters exhibit, when some important conditions are fulfilled, extremely interesting magnetic properties: they can act as single molecule magnets [26-28].

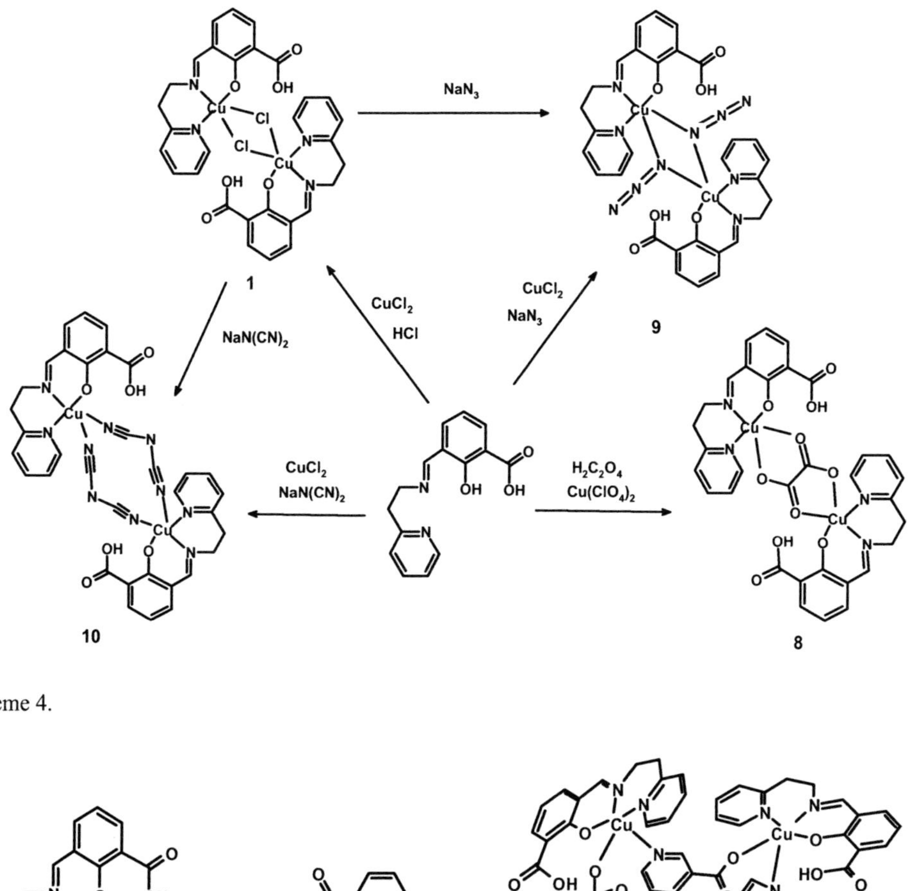

Scheme 4.

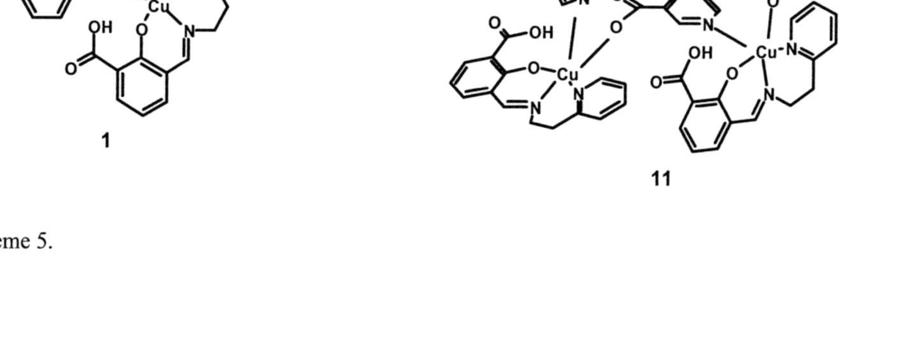

Scheme 5.

We rationalized that ligands L^1-L^4, which contain two dissimilar binding sites, would allow us to synthesize heterotrinuclear species, as schematically illustrated in Scheme 6.

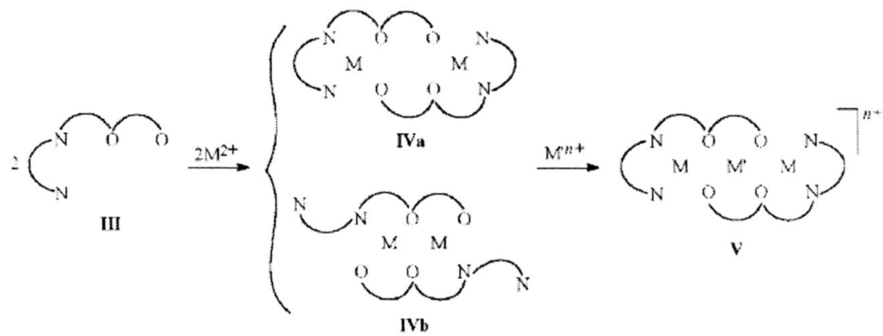

Scheme 6.

To prove that, oxophilic lanthanide ions and [{Cu(fsaaep)(H_2O)}$_2$] **2** were involved [20]. The reaction between **2** and various lanthanide nitrates affords heteronuclear 3d-4f complexes (Scheme 7). The crystal structure of the praseodymium derivative, [Pr_2Cu_4(fsaaep)$_4$(NO_3)$_6$] **12**, has been solved.

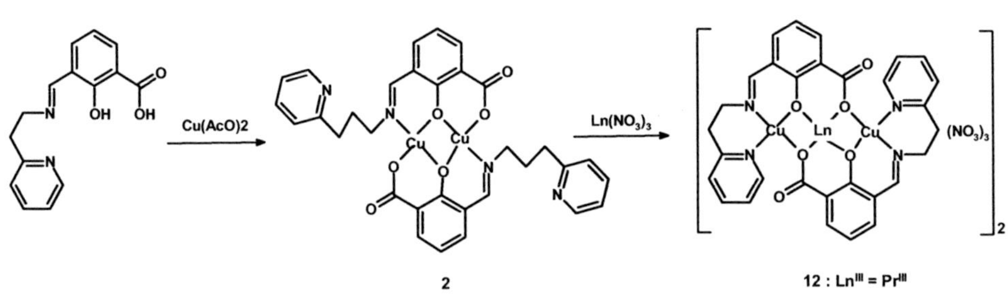

Scheme 7.

A simplified view of the structure of **12** is presented in Figure 11. The six metal atoms form a chair-shaped hexagon, with two praseodymium atoms located on both sides of a double layer containing the four copper atoms. The hexanuclear entity may be viewed as a pair of heterotrinuclear units. The copper atoms belonging to two {$PrCu_2$} units are bridged two by two by carboxylato groups. Within a trinuclear unit, each praseodymium atom is bridged to two atoms through one carboxylic and one phenolic oxygen atom. The Pr···Cu distances range from 3.400 to 3.428 Å.

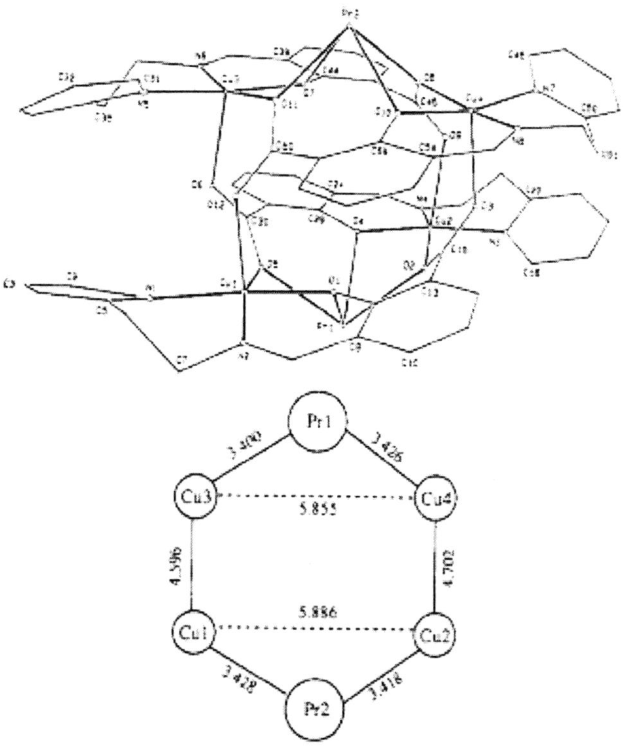

Figure 11. Crystal structure of the hexanuclear complex **12**

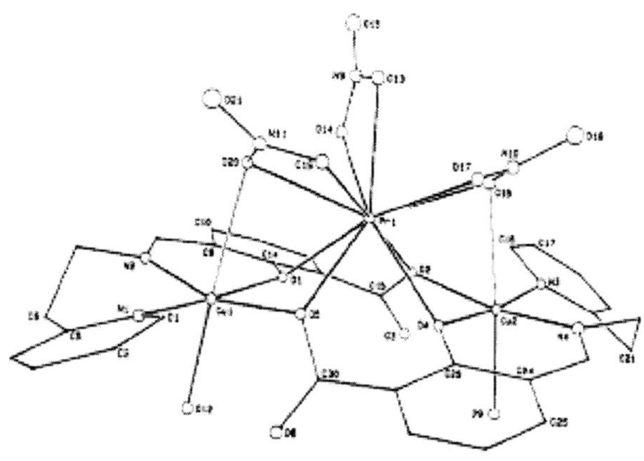

Figure 12. Perspective view of the trinuclear moiety in **12**

The copper environments may be considered as elongated distorted octahedral. The equatorial positions are occupied by two nitrogen and two oxygen atoms. One of the apical positions is occupied by the oxygen atom of the carboxylic group involved in the bridges between the layers with an average Cu-O bond length of 2.32 Å. The other apical position is occupied by an oxygen atom belonging to a nitrato ion already linked to a rare earth atom. In addition to the oxygen atoms belonging to the fsaaep^{2-} ligands, the rare earth atoms achieve their environment with six oxygen atoms coming from three bidentate nitrato ions, as shown in Figure 12. The coordination polyhedron around the praseodymium atom may be described as a bicapped square antiprism.

The isostructural [Cu_4La_2] and [Cu_4Gd_2] complexes have been also synthesized. They were very good model compounds for the study of the 3d-4f magnetic interaction. From the magnetic point of view, these clusters behave as two independent Cu(II)Ln(III)Cu(II) triads (Ln=La and Gd), since the Cu(1) – Cu(3) and Cu(2) – Cu(4) interactions are negligible (the magnetic orbital, d_{x2-y2}, is located in the equatorial plane of the elongated octahedron). The magnetic measurements show that the Cu(II)-Gd(III) interaction is ferromagnetic (J = + 6.0 cm^{-1}) [20]. The magnetic susceptibility data for the lanthanum derivative, [Cu_4La_2], with diamagnetic lanthanum(III), revealed a weak antiferromagnetic Cu···Cu interaction within a triad: J = - 3.13 cm^{-1}.

As an extension of the strategy of using 'metal complexes' as ligands we envisaged the possibility of preparing heterometallic 3d-3d complexes by using dimeric species as precursors [19b]. A heterobinuclear [CuCr] compound, **13**, has been synthesized by reacting the bis-oxalato chromium complex, [$NaCr(bipy)(C_2O_4)_2(H_2O)$].$2H_2O$ with the binuclear chloro-bridged copper(II) complex, $[Cu(Hfsaaep)Cl]_2$ 1. The chloro anion in the {Cu(Hfsaaep)Cl} moiety was substituted by the $[Cr(bipy)(C_2O_4)_2]^-$ unit, resulting in a neutral [CuCr] heterobinuclear complex, [$(bipy)(C_2O_4)Cr(\mu$-$C_2O_4)Cu(Hfsaaep)(H_2O)$]·$2H_2O$ (Figure 13). The $[Cr(bipy)(C_2O_4)_2]^-$ ion acts as a bidentate ligand toward the copper(II) ion, that is, one oxalato group acts simultaneously as a bidentate ligand toward chromium and copper. The distance between the Cr^{III} and Cu^{II} ions across the bridging oxalate is 5.506 Å. The copper atom is coordinated by two oxygen atoms arising from the oxalato bridge, one phenolic oxygen and two nitrogen atoms arising from the organic ligand (one nitrogen atom from the azomethine group, and the other one from the pyridyl ring) and one aqua ligand. The coordination polyhedron of the copper atom may be described as a distorted square-based bipyramid.

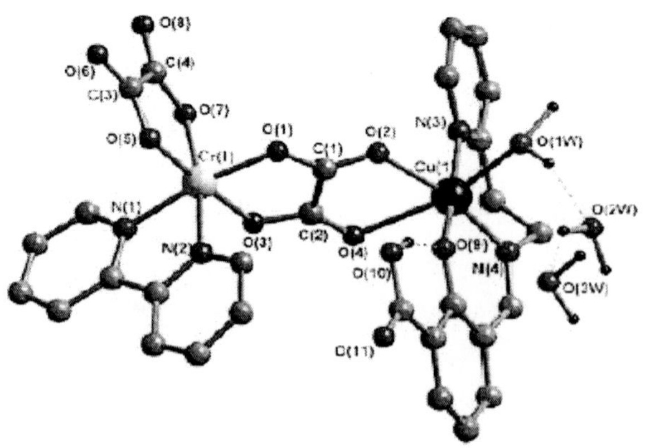

Figure 13. Perspective view of the binuclear unit in **13**

The apical positions are occupied by the aqua ligand [Cu(1) – O(1w) = 2.360(3) Å] and one oxygen atom from the bridging oxalato group, Cu(1) – O(4) = 2.504(2) Å. The phenolic group of the Schiff-base ligand is deprotonated, while the carboxylic group is protonated. The strong intramolecular hydrogen bond, O(10)-H···O(9), (Figure 13) binds these moieties and contributes to their planar arrangement. The magnetic coupling between Cr^{III} and Cu^{II} in **13** was found to be ferromagnetic (J = +1.4 cm^{-1}) – Figure 14. The nature of the ground state (S = 2) has been further confirmed by the magnetization versus field measurements. It is worthy to note that most of the μ-oxalato Cr(III)-Cu(II) complexes reported in the literature show either no, or antiferromagnetic interaction, although the Cr(III)-Cu(II) pair is a classical candidate for ferromagnetic couplings imposed by the orthogonality of the magnetic orbitals. For such Cr(III)-Cu(II) complexes the symmetry conditions are not fulfilled and the interaction is antiferromagnetic. Compound **13** is a rare example of oxalato-bridged Cu(II)-Cr(III) complex showing a ferromagnetic coupling. The copper $d_{x^2-y^2}$ orbital interacts with the σ type orbitals of the oxalato group, and, on the other hand, the t_{2g} metallic orbitals of the Cr^{III} ions interact with the π type orbitals of the oxalato bridge, leading to the classical example of ferromagnetic coupling due to the strict orthogonality of the magnetic orbitals.

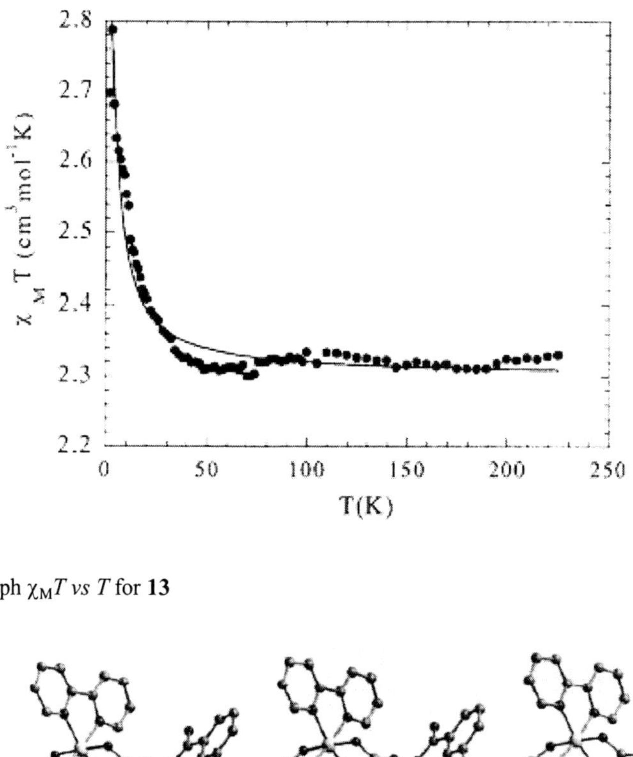

Figure 14. The graph $\chi_M T$ *vs* T for **13**

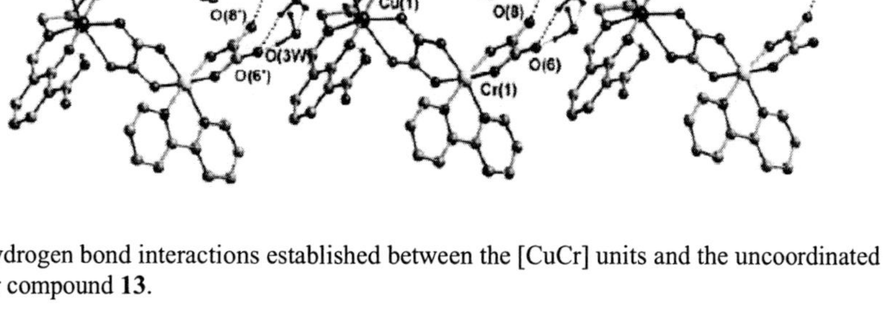

Figure 15. Hydrogen bond interactions established between the [CuCr] units and the uncoordinated water molecules for compound **13**.

The analysis of the packing diagram reveals interesting features. The heterobinuclear [CuCr] entities are interconnected through hydrogen-bond interactions involving the two water molecules, the oxygen atoms from the terminal oxalato group, and the aqua ligand coordinated to copper, resulting in infinite chains (Fig. 15). The crystal structure is further sustained by π–π stacking interactions (3.44 and 3.59 Å), which are established between the bipy ligands from neighboring chains.

In conclusion, we have shown that readily-prepared unsymmetrical Schiff-bases derived from 3-formyl-salycilic acid act as versatile ligands, and generate a quite rich coordination chemistry. Homometallic, as well as heterometallic complexes with interesting magnetic properties can be obtained. The employment of additional ligands (*e. g.* oxalato, azido, dicyanamido, nicotinato) enlarges the structural variety of the resulting polynuclear complexes, whose comparative magnetic properties are anticipated to lend valuable insights into the molecular magnetism field. The potential of these, or other unsymmetrical Schiff-base ligands in designing single-molecule magnets or molecule-based magnets remains to be further explored.

ACKNOWLEDGEMENTS

The synthesis of the ligands and of the corresponding complexes, excepting the zinc compounds (**4**, **6** and **7**), was accomplished by Dr. Gabriela Grasa, Mirela Pascu, Gabriel Pascu, Dr. Diana Visinescu, and Dr. Gabriela Marinescu. Their enthusiastic work is gratefully acknowledged.

REFERENCES

[1] Kahn, O., Molecular Magnetism, VCH, New York, 1993; (b) Kahn, O. *Adv. Inorg. Chem.* 1995, *43*, 179; (c) Miller, J. S.; Drillon, M., Magnetism: Molecules to Materials, Wiley-VCH, Weinheim, 2002.

[2] Magneto-Structural Correlation in Exchange Coupled Systems; Willet, R. D., Gatteschi , D., Kahn, O., Eds.; NATO ASI Series C140; Reidel: Dordrecht, The Netherlands, 1985; (b) Molecular Magnetic Materials; Gatteschi, D., Kahn, O., Miller, J., Eds.; NATO ASI Series E198; Kluwer: Dordrecht, The Netherlands, 1991; (c) Molecular Magnetism: From Molecular Assemblies to the Devices, Coronado, E., Delhaès, P., Gatteschi, D., Miller, J., Eds., NATO ASI Series E321; Kluwer: Dordrecht, The Netherlands, 1996.

[3] Neumann, R.; Dahan, M. *Nature* 1997, *388*, 353.

[4] (a)Holm, R. H.; Solomon, E. I. *Chem. Rev.* (Guest Editors), 1996, *96*, No 7; (b) Feig, A. L.; Lippard, S. J. *Chem. Rev.* 1994, *94*, 759.

[5] Crichton, R. R., Inorganic Biochemistry of Iron Metabolism; Horwood: New York, 1991.

[6] (a)Palmer, G.; Babcock, G. T.; Vickery, L. E. *Proc. Natl. Acad. Sci. U. S. A.* 1976, *73*, 2206; (b) Babcock, G. T.; Vickery, L. E.; Palmer, G. *J. Biol. Chem.* 1978, *253*, 2400.

[7] Lieberman, R. L.; Shrestha, D. B.; Doan, P. E.; Hoffman, B. M.; Stemmler, T. L.; Rosenzweig, A. C. *Proc. Natl. Acad. Sci. U. S. A.* 2003, *100,* 3820.

[8] (a)Brown, K.; Djinovic-Carugo, K.; Haltia, T.; Cabrito, I.; Saraste, M.; Moura, J. J. G.; Moura, I.; Tegoni, M.; Cambillau, C. *J. Biol. Chem.* 2000, *275*, 41133; (b) Brown, K.; Tegoni, M.; Prudêncio, M.; Pereira, A. S.; Besson, S.; Moura, J. J. G.; Moura, I.; Cambillau, C. *Nat. Struct. Biol.* 2000, *7*, 191.

[9] K. Wieghardt, *Angew. Chem.* 1994, *106*, 765; (b) Rabblee, J. H.; Messinger, J.; Cinco, R. M.; McFarlane, K. L.; Fernandez, C., Pizarro, S. A.; Sauer, K., Yachandra, V. K. *J. Am. Chem. Soc.* 2002, 124, 7459.

[10] J. M. Lehn, Supramolecular Chemistry Concepts and Perspectives, VCH, Weinheim, 1995.

[11] (a)Zanello, P.; Tamburini, S.; Vigato, P. A.; Mazzochim, G. A. *Coord. Chem. Rev.* 1987, *77*, 165; (b) Atkins, A. J.; Black, D.; Blake, A. J.; Marin-Becerra, A.; Parsons, S.; Ruiz-Ramirez, L.; Schröder, M. *Chem. Commun.* 1996, 457; (c) Vigato, P. A.; Tamburini, S. *Coord. Chem. Rev.* 2003, 248, 1717; (d) Gavrilova, A. L.; Bosnich, B. *Chem. Rev.* 2004, *104*, 349; (e) Brooker, S. *Coord. Chem. Rev.* 2001, *222*, 33; (e) Winpenny, R. E. P. *Chem. Soc. Rev.* 1998, 27, 447.

[12] Okawa, H.; Nishio, J; Ohba, M.; Tadokoro, M.; Matsumoto, N., Kolkawa, M., Kida, S., Fenton, D. E. *Inorg. Chem.* 1993, *32*, 2949.

[13] (a)Verani, C. N.; Rentschler, E.; Weyhermüller, T.; Bill, E.; Chaudhuri, P. *J. Chem. Soc., Dalton Trans.* 2000, 251; (b) Ikeda, K; Ohba, M.; Okawa, H. *J. Chem. Soc., Dalton Trans.* 2001, 3119; (c) Visinescu, D.; Andruh, M.; Müller, A.; Schmidtmann, M.; Journaux, Y. *Inorg. Chem. Commun.* 2002, *5*, 42; (d) Visinescu, D.; Pascu, G. I., Andruh, M.; Magull, J.; Roesky, H. W. *Inorg. Chim. Acta* 2002, *340*, 201.

[14] Grasa, G.; Tuna, F.; Gheorghe, R.; Leznoff, D. B.; Rettig, S. J.; Andruh, M. *New J. Chem.* 2000, *24*, 615; (b) Pascu, M.; Andruh, M.; Müller, A.; Schmidtmann, M. *Polyhedron* 2004, *23*, 673.

[15] Thompson, L. J.; Mandal, S.K.; Tandon, S.S.; Bridson, J.N.; Park, M.K. *Inorg. Chem.* 1996, *35,* 3117.

[16] Kahn, O.; Galy, J.; Journaux, Y.; Morgenstern-Badarau, I. *J. Am. Chem. Soc.* 1982, *104*, 2165.

[17] (a)Tuna, F.; Patron, L.; Journaux, Y.; Andruh, M. *Rev. Roum. Chim.* 1997, *42*, 579; (b) Tuna, F.; Grasa, G.; Patron, L.; Andruh, M. *Rev. Roum. Chim.* 1997, *42*, 793; (c) Tuna, F.; Patron, L.; Andruh, M. *Synth. React. Inorg. Met.-Org. Chem.* 1998, *28*, 13; (d) Tuna, F.; Grasa, G.; Patron, L.; Campeanu, V.; Meghea, A.; Andruh, M. *Rev. Roum. Chim.* 1999, *44*, 881; (e) Tuna, F.; Patron, L.; Lazarescu, A.; Andruh, M. *Rev. Roum. Chim.* 2000, *44*, 795.

[18] Tuna, F.; Patron, L.; Journaux, Y.; Andruh, M.; Plass, W.; Trombe, J.C. *J. Chem. Soc., Dalton Trans.* 1999, 539.

[19] (a)Tuna, F.; Pascu, G. I.; Sutter, J. P.; Andruh, M.; Golhen, S.; Guillevic, J.; Pritzkow, H. *Inorg. Chim. Acta* 2003, *342*, 131; (b) Marinescu, G.; Visinescu, D.; Cucos, A.; Andruh, M.; Journaux, Y.; Kravtsov, V.; Simonov, Y. A.; Lipkowski, J. *Eur. J. Inorg. Chem.* 2004, 2914.

[20] Andruh, M.; Ramade, I.; Codjovi, E.; Guillou, O.; Kahn, O.; Trombe, J.C. *J. Am. Chem. Soc.* 1993, *115*, 1822.

[21] Marsh, W. E.; Hatfield, W. E.; Hodgson, D. J. *Inorg. Chem.* 1982, *21*, 2679.

[22] Erxleben, A. *Inorg. Chem.* 2001, *40*, 208.

[23] (a)Koner, S.; Saha, S.; Okamoto, K. I.; Tuchagues, J. P. *Inorg. Chem.* 2003, *42*, 4668; (b) Goher, M. A. S.; Maunter, F. A. *J. Chem. Soc., Dalton Trans.* 1999, 1535.

[24] Whittaker, M. M.; Whittaker, J. W.; *Biophys. J.* 1993, *64*, 762; (b) Solomon, E. I.; Sundaram, U. M.; Machonkin, T. E. *Chem. Rev.* 1996, *96*, 2563.

[25] Pascu, M.; Tuna, F.; Andruh, M. unpublished results.

[26] (a)Sessoli, R.; Tsai, H. L.; Schanke, A. R.; Wang, S.; Vincent, J. B.; Folting, K.; Gatteschi, D.; Christou, G.; Hendrinckson, D. N. *J. Am. Chem. Soc.* 1993, *115*, 1804; (b) Sessoli, R.; Gatteschi, D.; Caneschi, A.; Novak, M. A. *Nature* 1993, *365*, 141.

[27] Gatteschi, D.; Sessoli, R. *Angew. Chem. Int. Ed.* 2003, *42*, 268; (b) Sokol, J. J.; Hee, A. G.; Long, J. R. *J. Am. Chem. Soc.* 2002, *124*, 7656.

[28] Aliaga-Alcade, N.; Edwards, R. S.; Hill, S. O.; Wernsdorfer, W.; Folting, K.; Christou, G. *J. Am. Chem. Soc.* 2004, *126*, 12503; (b) Oshio, H.; Hoshino, N.; Ito, T.; Nakano, M. *J. Am. Chem. Soc.* 2004, *126*, 8805; (c) Brechin, E. C.; Cador, O.; Caneschi, A.; Cadiou, C.; Harris, S. G.; Parson, S.; Vonci, M.; Winpenny, R. E. P. *Chem. Commun.* 2002, 1860; (d) Milios, C. J.; Raptopoulou, C. P.; Terzis, A.; Lloret, F.; Vicente, R.; Perlepes, S. P.; Escuer, A. *Angew. Chem. Int. Ed.* 2004, *43*, 210.

In: Focus on Organometallic Chemistry Research
Editor: Marin A. Cato, pp. 167-195
ISBN 1-59454-495-6

Chapter 7

SYNTHESIS, CHEMISTRY AND APPLICATIONS OF α-AMINOPHOSPHONIC AND α-AMINOPHOSPHONOUS ACIDS DERIVATIVES BEARING FURAN MOIETY

Jarosław Lewkowski*
Department of Organic Chemistry, University of Łódź, Narutowicza 68, 90-136 Łódź, Poland

ABSTRACT

This chapter describes aminophosphonic and aminophosphonous acids and esters, which bear furan moiety substituted in the '2' or '2' and '5' positions.

It demonstrates the importance of aminophosphonic acids and esters basing on the most important reviews treating with them – especially reviews by Pudovik and Konovalova or by Drabowicz and Mikołajczyk. I made the attempt to present the furan-containing drugs in pharmacy in order to demonstrate the biological importance of furan compounds.

This chapter describes all achievements concerning the chemistry of aminophosphonates bearing furan moiety in the field of synthesis, chemical conversions as well as the applications in the chemistry of polymers and crop protection. The literature is covered since 1952 (Ellis Fields' work) till today and works of such distinguished specialists in organophosphorus chemistry as Boduszek, Kraicheva or Finnocchiaro are described. The stereochemical aspect of their synthesis is also discussed.

The importance and methods for synthesis of aminophosphonous acids is also described and such papers as Baylis' et al. one, Grobelny's one or Xiao et al. one are quoted. The synthesis of title furan-derived aminophosphonous acids and formation of their ester with adenosine and cholesterol is discussed, the discussion covers also the stereochemical aspect of these syntheses.

This topic has never been described and no review has been written about furyl aminophosphonic derivatives.

---

* Phone: (+48) 42.6.78.47.31; Fax: (+48) 42.6.78.65.83; E-mail: *JLEWKOW@uni.lodz.pl*

1. Introduction – Some Words about Aminophosphonates and Furans

The great interest of aminophosphonic acids comes from late 40's of the 20th century. It was then Egelmann and Pikl [1] as well as Chavanne [2] performed for the first time the synthesis of these analogs of natural amino acids.

At the beginning of 50's, independently, Fields [3] and Kabachnik [4] worked out the method for aminophosphonic acid synthesis, which influenced much on the development of studies on the chemistry of these compounds. After their papers had been published, studies on these compounds moved on, when Horiguchi *et al.* [5] reported the isolation of 2-aminoethanephosphonic acid from biological stuff. But the most important development of this chemistry was observed in 70's, when Franz [6] patented the synthesis and properties of N-phosphomethyleneglycine (1). This compound is the main component of the commercially available very efficient herbicide, well known as Glyphosate. As Glyphosate was a really excellent herbicide, it is not strange that a variety of its analogs have been synthesized. They are: N,N-bis-(phosphonomethylene)glycine (2) (Glyphosine), Azaglyphosate (3) and Aminoglyphosate (4) or diethyl phosphonomethyleneaminomalonate [7,8]. It is also to mention the pharmacological application of aminophosphonic acids and the aminophosphonic antibiotic of a largely described efficacy – alaphosphaline (5) [9].

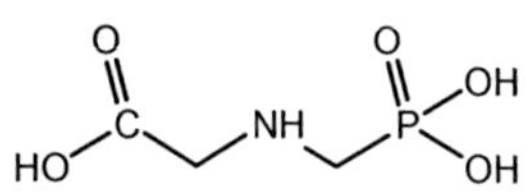

1

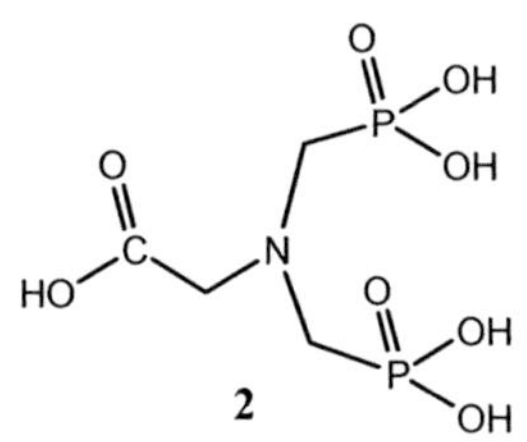

2

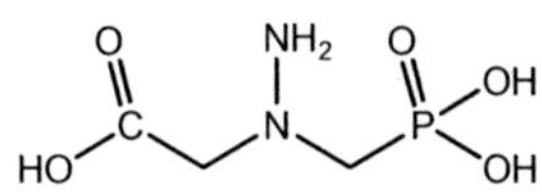

3

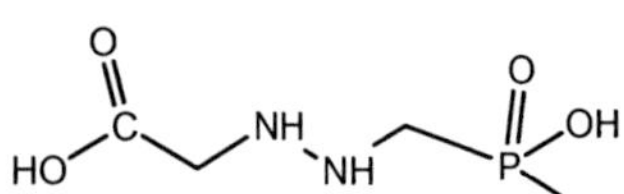

4

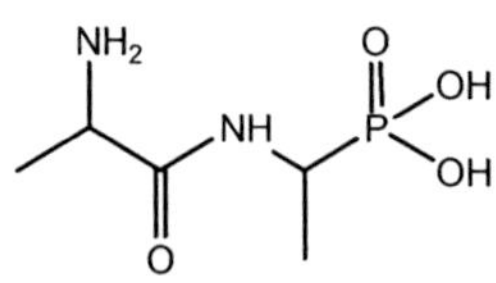

5

Numerous papers describing studies on the synthesis of aminophosphonic acids have been published and works of Mastalerz's team from Wrocław Technical University belong to most important ones. During last 25 years they dealt with the general methods for synthesis [10-15] synthesis of several important phosphono-analogues of natural amino acids [16-21] or the application of aminophosphonic acids [22-24].

Until today, over 5000 papers about aminophosphonic acids have been published describing methods for their synthesis, their chemical properties or their biological activity in the largest sense of this word. I do not want even try to summarize all of them, especially that there are several large reviews, which deeply analyzed the topic. They were written by such outstanding chemists as Pudovik [25], Mikołajczyk and Drabowicz [26], who described methods for synthesis of aminophosphonic acids, Kafarski and Lejczak [27] or Maier [28], who reviewed the application and the biological properties of this compounds.

I would rather try to confine my paper to review achievements of studies on furan-derived aminophosphonic acid derivatives. Some could ask the question: why furan, why not other molecules? Let the answer to this question be a short summary of the application of furan derivatives in medicine.

Some time ago, there was a popular anti-tuberculosis drug – isoniazide [29], which was a hydrazide of isonicotinic acid. Its action was based on that bacterium cells recognize isonicotinic acid as nicotinic acid. Isonicotinic acid forms as a result of the decomposition of isoniazide. When bacterium cells recognize isonicotinic acid as nicotinic acid, they included it to the synthesis of NAD i.e. nicotine-adenine dinucleotide. In such a way the "false" NAD forms, which causes the disturbance of respiration cycles [29]. But isoniazide was not a perfect drug – its failure was a high degree of toxicity. However, when the furan molecule was introduced to isoniazide to obtain a hydrazone – furfuryl-methyl ketone isonicotinoylhydrazone, a new drug was obtained of a commercial name – furilazone [30] (6). It is a drug of much lowered toxicity, better absorption and prolonged action.

Diloxanide [31], which chemically is hydroxy-N-methyl-dichloroacetanilide (7) was discovered to be rather efficient in anti-amoebic therapy. Further clinical studies revealed that this drug is not so efficient as it is required, but introduction of a furan ring by the ester formation allowed to achieve furamide [31] – 2-furoyloxy-N-methyl-dichloroacetanilide (8). Furamide is much more efficient in the anti-amoebic therapy.

Furosemide is a strong diuretic agent and hypotensive drug belonging to the sulfonamide group. Its action is based on the inhibition of sodium and chloride ions and water absorption [32].

It is worthwhile mentioning the group of drugs derived from nitrofurfural. Its action is based on the redox biotransformation of nitrofurfural, which is reduced with nitroreductase to an active radical responsible for the antibacterial and anti-protozoal action of nitrofurfurals [33].

To this group of drugs, they belong preparations against inflammatory states of alimentary canal and against inflammatory states of uric system, antiseptics for external use only or even agents for Chagas' disease treatment such as anti-*Trypanosoma cruzi* nifurtimox (9) [34]. The first and used up till now drug is nitrofurazone (10) [35], the agent in the anti-anginal treatment. Good antiseptics for uric system diseases treatment are nitrofurantoin (11) [36], furagin (12) [37] and furaltadone (13) [38]. Quite efficient agents in the alimentary canal inflammations treatment are furazolidone (14) [39] and nifuroxazide (15) [40]. The latter is even more efficient than sulfaguanidine.

To complete the list, let me be allowed to mention drugs such as two paramimetosympathics furthretonium (16) [41] and methfurthetonium (17) [41], which demonstrate the large structural similarity to muscarine (18).

It is also to mention ranitidine (19), which is the antihistamine agent blocking the H_2 receptors and which is the efficient drug in the treatment against a digestive ulcer [42].

At the end of this laconic overview, I would like to mention the new generation drugs. The first is prazosin, which chemical name is: 2-[4-(2-furoyl)-piperazine-1-yl-4-amino-6,7-dimethylquinoline hydrochloride (20) and which is the α_1-adrenolytic agent. Its hypotensive action is based on the phosphorodiesterase inhibition [43].

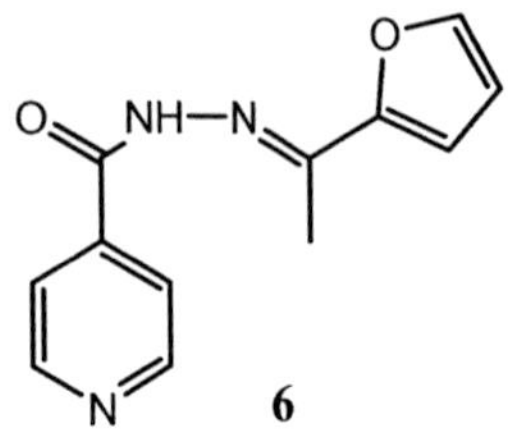

6

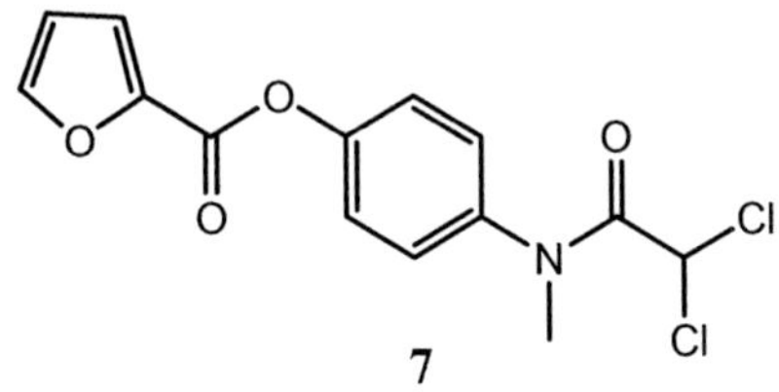

7

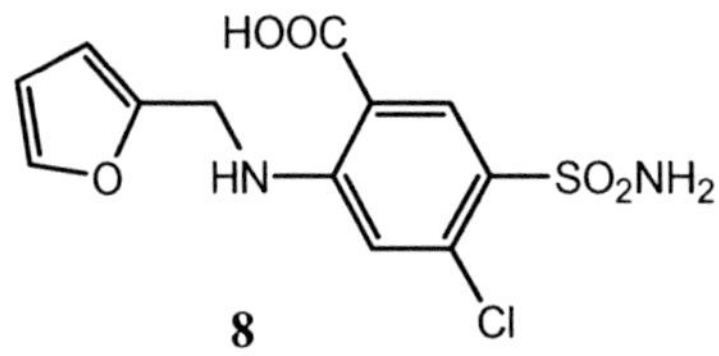

8

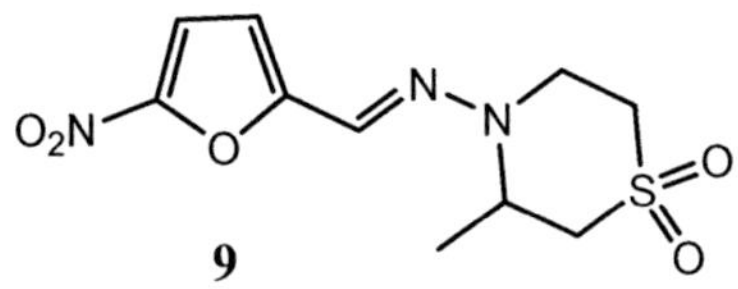

9

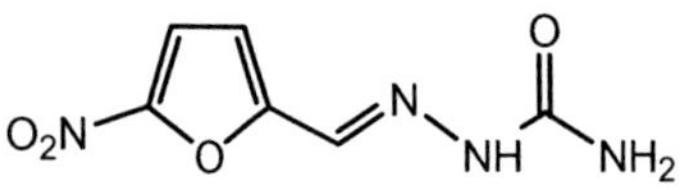

10

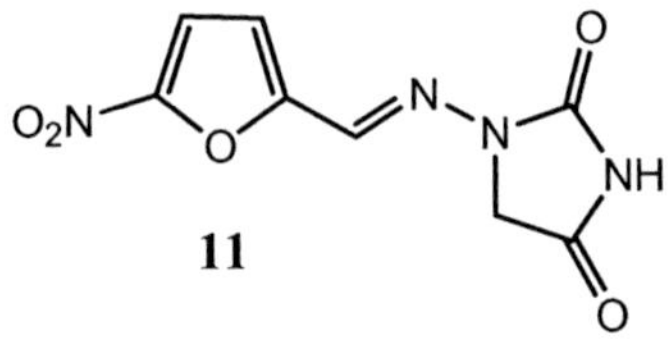

11

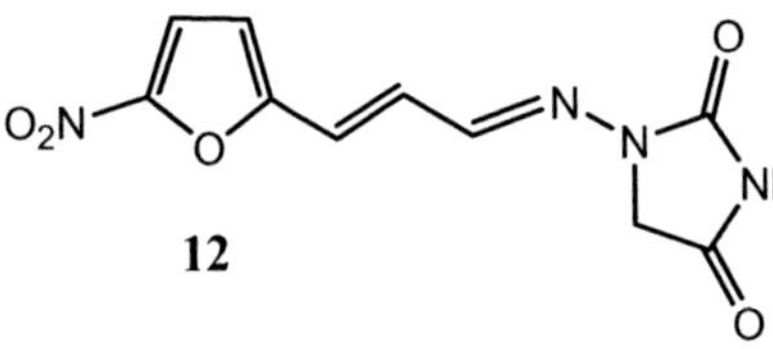

12

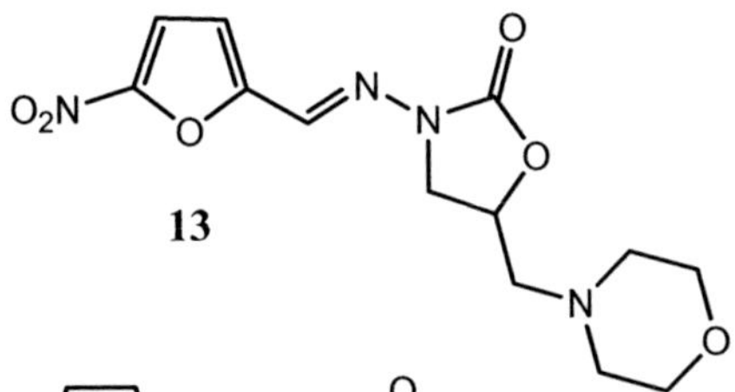

13

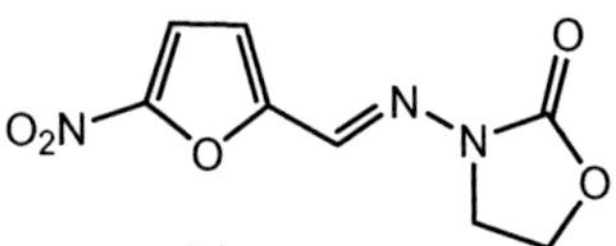

14

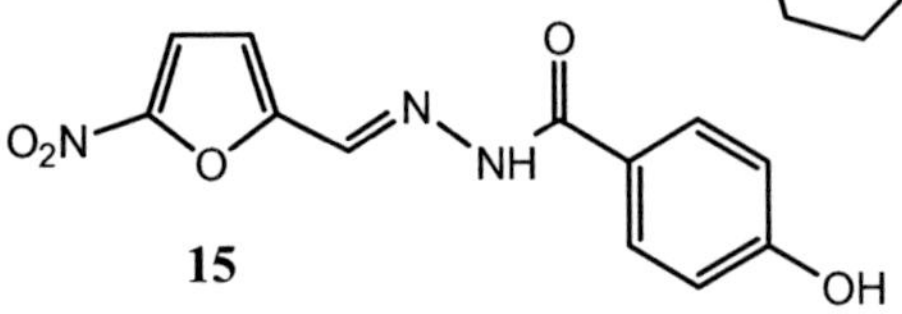

15

It is also to mention antibiotics bearing the furan moiety: furbenicillin (21) [44] belonging to the penicillin 5 group, cefuroxime (22) [45], which is the second generation cephalosporin and the still experimental group of anti-*Pneumocystis carnii* agents deriving from amidinediarylfuran (23) [46]. The latter group forms the weak bonding with the DNA by the intercalation, which disrupts normal functions of a bacterium cell.

Considering below, it does not seem strange to me that chemists have decided to conjoin the biochemical and pharmaceutical properties of both groups, i.e. aminophosphonic acids and furans, and to create the new series of α-aminophosphonic acids and esters bearing the furan moiety substituted in the '2' position. Did they all really think about it in this way, one does not know for sure, but it would be quite possible, if so.

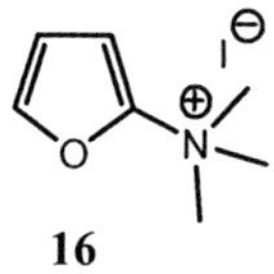

16

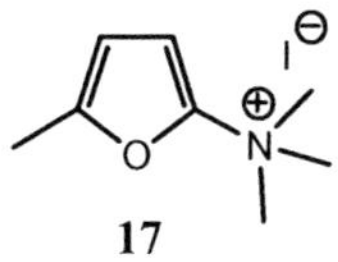

17

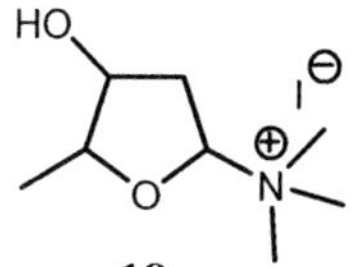

18

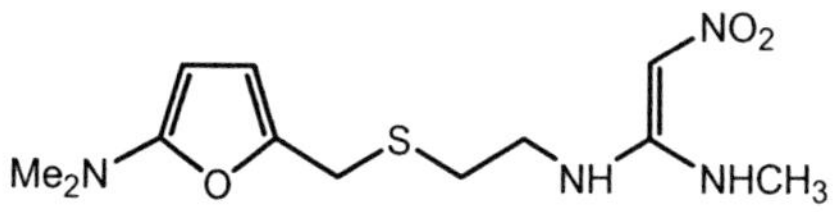

19

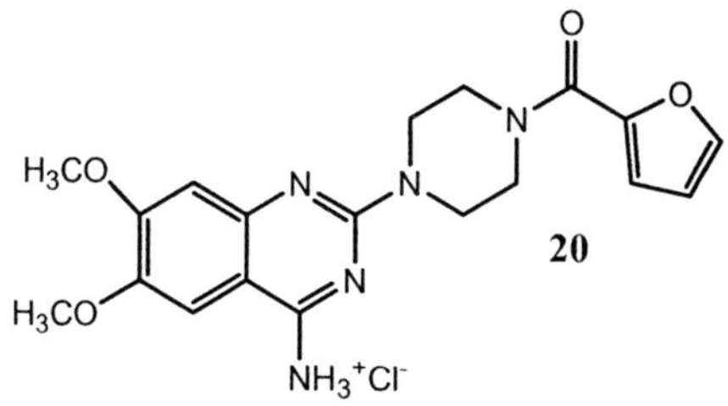

20

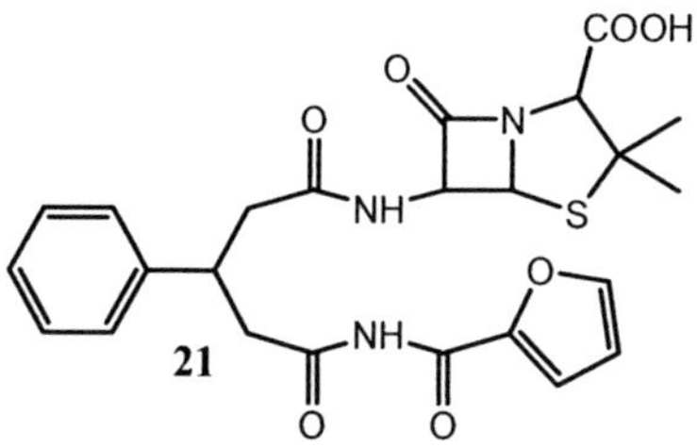

21

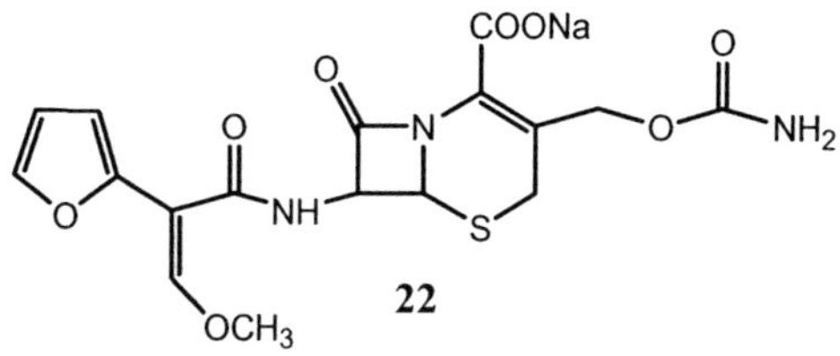

22

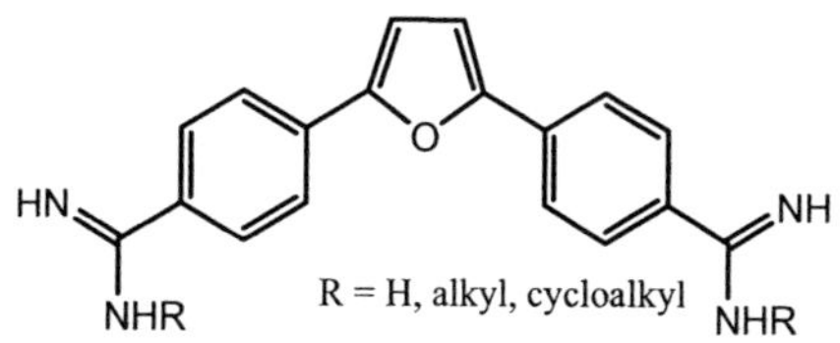

23

2. Furan Derived Aminophosphonic Acids and Esters

The first was Ellis Fields starting up with his pioneer work from 1952 [3], where he described, among others the synthesis of two furan derived diethyl aminophosphonates: diethyl (2-furyl)-N-ethylaminomethane phosphonate (24a) and diethyl (2-furyl)-N,N-diethylaminomethane phosphonate (24b), obtaining each of then by the use of different method. N,N-Diethyl derivative was obtained by the direct reaction of furfural, diethylamine and phosphite, and the N-monoethyl derivative was achieved by the two step reaction: the first step was the imine synthesis and the second one – the phosphite addition (*Scheme 1*).

Lugovkin and others [47] performed the synthesis of the series of diethyl (2-furyl)aminophosphonates applying variously substituted aminopyridines and aminoquinolines as amines. The synthesis of Schiff bases followed by the addition of diethyl phosphite to an azomethine bond led to the formation of following derivatives: N-(α-pyridinyl) (25), N-(γ-pyridinyl) (26), N-[2-(2-, 3- and 4-methyl)pyridinyl (27), (28) and (29) respectively, as well as the N-2-quinolinyl derivative (30). (*Scheme 2*) One can ask the question: why, in such a nice set of derivatives, there is no the N-β-pyridinyl derivative? Is that caused by the difficulties with the synthesis of corresponding Schiff base? I found nothing about it in the article by Lugovkin [47].

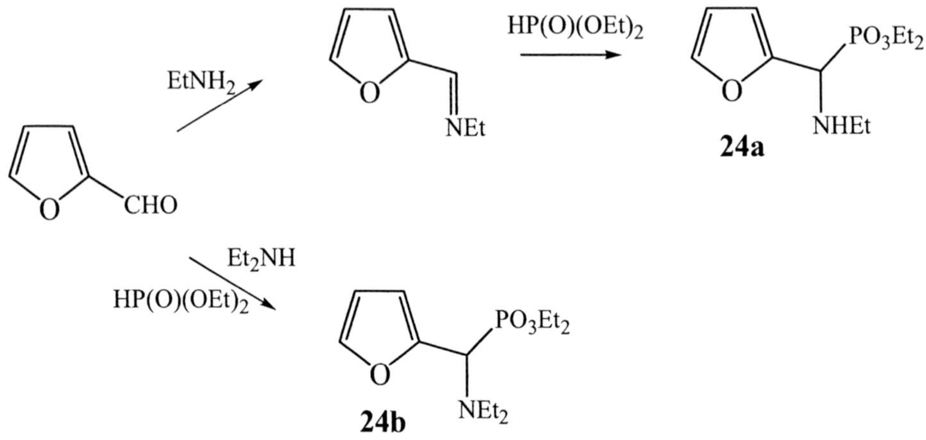

Scheme 1

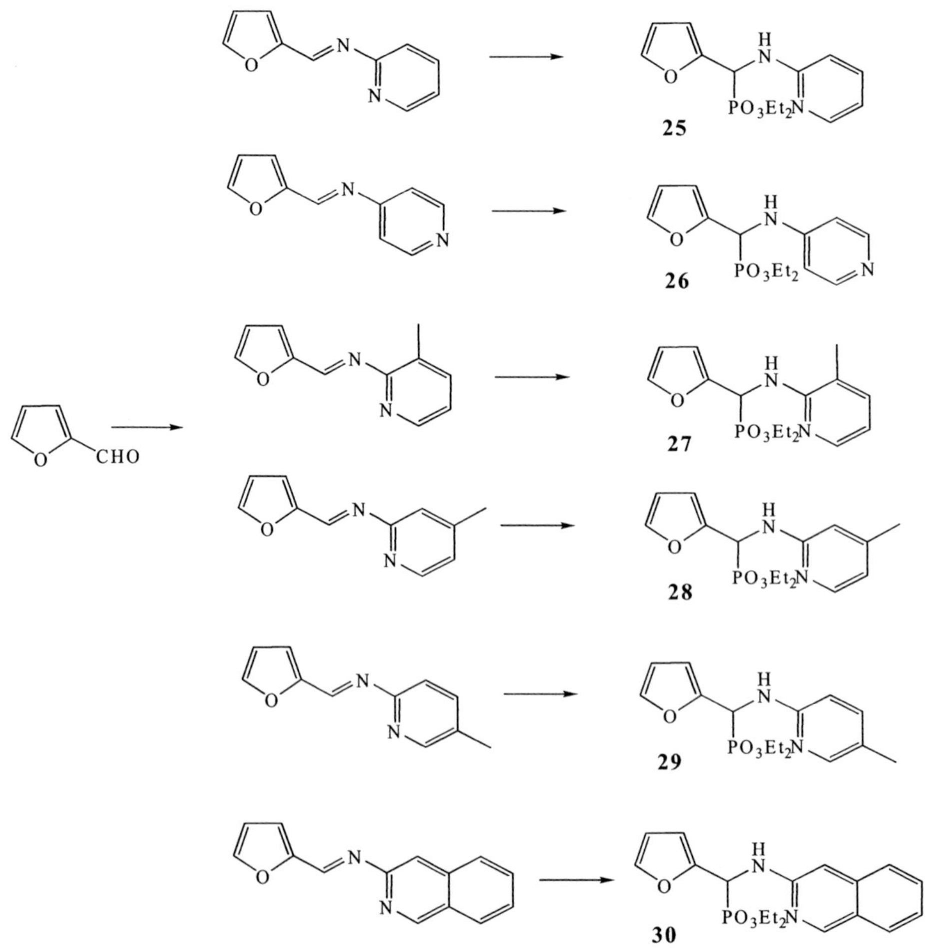

Scheme 2

Osipova with coworkers [48] performed with success the modified Arbuzov reaction of furfuraldoxime with triethyl phosphite. This reaction resulted in the formation of diethyl (2-furyl)-N-ethoxyaminomethanephosphonate (31) and 2-cyanofuran as a side-product. (*Scheme 3*)

Demir *et al.* [49] have also worked with oxime derivatives. As a result of the Arbuzov reaction between furan-2-carboxylic acid chloride and triethyl phosphite, they obtained diethyl α-ketophosphonate. The latter, after the reaction with hydroxylamine hydrochloride, was converted into the corresponding α-(hydroxyimino)phosphonate, which was then undergone the reduction with sodium borohydride catalyzed by molybdenum trioxide leading to diethyl (2-furyl) aminomethanephosphonate (32). Subsequently, they applied the rather "exotic" method for hydrolysis of aminophosphonates, i.e. heating diethyl (2-furyl) aminomethanephosphonate (32) in ethanol adding 1ml of propylene oxide, which allowed them to achieve (2-furyl)aminomethanephosphonic acid (33) in 86% yield. (*Scheme 3*) I was puzzling very much about this method: how it was possible that the aminophosphonate cleaved to aminophosphonic acid without any cleaving agent. Indeed, propylene oxide is not the reagent for hydrolysis, it is used after the cleavage to facilitate the precipitation by reacting with water. I suppose that authors [49] forgot to mention the cleaving agent they used.

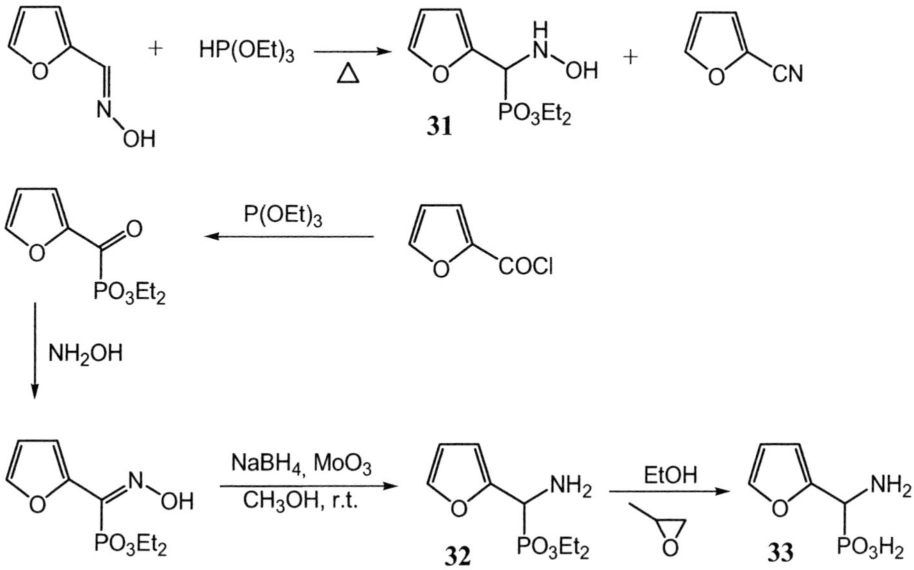

Scheme 3

Bogdan Boduszek has a great contribution to studies on aminophosphonic acids and esters bearing heteroaromatic moieties. It is then obvious that he investigated furan-derived aminophosphonates too. Boduszek [50] converted furfural into its Schiff basis – N-furfurylidenebenzylamine (34), which was subsequently undergone the addition of diethyl and diphenyl phosphite. This reaction led to the formation of corresponding diethyl and diphenyl (2-furyl)-N-benzylaminomethanephosphonate (35) and (36) respectively. Diethyl ester 35 was then hydrolysed with concentrated hydrobromic acid to obtain (2-furyl)-N-benzylaminomethanephosphonic acid (37) in 25% yield, which, according to Boduszek [50] was caused by the partial decomposition of the furan ring in acidic medium.

Both esters [50] 35 and 36 was undergone the hydrogenolysis with palladium-on-charcoal which led to the formation of diethyl (2-furyl) aminomethanephosphonate (38) and diphenyl (2-furyl) aminomethanephosphonate (39). (*Scheme 4*)

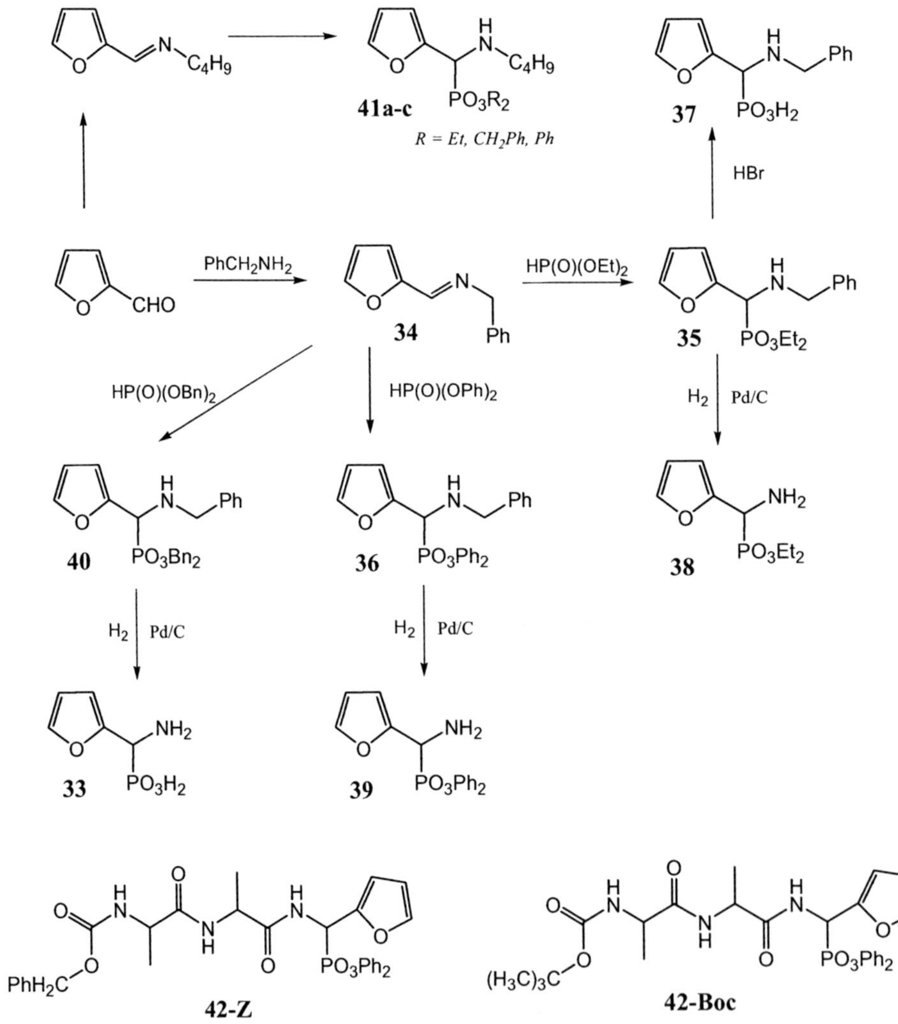

Scheme 4

In his next paper [51], Boduszek reported the addition of dibenzyl phosphite to the azomethine bond of the Schiff basis 34, which led to dibenzyl (2-furyl)-N-benzylaminomethanephosphonate (40). The hydrogenolysis of aminophosphonate 40 yielded (2-furyl)aminomethanephosphonic acid (33). He also presented [52] the synthesis of N-butyl (2-furyl)aminophosphonates (41a-c) and some attempts [52] of the preparation of oligopeptides with aminophosphonic monomers bearing the furan ring. Boduszek applied a variety of protecting groups using a *Boc* (tert-butoxycarbonyl) group as well as a *Z* (benzyloxycarbonyl) group. The dipeptide alaninyloalanine, protected in one of ways, was undergone the condensation with diphenyl (2-furyl) aminomethanephosphonate (39) in the presence of dicyclohexylcarbodiimide (DCC). It allowed to obtain two mixed tripeptides 42-Boc and 42-Z. Further biochemical studies allowed to state the ability of both peptides to the irreversible inhibition of chymotrypsin, human neutrophilic elastase and pig pancreas elastase. Moreover, biological studies of esters demonstrated that diphenyl (2-furyl)-N-benzylaminomethanephosphonate (35) had the herbicidal action inhibiting the root growth of such plants as: *Lepidium sativum L.* or *Cumus stivus*. (*Scheme 4*)

Boduszek interested also in furfural derivatives substituted in the position '5'. In his work from 2001, he presented [53] the synthesis of [2-(5-nitrofuryl)]-N-benzylaminomethane-phosphonic acid (43). This acid was obtained by the reaction of N-(5-nitrofurylidene)benzylamine (44), trimethylsilyl bromide and triethyl phosphite leading to the formation of unstable silyl derivative (45), which hydrolysed quickly to the acid 43 when treated with aqueous methanol. (*Scheme 5*)

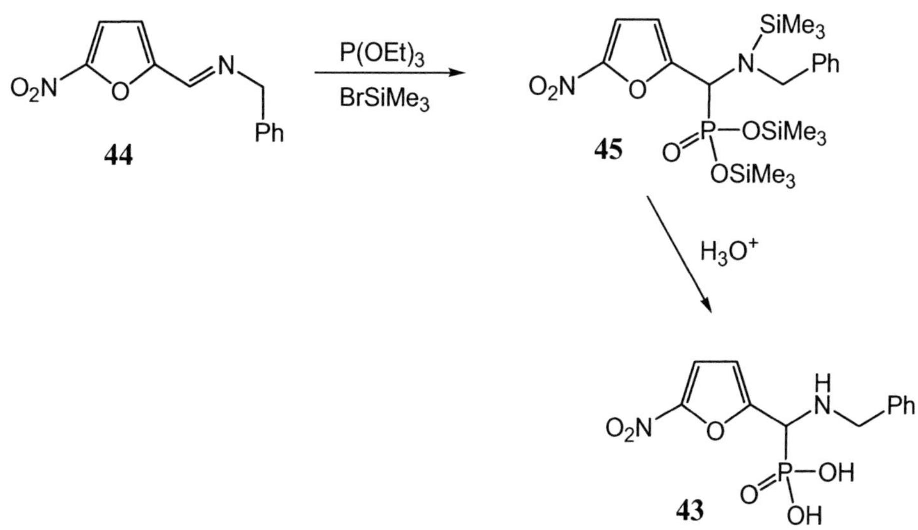

Scheme 5

As for the studies on the biological or even pharmacological activity, it is to mention the work by Alonso *et al.* [54] who performed the synthesis of aminophosphonic esters, which are the potential protease HIV virus inhibitors. They used the information that this protease is in fact aspartic proteinase and they stated that the potential protease inhibitor should be an analogue of the active complex, which binds protease in a stable way. α-Aminophosphonic esters and α-aminophosphine oxides were discovered to be such compounds. In both cases, furan derivatives were considered to be potentially highly active towards the protease HIV virus.

Authors [54] performed the synthesis of these furan derived aminophosphonates in three steps. First one included the formation of leucine ester with benzyl alcohol in the presence of p-toluenesulfonic acid. The ester obtained in this way, was then undergone the condensation with furfural, which led to the formation of the Schiff base 46. The subsequent addition of dimethyl phosphite allowed to achieve (2-furyl)aminophosphonate 47 in 86% yield. (*Scheme 6*)

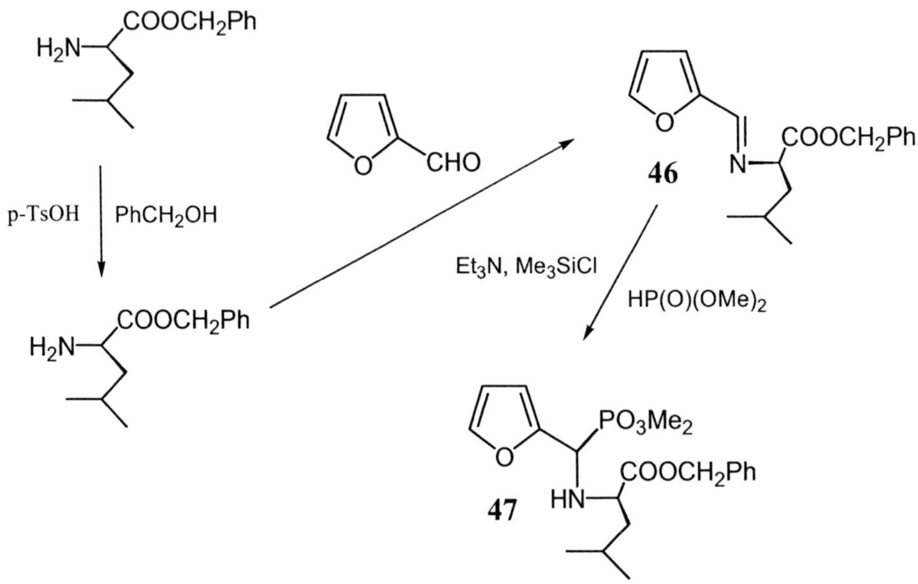

Scheme 6

The very interesting paper was written in 1992 by Laschat and Kunz [55]. They reported a new method for the stereoselective synthesis of aminophosphonic esters, among others diethyl (2-furyl)aminomethanephosphonate (32). They used aminosugars as the chiral auxiliaries, and more precisely galactosylamine and fucopyranosylamine both protected by pivaloyl groups (Piv), which were undergone the reaction with furfural to obtain corresponding Schiff bases (48) and (49) subsequently undergone the addition of diethyl phosphite in the presence of tin tetrachloride as a catalyst. In this way, authors [55] obtained (2-furyl)-N-galactosylaminomethanephosphonate (50) and (2-furyl)-N-fucopyranosylamino-methane-phosphonate (51). Both esters were also obtained in the course of a one pot reaction of an aminosugar, aldehyde and phosphite in the presence of tin tetrachloride as a catalyst. Yields were very high up to 90%, as well as stereoselectivity. The major product formed in 70%, and it was either the β-S or the β-R isomer, dependently on the applied sugar. Derivatives of the α-anomeric isomer formed evidently in minority and did not exceed overall yield (αS + αR) 15%. The simple hydrolysis of N-glycosyl esters led to the formation of chiral (2-furyl)aminophosphonic acids (32). But authors [55] did not succeed in separating diastereoisomers. (*Scheme 7*)

Works of Ivanka Kraicheva from the Bulgarian Academy of Science in Sofia seem to me extremely interesting in the point of view of application. She performs studies on the effect of the presence of the furan-containing aminophosphonates on properties of various polymers. Her preliminary investigation demonstrated that such polymers are more thermo-resistant and more fire-resistant.

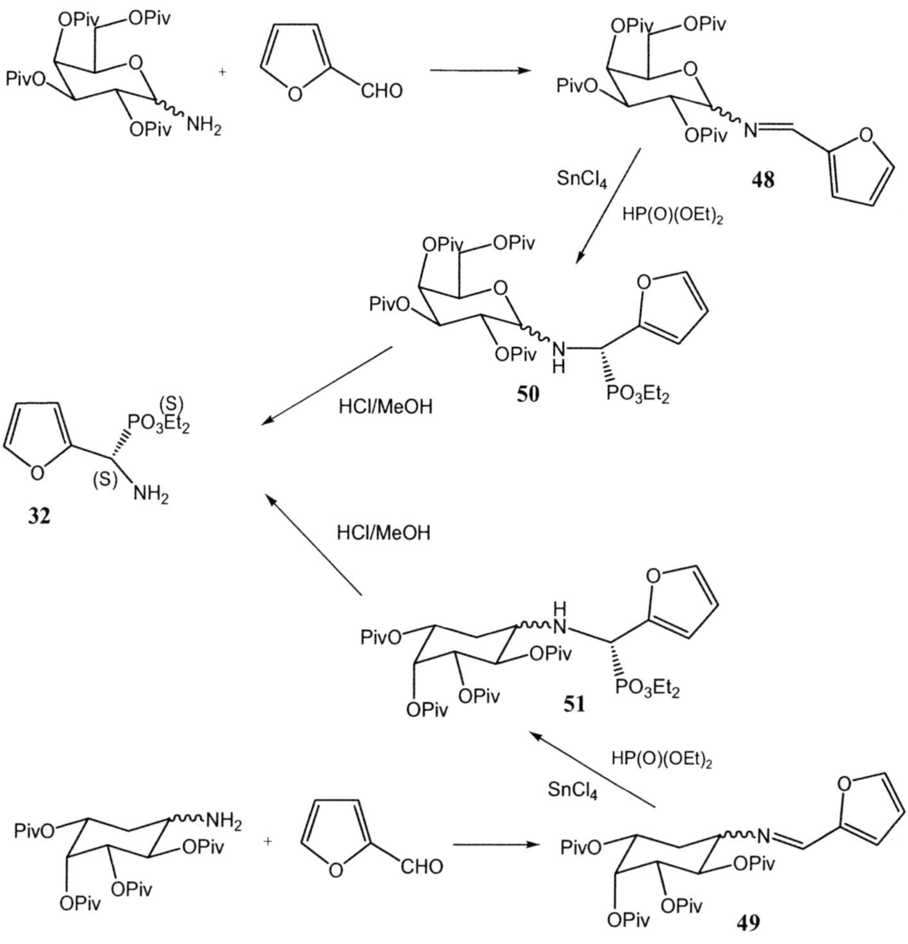

Scheme 7

Kraicheva performed the synthesis of the variety of furfural-derived bis-aminophosphonates using as diamines: diaminodiphenyl ether [56,59,60], diaminodiphenylmethane [56.59,60], benzidine [61] and 1,4-phenylenediamine [56,59,60]. She used such phosphites as: diethyl, dimethyl and di-isopropyl phosphite [56,59,61] or even dibutyl one [60]. Reactions were based on the conversion of furfural to corresponding Schiff bases with above mentioned amines and the subsequent addition of a phosphite in the presence of an alkoxide. This is important that Kraicheva always used an alkoxide appropriately to a phosphite, i.e. ethoxide in a case of diethyl phosphite, methoxide in a case of dimethyl phosphite etc. It was a protection from any uncontrolled and undesirable exchange reactions. (*Scheme 8*)

Kraicheva performed also the detailed NMR analysis of the above compounds [59-61] varying deuterated solvents ($CDCl_3$, DMSO-D_6) applying shaking with D_2O. She performed even the 2-D correlation H,H-COSY spectrum [60].

The esters were subsequently introduced to a polymer [56] in the course of the reaction with 2,4-toluilenediisocyanate, which allowed to obtain polycondensation products 52.

Thermogravimetric measurements allowed to state that polymers belonging to this group are stable at the temperature above 500°C and moreover, they are fireproof.

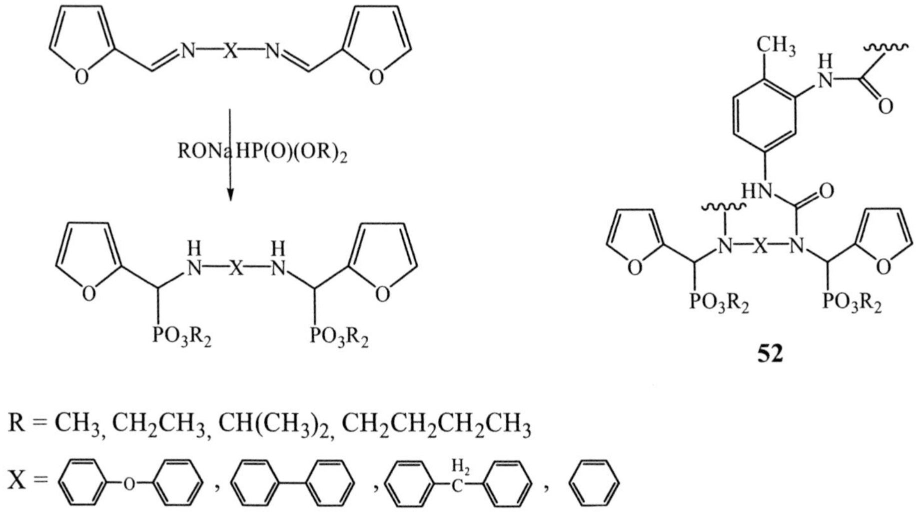

Scheme 8

In a search for new additives to polymers, which would increase polymer resistance, Kraicheva underwent the synthesized esters the Diels-Alder reaction [57,58]. Furan rings in aminophosphonates 53 reacted as dienes with maleic anhydride as dienophile. It allowed to achieve interesting compounds 54, which would be introduced to polymers to give them new interesting properties. (*Scheme 9*)

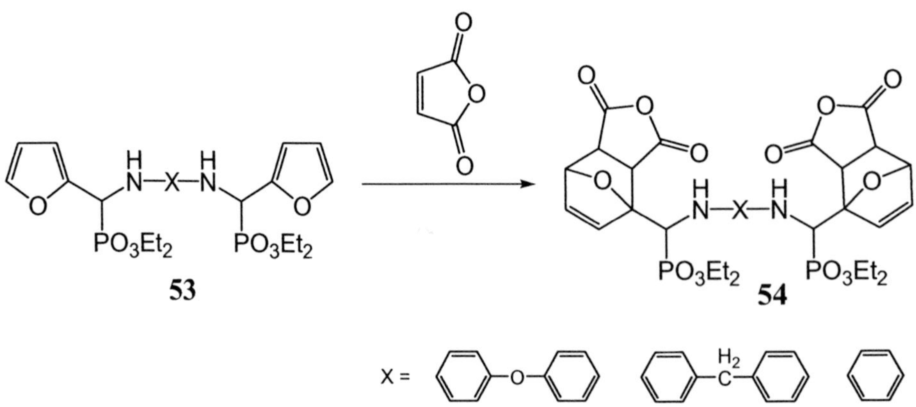

Scheme 9

In one of her most recent works [62], Kraicheva synthesized different phosphonic derivatives of furan, i.e. furfural phosphonohydrazone (55) and 2-acetylfuran phosphonohydrazones (56). (*Scheme 10*) This group of compounds may be applied as fluorescent dyes for polymeric fibres or cytostatic agents.

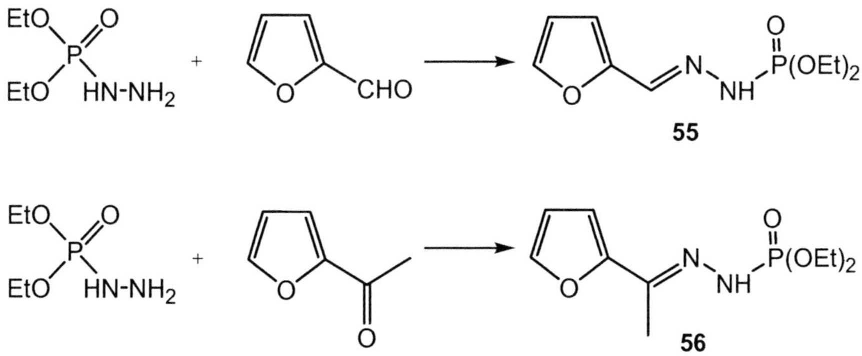

Scheme 10

Consiglio and coworkers [63] performed the synthesis of two complex (2-furyl) aminophosphonates: diphenyl [2-(5-(2-chlorophenyl))furyl]-N-benzyloxycarbonylamino-methanephosphonate (57) and diphenyl [2-(5-(acetylmethyl))furyl]-N-benzyloxycarbonyl-aminomethane phosphonate (58). They presented also the detailed NMR characterization of these compounds. (*Scheme 11*)

The reaction applied by Consiglio and coworkers [63] was the one pot reaction carried out by simple mixing the reagents: an aldehyde, triphenyl phosphite and benzyl carbamate. This method was worked out by Mastalerz and coworkers [64].

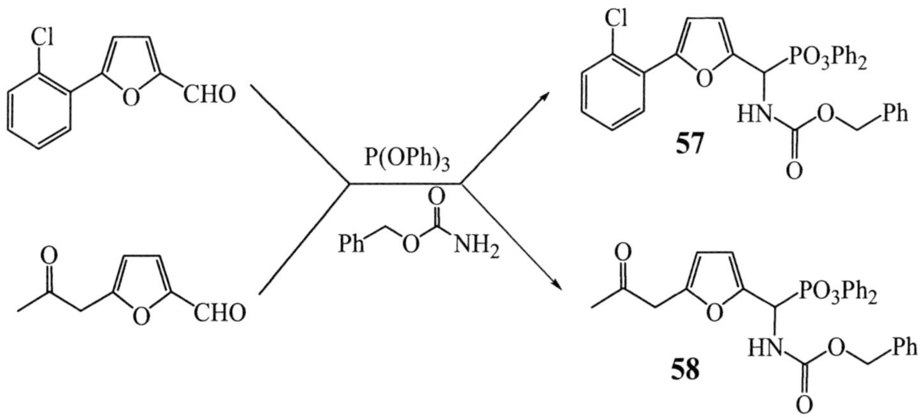

Scheme 11

Prishchenko *et al.* [65] also synthesized N-acyl derivatives of (2-furyl)aminophosphonic esters, but their reactions were performed by the acylation reaction of diethyl (2-furyl)-N-methylaminomethane phosphonate (59). In such a way, they obtained the series of compounds, among which there are: diethyl (2-furyl)-N-formyl-N-methylaminomethane phosphonate (60), the N-acetyl (61) and the N-cyclopropanecarbonyl (62) derivative and diethyl (2-furyl)-N-sulfonyl-N-methylaminomethane phosphonate (63). They prepared also ethyl (2-furyl)-N-sulfonyl-N-methylaminomethane-methane-phosphinate (64). (*Scheme 12*)

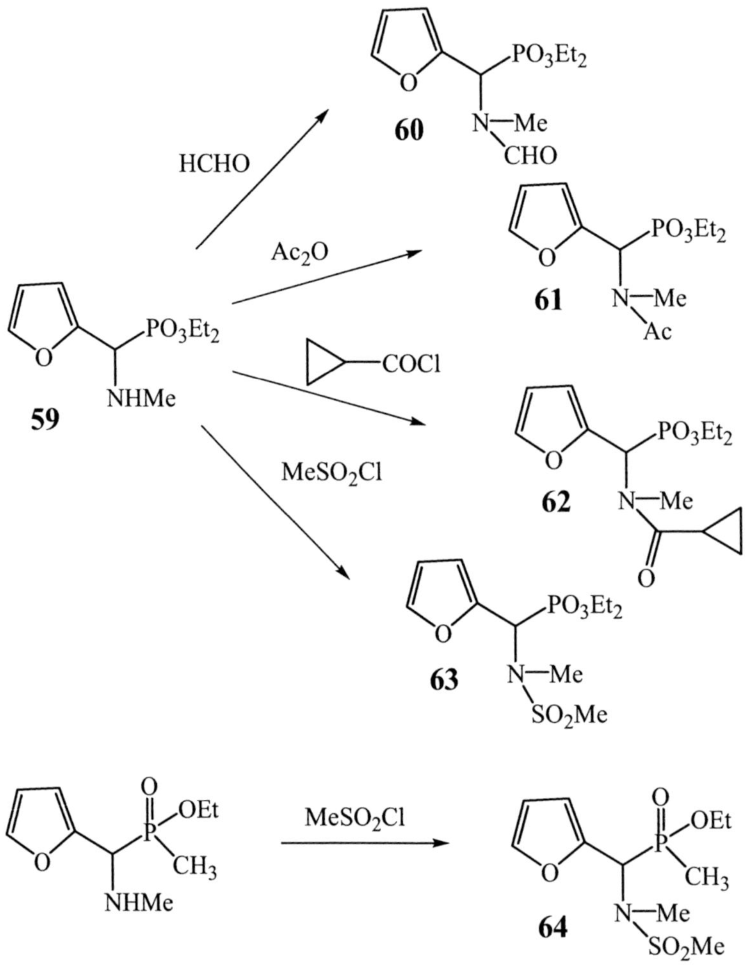

Scheme 12

Now, I wish to present three papers describing new methods for the synthesis of the title compounds. First is the paper by Qian and Huong [66], who applied ytterbium trifloromethanesulfonate $Yb(OTf)_3$ as a catalyst. They tested both the addition of diethyl phosphite to an azomethine bond of Schiff bases and the classical Kabachnik-Fields method. They performed tests to answer the question which catalyst is the most efficient. Thus they investigated lanthanum trifloromethanesulfonate, samarium trifloromethanesulfonate and even tin tetrachloride but the best results were obtained when the reaction was carried out in dichloromethane in the presence of ytterbium trifloromethanesulfonate and molecular sieves. In such a way, authors of the discussed paper [66] obtained among others diethyl (2-furyl)-N-anilinomethanephosphonate (65). (Scheme 13)

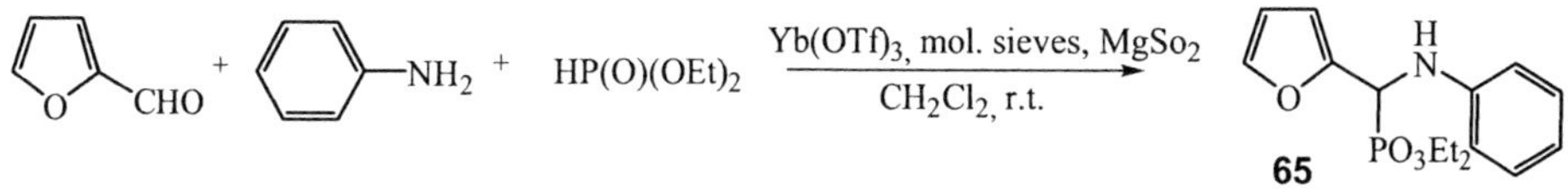

Scheme 13

The next paper that I would like to discuss here is the work by Manabe and Kobayashi [67]. They applied a surfactant as a catalyst and they called it "Lewis acid-surfactant-combined catalyst", which abbreviation is LASC. They used scandium tris-(dodecyl sulfate) as the LASC and the method required the use of triethyl phosphite as while applying diethyl phosphite, the reaction did not occur. The reaction was carried out in water at a room temperature and it was in full accord with authors' intentions, that is why they used a surfactant for this reaction. The suggested mechanism was as follows: the reaction of an amine with an aldehyde resulted in an imine, which underwent subsequently the attack of triethyl phosphite to form the intermediate (66). This intermediate underwent the hydrolysis, which allows to obtain among others diethyl (2-furyl)-N-diphenylmethylaminomethane phosphonate (67) in 80% yield. (Scheme 14)

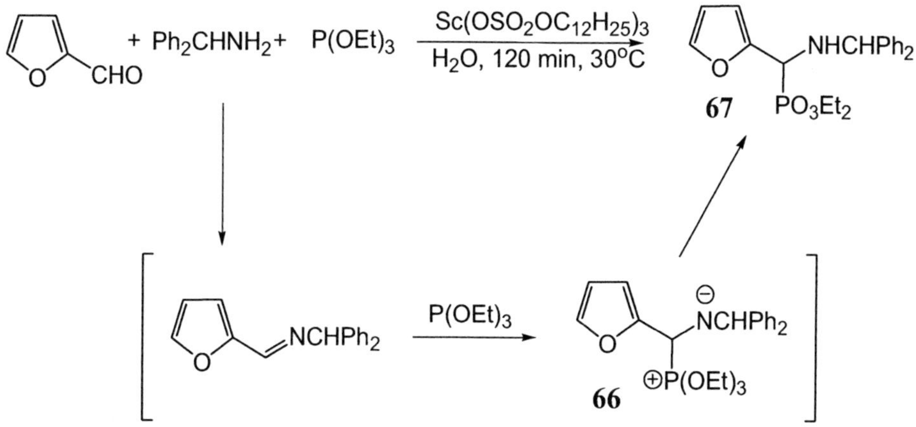

Scheme 14

Kaboudin and Nazari [68] applied the Kabachnik-Fields method modifying it slightly and they obtained diethyl (2-furyl) aminomethanephosphonate (32). There were three main modifications of this method, first ammonium formate was used as the "source" of amine group, the second was based on applying acidic aluminum oxide as a catalyst for the microwave stimulation of the reaction. The use of microwaves was the third modification and permitted to obtain the ester 32 in high yield. (Scheme 15)

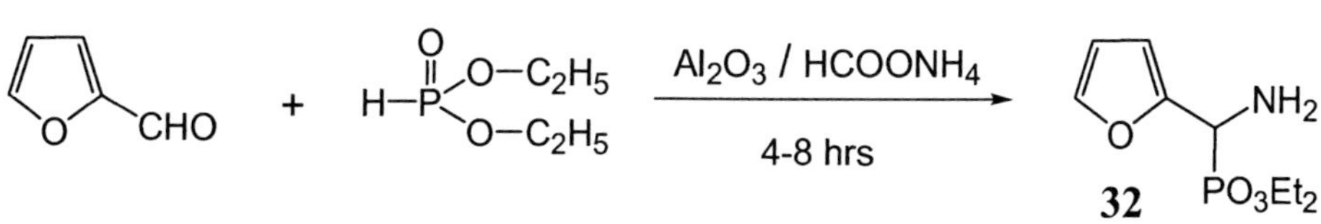

Scheme 15

A similar reaction was performed seven years before by Takahashi *et al.* [69] who applied molecular sieves instead aluminum oxide and ammonium acetate instead ammonium formate. Using the Takahashi's method, I obtained diethyl (2-furyl) aminomethanephosphonate (32) in 55% yield [70].

The author of this review also contributed to this topic continuing his study on 5-hydroxymethylfurfural [71]. I performed the condensation of this aldehyde (HMF) with various amines to obtain corresponding Schiff bases 67a-d [72,73]. The Schiff bases of 5-hydroxymethylfurfural 67a-d were undergone the addition of diethyl and dibenzyl phosphites in the presence of catalytic amount of trifluoroacetic acid to obtain eight [2-(5-hydroxymethylfuryl)]aminomethanephosphonates 68a-h [72,73]. (Scheme 16)

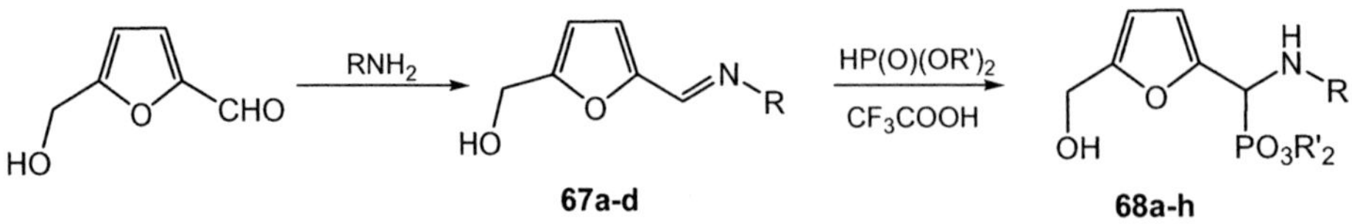

R = CH_2Ph, CH_2Fur, $C(CH_3)_3$, $CH(CH_3)Ph$
R' = CH_2CH_3, CH_2Ph

Scheme 16

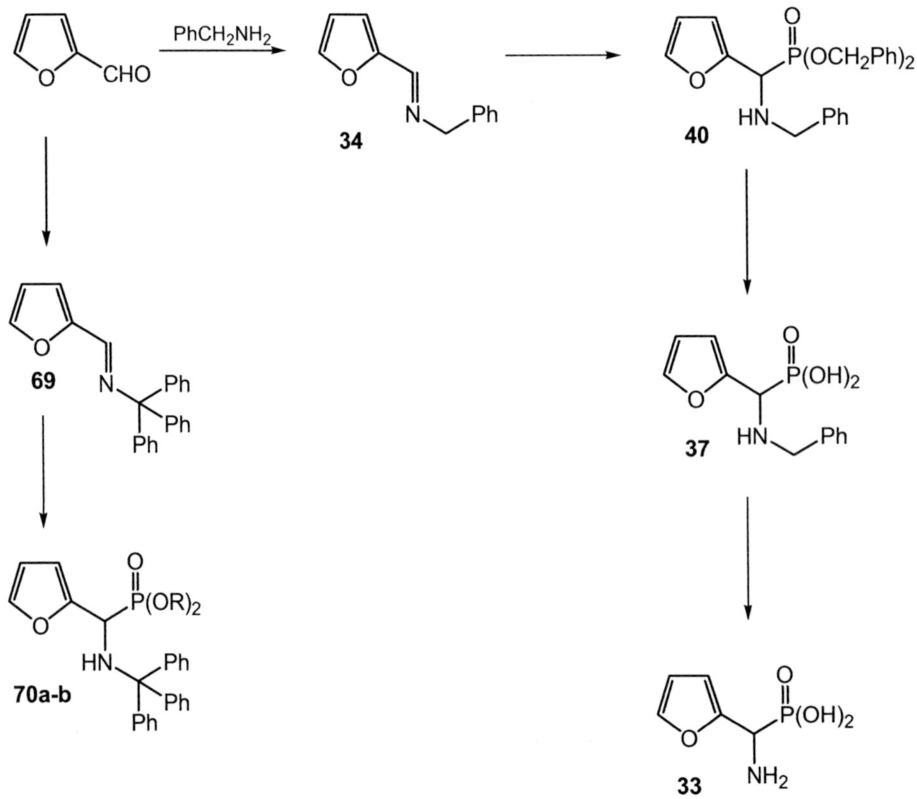

Scheme 17

The series of conversion of furfural derived aminophosphonates were also made by our team [74]. Furfural was converted into its N-benzyl Schiff base 34, which was subsequently undergone the addition of dibenzyl phosphite resulting in dibenzyl (2-furyl)-N-benzylaminomethanephosphonate (40). The aminophosphonate 40 has been undergone the two-step hydrogenolysis, which at its first step gave (2-furyl)-N-benzylaminomethanephosphonic acid (37) and at the second one yielded (2-furyl)-aminomethanephosphonic acid (33) [74]. (Scheme 17)

Moreover, furfural was converted into its N-trityl Schiff basis 69 and, after the addition of diethyl and dibenzyl phosphites, two aminophosphonates were obtained: diethyl (2-furyl)-N-tritylaminomethanephosphonate (70a) and dibenzyl (2-furyl)-N-tritylaminomethanephosphonate (70b) [74]. Unfortunately, the synthesis of diphenyl (2-furyl)-N-tritylaminomethanephosphonate (70c) was not successful, despite attempts, we were not able to obtain this compound. It is a great pity, because, according to the letter information from the National Cancer Institute (NCI), the aminophosphonate 70c could have some cytostatic action [70]. It is to early to suggest why the synthesis did not occurred, but the careful analysis of all concerning data allows to suppose the sterical hindrance. (Scheme 17)

The series of twelve furan-derived aminophosphonates 72a-l were synthesized by the author of this review [75]. We applied the classical addition of three phosphites: diethyl, dibenzyl and diphenyl one to azomethine bonds of Schiff bases 71a-d and (2-furyl)aminophosphonates 72a-l were obtained in fair yields [75]. (Scheme 18)

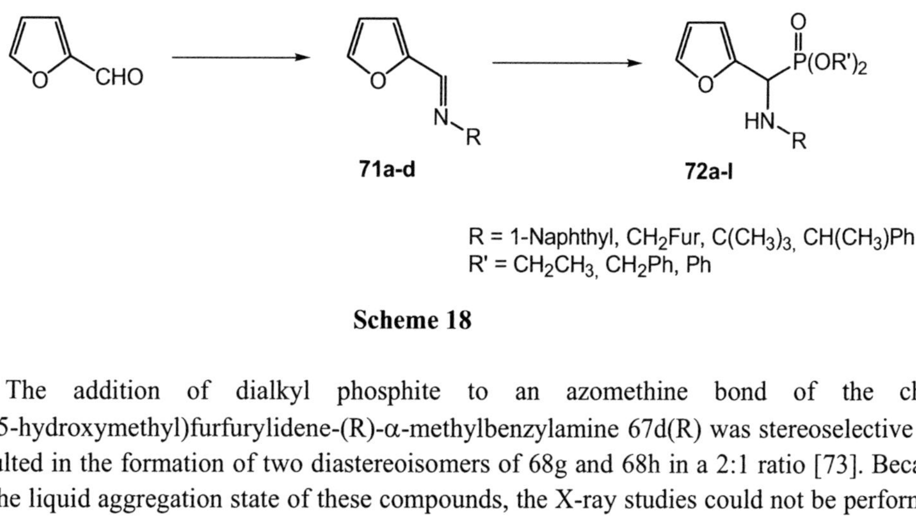

Scheme 18

The addition of dialkyl phosphite to an azomethine bond of the chiral N-(5-hydroxymethyl)furfurylidene-(R)-α-methylbenzylamine 67d(R) was stereoselective and resulted in the formation of two diastereoisomers of 68g and 68h in a 2:1 ratio [73]. Because of the liquid aggregation state of these compounds, the X-ray studies could not be performed, so attempts were made to establish the absolute configuration by the qualitative analysis of the Cram model [76]. This analysis led to the supposition that when the Schiff base of the R configuration was used, the major diastereoisomer was the aminophosphonate of the RR configuration [73]. (Scheme 19)

The analysis of ^{1}H NMR spectra was also preformed [73] basing on the method described by Latypov and Riguera [77], where each diastereoisomer was undergone the analysis of the adopted conformation and where it was predicted which proton found itself in a magnetic field induced by the stream of π-electrons, which caused its shielding [73]. In a result of the above analysis, the supposition has been made that the major diastereoisomer was the aminophosphonate of the RR configuration. (Scheme 20)

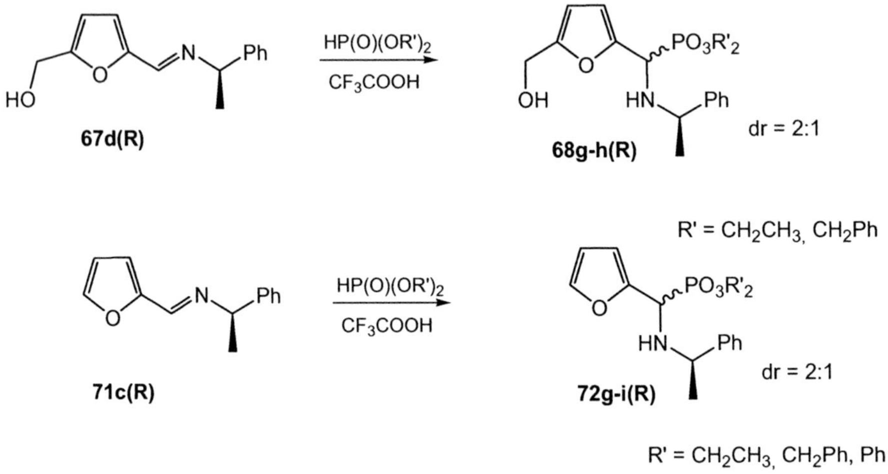

Scheme 19

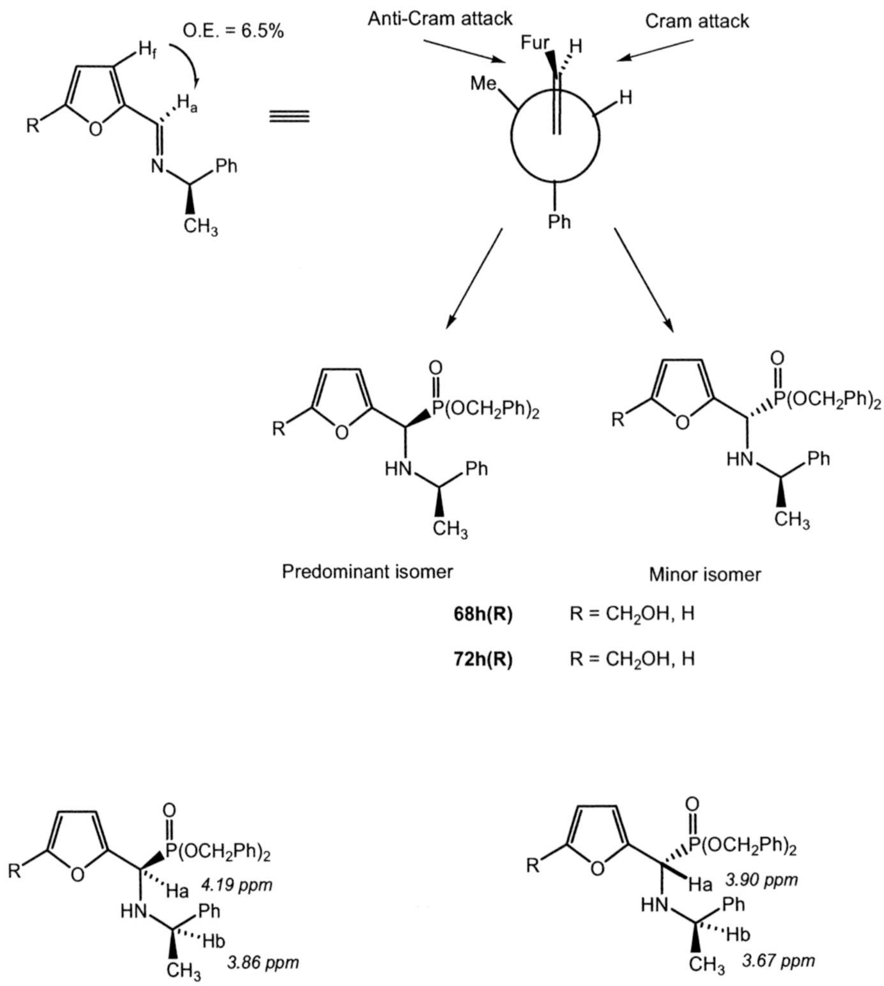

Scheme 20

The similar results were obtained in case of furfural derivatives. The addition of phosphites to the chiral Schiff base – N-furfurylidene-(R)-α-methylbenzylamine 71c(R) was stereoselective to the certain degree leading to aminophosphonates 72g-i with the diastereoisomers distribution 2:1 [75]. (Scheme 19) The separation of diastereoisomers was successful only in a case of the dibenzyl derivative 72h, other two derivatives 72g and 72i were characterized by too close R_f coefficients, which made the resolution impossible. However, the aggregation state of both diastereoisomers of the ester 72h (oily liquids) prevented successfully from performing the X-ray study [75]. Generally, it was to state that the stereochemistry of the phosphite addition to an azomethine bond of furfural Schiff bases was very similar to the case of 5-hydroxymethylfurfural ones.

Despite all previous attempts to explain the stereochemistry of this stereochemistry [73,78], the problem was undergone the theoretical qualitative analysis, which led to the more general outlook and demonstrated some failures of taken assumptions. We thought that, if it was taken that the reaction should have proceeded according to the Cram model modified by Yamamoto [76], the a priori assumption of the chiral imine conformation is false. Of course, the ^{1}H NMR Overhauser effect measurements demonstrated the signal intensity increase of around 7%, but this value was too small to state the strong predominance of the "cisoid" conformation. For the same reason, i.e. the lack of certitude as for the predominant conformation of the Schiff base 67d(R) and 71d(R), effects in the ^{1}H NMR spectroscopy described above on the base of Riguera's studies [77] became doubtful, as for these investigation assuming the conformation of the esters 68h and 72h was necessary.

Regarding all above, our team applied theoretical semi-empirical calculations, in order to get the knowledge about the most predominant conformation of both, Schiff bases 67d(R) and 71d(R) and aminophosphonates 68h and 72h [79]. The geometrical structure of the Schiff base was obtained by the quantum mechanics ab initio computations in the 3-21 G function base. Results demonstrated that the most predominant conformation is the "transoid" form for the azomethine C=N electrophilic center in refer to the furan ring and that the azomethine bond itself had the configuration Z. Because of hardware limitations, a transition state could not be investigated by any ab initio method, we presented the approximate qualitative picture of a transition state [79]. (Scheme 21) The attack of a phosphite molecule came from the least sterically hindered the pro-S side to form the RS diastereoisomer, which conformed with the Cram-Felkin-Ahn model [80].

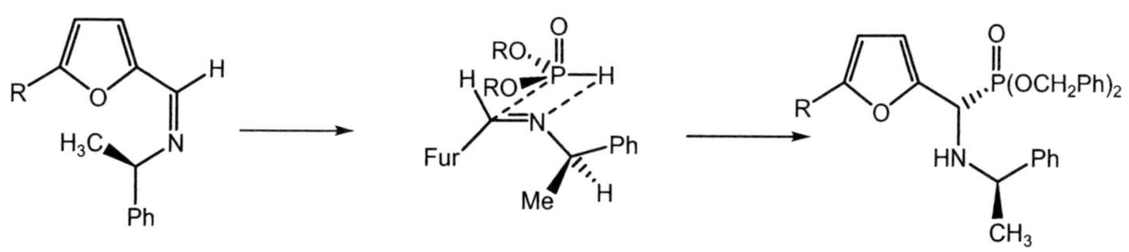

68h(R) R = CH_2OH, H

72h(R) R = CH_2OH, H

Scheme 21

Diastereoisomers were studied using the semi-empirical AM1 computations and the results demonstrated that the most energetically convenient diastereoisomer of the esters 68h and 72h was the isomer of the RS configuration, so the same, which was predicted in the course of the Cram-Felkin-Ahn model analysis. The difference of energy was 1.70 kJ/mol which allowed to predict the diastereoisomeric ratio 1:2 [79]. The 3-D picture of both diastereoisomers of 68h demonstrated that the shielding of protons caused by the magnetic field induced by the π-electron stream took place in the RR-isomer and this isomer formed in minority [79]. All these phenomena observed on NMR spectra of both diastereoisomers of the aminophosphonate 68h found their reflection in the charge distribution according to the Mulliken population analysis [79].

These results [79] made me to think it over and demonstrated that the previous analysis of the problem was false and drove to the conclusion that, in the course of the addition of dialkyl phosphite molecule to N-(5-hydroxymethyl)furfurylidene-(R)-α-methylbenzylamine 67d(R), the RS diastereoisomer formed in majority. The findings of our work [79] were in accordance with the preference in stereochemistry of a structurally similar Strecker-like synthesis of furfural derivatives using chiral alkylammonium hypophosphite. But this will be the subject of the next chapter.

3. Some Words about Aminophosphonous Acids

Aminophosphonous acids were stated to be compounds of the structure even closer to natural amino acids than aminophosphonic ones [81]. If one considers the structures of both compounds, this thesis seems obvious even by the visual examination. (Scheme 22).

But the first isolation of phosphonothricine (73) [81] and discovering its herbicidal properties [82] prompted chemists to perform studies on this group of compounds. Since this date, almost all aminophosphonous analogues of natural amino acids have been synthesized [81]. Biochemical investigations have been also performed discovering among others that phosphinic creatine analogue (74) is the best precursor for the creatine kinase synthesis [83].

So, it is not weird that several interesting works appeared describing novel methods for the synthesis of aminophosphonous acids. Chronologically the first was the method applying the ammonolysis of chloromethanephosphonous acids [81]. But this method served only for obtaining phosphonous analogue of glycine (75) and attempts of an ammonolysis of 5-chloro-1-amino-1-pentanephosphonous acid and 4-chloro-1-amino-1-butanephosphonous acid led to the cyclization products. But, however these reactions allowed to obtain α-piperidynephosphonic acid (76) and phosphonous analogue of proline (77). [81] (Scheme 22)

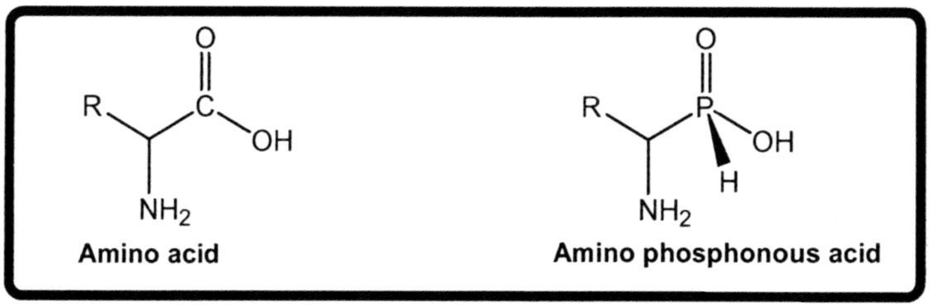

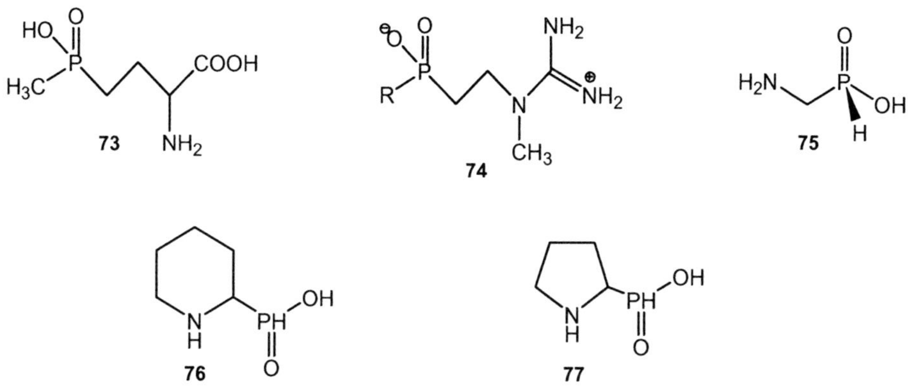

Scheme 22

The alternative method for the synthesis of aminophosphonous acids based on addition to an azomethine bond of imines was first performed by Schmidt [84] and improved by Baylis and coworkers [81], who proposed the laboratory methods for synthesis of aminophosphonous acids in three variants. First was the alteration of the Strecker synthesis of amino acids, and was based on simple mixing an aldehyde, an amine and hypophosphorous acid (Variant A), the second one was the addition of hypophosphorous acid to an azomethine bond of previously prepared Schiff bases (Variant B). (Scheme 23) Both methods seem to be equally efficient.

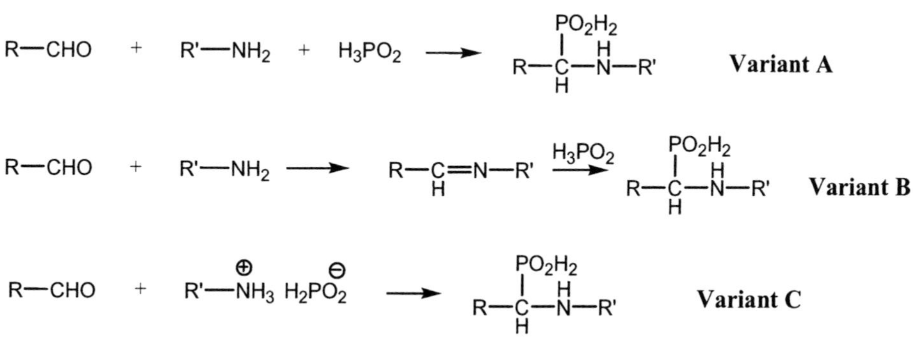

Scheme 23

The third method described by Baylis [81] (Variant C) was the reaction of an aldehyde with a previously prepared salt – alkylammonium hypophosphite. This reaction was also very efficient and was used in a case when both former ones failed. This was connected with two possible tendencies of the reagents: the affinity of an amine and hypophosphorous acid to form salts insoluble in solvent of the reaction, (dioxin or ethanol) which caused its precipitating off the reaction medium and quenching the reaction. The second one is an affinity of Schiff bases to form ammonium salts with hypophosphorous acid, which precipitates off the reaction medium and stops the reaction. But when the salt is introduced to the reaction medium, it reacts easily with an aldehyde. (Scheme 23)

The next method for the synthesis of aminophosphonous acids was proposed by Grobelny [85]. He reported the addition of di-(trimethylsilyl) hypophosphite to the azomethine bond of N-diphenylmethyl substituted Schiff bases, which after the acidic cleavage of silyl ester groups allowed to obtain corresponding N-diphenylmethylaminomethane phosphonous acids. (Scheme 24)

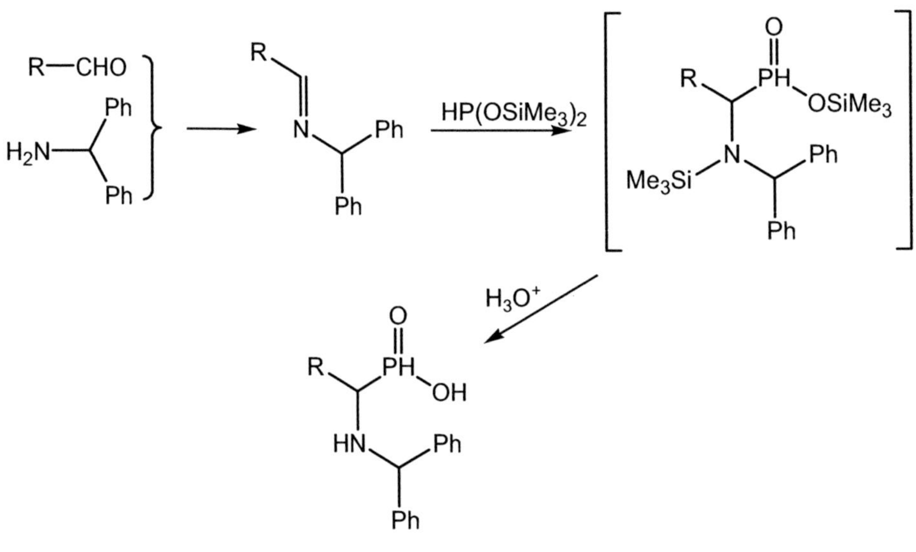

Scheme 24

Xiao et al. [86] reported the method a bit similar to Grobelny's [85] one, as they also performed the addition of di-(trimethylsilyl) hypophosphite to the azomethine bond of Schiff bases. But the idea of the Schiff base choice was different, as they have chosen N-triphenylmethyl (trityl) substituted imines. The advantage of this method in comparison to Grobelny's one [85] is that the acidic hydrolysis following the addition led to the formation of N-non substituted α-aminophosphonous acid. Conditions of acidic hydrolysis turned out to be strong enough to remove the easily leaving triphenylmethyl group. (Scheme 25)

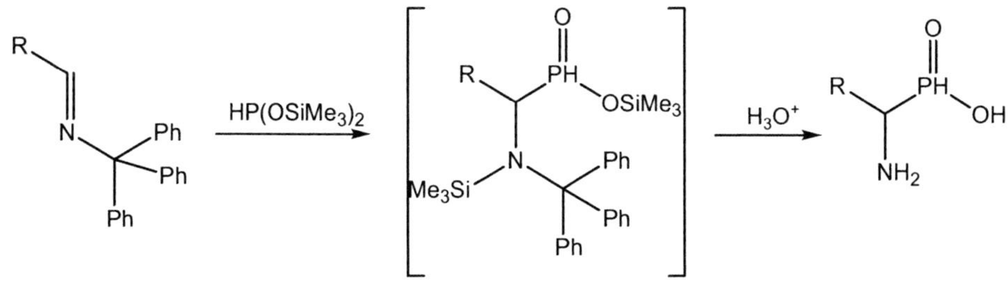

Scheme 25

The method worked out by Xiao et al. [86] is, in my opinion, the method of the great significance, because it opens a way for the synthesis of N-non substituted α-aminophosphonous acid. The largely used method for removal of the N-benzyl substituent, i.e. the hydrogenolysis on palladium on charcoal is useless in a case of aminophosphonous acids, because the phosphonous group deactivates the catalyst by poisoning it, as it is described by Baylis et al. [81] And the synthesis of aminophosphonous acids may be the first step for obtaining their aminophosphonic derivatives (in case when it is difficult to do directly) because α-aminophosphonous acid oxidize easily to their aminophosphonic analogues by the action of mercury (II) chloride or of bromine water [81]. (Scheme 26)

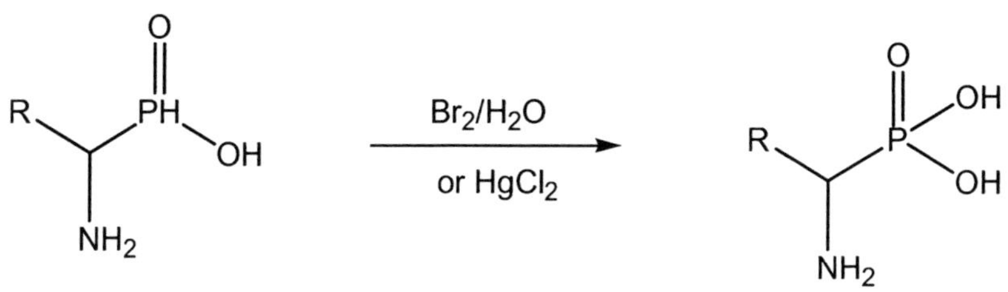

Scheme 26

One of the most important papers, which appeared in this area, is the work by Hamilton, Walker and Walker from 1995 [87]. They reported that the reaction of the chiral (R)-α-methylbenzylammonium hypophosphite with an aldehyde led to the formation of the exclusively one diastereoisomer. Studies of the post-reaction mixture drew to the conclusion that it is the only product of this reaction, so the reaction is stereoselective to 100%. Authors [87] tested five aldehydes and obtained chiral aminophosphonous acids were oxidized to corresponding aminophosphonic acids of a known configuration. In this way, they established that the reaction of (R)-α-methylbenzylammonium hypophosphite with an aromatic or aliphatic aldehyde leads to the formation of N-(R)-α-methylbenzylaminomethane-(S)-phosphonous acid. (Scheme 27)

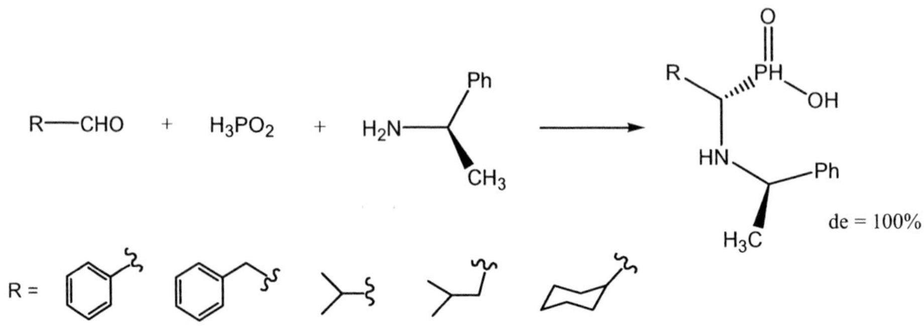

Scheme 27

4. Furan Derived Aminophosphonous Acids

It is to state that to the knowledge of the author of this review, he is the only one, who dealt with aminophosphonous acids bearing furan moiety. That is why, I will try to describe the topic as concisely as possible.

To synthesize variously N-substituted (2-furyl)aminomethanephosphonous acids 78a-e, we applied first the addition of hypophosphorous acid to the azomethine bond of Schiff bases of furfural [78,88-89,96] The reaction carried out in dioxane allowed to obtain the N-benzyl derivative 78a in 60% yield, but (2-furyl)-N-furfurylaminomethanephosphonous acid 78b was obtained in only 40%, when this method was applied [89]. But the application of the method described by Baylis [81], i.e. the reaction between furfural, hypophosphorous acid and a corresponding amine allowed to obtain four acids 78a-d in fair yields. As for the N-benzyl derivative, when synthesized by this method, its yield raised from 60% to 70% [89]. (Scheme 28)

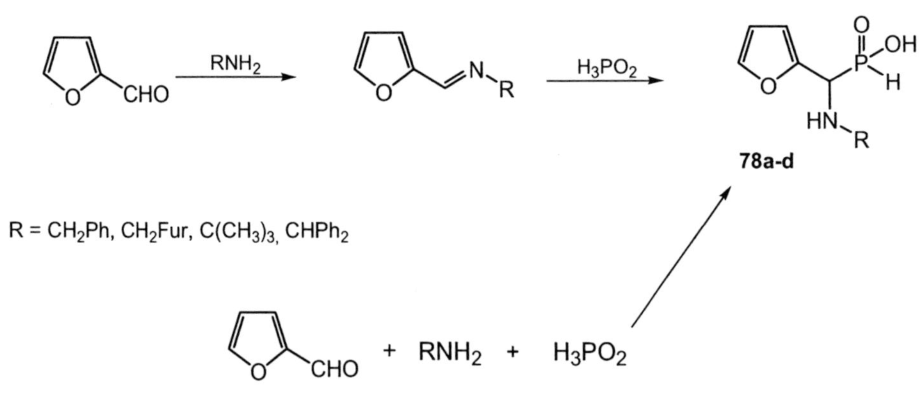

Scheme 28

The cause for that the method of the addition of hypophosphorous acid to the azomethine bond of Schiff bases failed, is the concurrence reaction of the formation hypophosphorous salts with imines. Such a salt is poorly soluble in dioxane, so it precipitates off the reaction mixture and the reaction does not occur.

Using the method described by Hamilton et al. [87], we performed the synthesis of (2-furyl)-N-(R)-α-methylbenzylaminomethane-(S)-phosphonous acid (78e). We have also performed the X-ray studies of this compound to ensure what its absolute configuration is [89]. This study demonstrated the existence of the S configuration around the α-carbon atom linked to phosphorus and nitrogen atoms [89]. (Scheme 29)

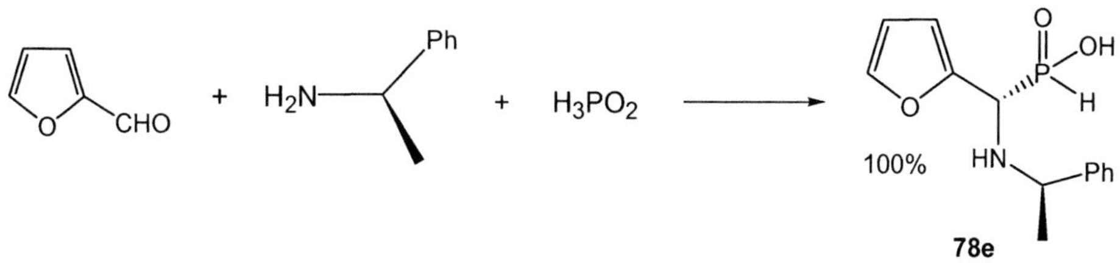

Scheme 29

The next step was to find the way for the ester formation of acids 78a-e with cholesterol and adenosine, in hope that they have any herbicidal activity.

Biological substances often bear in their structure a phosphorylated alcohol group. These are phospholipids, oligonucleotides or carbohydrate phosphates [90]. The compounds of phosphorous or aminophosphonic acids with cholesterol or adenosine were not largely described. Kashman [91] presented the synthesis of O-cholesteryl, O-methyl phosphite by the reaction of dimethyl phosphite with cholesterol. Riess et al. [92] reported the preparation of mono- or dicholesteryl phosphites by the reaction of triethyl phosphite with cholesterol. Gibbs and Larsen [93] and Knerr et al. [90] proposed the preparation of monophosphites of cholesterol and uridine and thymidine, Puech et al. [94] of adenosine.

Stec et al. [95] reported the synthesis of mono-adenosine and mono-uridine esters of 1-aminoethane phosphonic acid using mesitylenesulfonyl triazole as a condensing agent and stated that dicyclohexylcarbodiimide is not a suitable condensing agent in this case. The author of this review has never met any paper describing the adenosinyl or cholesteryl esters of α-aminophosphonous acids.

I reported the formation of variously N-substituted (2-furyl)aminomethanephosphonous acids 78a-e with protected adenosine and cholesterol carried out in dichloromethane in the presence of dicyclohexylcarbodiimide [96]. Reactions lasted several days and were carried out at the boiling temperature and permitted to obtain cholesteryl (2-furyl)aminomethane phosphonites 79a-e in 70-85% yields and O,O'-isopropylidene-adenosinyl (2-furyl)aminomethane phosphonites 80a-e in 53-63% yields. (Scheme 30) In a case of optically inactive acids 78a-d, which were used in the form of racemic mixtures, four diastereoisomeric of cholesteryl esters 79a-d formed in 1:1:2:2 up to 1:1:5:5 ratios, which was in full accord in our expectations. The similar situation was observed in a case of adenosinyl esters [96]. (Scheme 30)

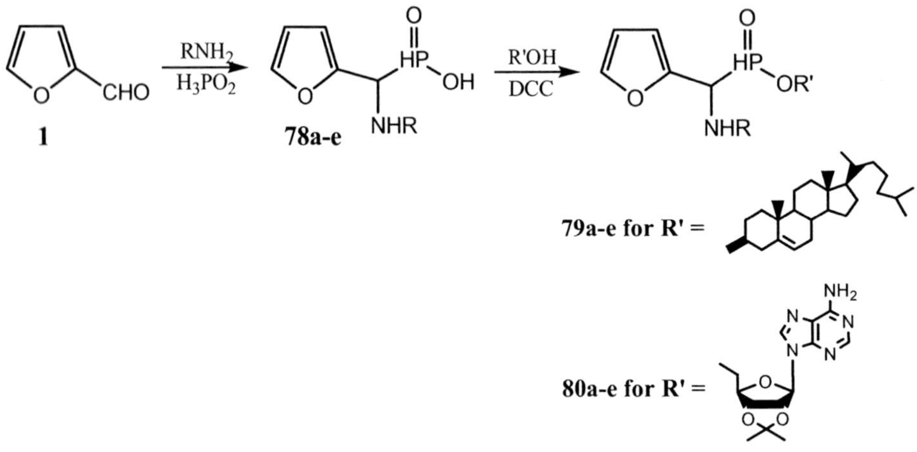

Scheme 30

The condensation of the optically active (S)-(2-furyl)-N-(R)-α-methylbenzylaminomethane phosphonous acid 78e with cholesterol or the O,O'-isopropylidene adenosine led to the formation of two diastereoisomers of 79e and 80e respectively. Their resolution however turned out to be very difficult to do and we did not succeed in separating them.

5. CONCLUSIONS

In this short review, I suppose that I managed to show how interesting are and how important may be furan derived aminophosphonic and aminophosphonous derivatives. Although results of studies on their biological activity did not demonstrate something especially astounding, it is to say that Boduszek's results [53] are promising. After all, no one said one's last word, according to screening studies made in the NCI, some of them may have their cytostatic action. And, when it is well known that nitrofurfural derivatives have strong antibacterial and anti-protozoal action, why nitrofuran-bearing aminophosphonates should not have any? This is the matter, which should be investigated carefully and deeply.

6. REFERENCES

[1] Egelman, M.; Pikl, J., *US Pat*, 2304156 (1942); C.A. 1943, *37*, 3261.
[2] Chavane, V.*Bull. Soc. Chim. Fr.*, 1948, 774
[3] Fields, E.K. *J. Am. Chem. Soc.* 1952, *74*, 1528.
[4] Kabachnik, M.I.; Medved, M.Ya. *Dokl. Akad. Nauk SSSR* 1952, *83*, 689.
[5] Horiguchi, M.; Kandatsu, M. *Nature* 1959, *184*, 901.
[6] Franz, J.E. *US Pat*, 3997860 (1976); C.A. 1976, *86*, 43812.
[7] Diel, D.J.; Maier, L. *Phosphorus Sulfur* 1987, *29*, 201.
[8] Maier, L. *Phosphorus Sulfur*1988, *36*, 1.
[9] Atherton, F.R.; Hassall, C.H. *Nature* 1978, *272*, 56.

[10] Lukszo; J., Tyka; R. *Pol. J. Chem.*, 1978, *52*, 321-327.
[11] Lukszo; J., Tyka; R. *Synthesis*, 1977, 239-240.
[12] Tyka; R., *Tetrahedron Lett.* 1970, 677-680.
[13] Kowalik; J., Kupczyk-Subotkowska; L., Mastalerz; P. *Synthesis*, 1981, 57-58.
[14] Zygmunt; J., Mastalerz; P. *Pol. J. Chem.*, 1978, *52*, 2271-2273.
[15] Soroka; M., Mastalerz; P. *Tetrahedron Lett.*, 1973, 5201-5202.
[16] Subotkowski; W., Tyka; R., Mastalerz; P. *Pol. J. Chem.*, 1980, *54*, 503-505.
[17] Nelson; V., Mastalerz; P. *J. Pharm. Sci.*, 1984, *73*, 1844-1846.
[18] Kowalik; J., Zygmunt; J., Mastalerz; P., *Pol. J. Chem.*, 1981, *55*, 712-715.
[19] Zygmunt; J., Mastalerz; P., *Pol. J. Chem.*, 1981, *55*, 411-414.
[20] Oleksyszyn; J., Gruszecka; E., Kafarski; P., Mastalerz; P., *Monatsh. Chem.*, 1982, *113*, 59-72.
[21] Lejczak; B., Kafarski; P., Soroka; M., Mastalerz; P., *Synthesis*, 1984, 577-240.
[22] Lejczak; B., Kafarski; P., Sztajer; H., Mastalerz; P., *J. Med. Chem.*, 1986, *29*, 2212-2217.
[23] Kafarski; P., Lejczak; B., Mastalerz; P., Dus; D., Radzikowski; C., *J. Med. Chem.*, 1985, *28*, 1555-1558.
[24] Garncarz; R., Kafarski; P., Lejczak; B., Mastalerz; P., Wieczorek; J.S., *Phosphorus Sulfur*, 1983, *18*, 373-376.
[25] Pudovik, A.N.; Konovalova, I.V. *Synthesis*, 1979, 81.
[26] Mikolajczyk, M.; Drabowicz, J. Formation of C-P Bond., Houben-Weyl Ed.; Chapter 8, Vol.21e 1995.
[27] Kafarski, P.; Lejczak, B. *Phosphorus Sulfur Silicon* 1991, *63*, 193.
[28] L. Maier, *Phosphorus Sulfur Silicon* 1991, *61*, 65.
[29] Chung, G.A.C.; Aktar, Z.; Jackson, S.; Duncan, K., *Antimicrob. Agents Chemother.*, 1995, *39*, 2235-2238.
[30] Schaberg, T; Lode, H., *Z. Gesamte Inn. Med.*, 1991, *46*, 276-279.
[31] Parthasarathy, P.C.; Ananthan, L.; Likhate, M.A.; Manjunatha, S.G.; Kalyanam, N., *Eur. Med. Chem. Chim. Ther.*, 1993, *28*, 195-199.
[32] Yagi, N.; Kiuchi, T.; Hayato, S.; Terashima Y.; Kenmotsu, H.; Sekikawa, H.; Takada, M., *Biol. Pharm. Bull.*, 1996, *19*, 616-622.
[33] Cerecetto, H.; Di Maio, R.; Gonzalez, M.; Risso, M.; Sagrera, G.; Seoane, G.; Denicola, A.; Peluffo, G.; Quijano, C.; Stoppani, A.O.M.; Paulino, M.; Olea-Azar, C.; Basombrio, M.A., *Eur. J. Med. Chem.*, 2000, *35*, 343-350.
[34] Morello, A., *Comp. Biochem. Physiol.*, 1988, *90C*, 1-12.
[35] Hoener, B.A.; Krueger, T.R., *J. Pharm. Sci.*, 1984, *73*, 1669-1671.
[36] Cadwallader; D.E., Jun; H.W., *Anal. Profiles Drug Subs.*, 1976, *5*, 345-373.
[37] Slapsyte, G.; Jankauskiene, A.; Mierauskiene, J.; Lazutka, J.R., *Mutat. Res.*, 2001, *491*, 25-30.
[38] Hoogenboom, L.A.P.; Polman, Th.H.G.; Lommen, A.; Huveneers, M.B.M.; van Rhijn, J., *Xenobiotica*, 1994, *24*, 713-728.
[39] Rossi, L.; Angelis, I.; Pedersen, J.Z.; Marchese, E.; Stammati, A., *Mol. Pharmacol.*, 1996, *49*, 547-555.
[40] Thabaut, A.; Durosoir, J.L., *Gazz. Med. Fr.*, 1978, *85*, 4516-4524; *Chem. Abstr.*, 1979, *90*, 1487.
[41] Grana, E.; Zonta, F.; Lucchelli, A., *Farmaco Ed. Sci.*, 1984, *39*

[42] Palop, D.; Agut, J.; Marquez, M.; Conejo, L.; Sacristan, A.; Ortiz, J.A., *Arzneim. Forsch.*, 1997, *47*, 439-446.

[43] Sironi, G.; Colombo, D.; Poggesi, E.; Leonardi, A.; Testa, R.; Rampin, O.; Bernabe, J.; Giuliano, F., *J. Pharmacol. Exp. Ther.*, 2000, *292*, 974-981.

[44] Wang, R.; Sun, X.D.; Cai, Q.M., *Chin. Med. J.*, 1989, *102*, 313-318.

[45] Wozniak; T.J., Hicks; J.R., *Anal. Profiles Drug Subs.*, 1991, *20*, 209-236.

[46] Bodkin, D.W.; Kumar, A.; Xiao, G.; Wilson, W.D.; Bender, B.C.; McCurdy, D.R.; Hall, J.E.; Tidwell, R.R. *J. Med. Chem.* 1998, *41*, 124.

[47] Lugovkin, B.P. et al., *Zh.Obshch.Khim.* 1976, *46*; 555.

[48] Osipova, M.P.; Kuz'mina, L.V.; Kukhtin, V.A. *Zh.Obshch.Khim.* 1982, *52*, 448.

[49] Demir, A.S.; Tanyeli, C.; Sesenoglu, O.; Demic, S.; Evin, O.O. *Tetrahedron Lett.* 1996, *37*; 407-410.

[50] Boduszek, B. *Phosphorus, Sulfur Silicon Relat. Elem.* 1995, *104*, 63-68.

[51] Boduszek, B. *Phosphorus, Sulfur Silicon Relat.Elem.* 1996, *113*, 209-218.

[52] Boduszek, B. *Phosphorus, Sulfur Silicon Relat.Elem.* 1999, *144*, 433-436.

[53] Boduszek, B.; Vegh, D.; Korenova, A.; Uher, M. *Pol. J. Chem.* 2001, *75*, 1271-1276.

[54] Alonso, E.; Alonso, E.; Solis, A.; del Pozo, C. *Synlett*, 2000, 698-700.

[55] Laschat, S.; Kunz, H. *Synthesis*, 1992, 90-95;

[56] Kraicheva, I.; Liogonkii, B.I.; Stefanova, R.; Borisov, G. *Phosphorus Sulfur* 1988, *40*, 145-148.

[57] Kraicheva, I.; Liogonkii, B.I.; Stefanova, R.; Borisov, G. *Eur. Polym. J.* 1988, *24*, 1167-1172.

[58] Kraicheva, I.; Stefanova, R.; Borisov, G. *Phosphorus, Sulfur Silicon Relat.Elem.* 1993, *79*, 107-112.

[59] Kraicheva, I. *Phosphorus, Sulfur Silicon Relat.Elem.* 1996, *118*, 21-29.

[60] Kraicheva, I. *Phosphorus, Sulfur Silicon Relat.Elem.* 1998, *134/135*, 287–294.

[61] Kraicheva, I. *Phosphorus, Sulfur Silicon Relat.Elem.* 1999, *155*, 127-136.

[62] Kraicheva, I. *Phosphorus, Sulfur Silicon Relat.Elem.* 2001, *173*, 234-252.

[63] Consiglio, G.A.; Failla, S.; Finocchiaro, P.; Siracusa, V. *Phosphorus, Sulfur Silicon Relat.Elem.* 1998, *143*, 159-166.

[64] Oleksyszyn, J.; Subotkowska, L.; Mastalerz, P. *Synthesis*, 1979, 985.

[65] Prishchenko, A.A.; Novikova, O.P.; Livantsov, M.V.; Kustrya, D.N.; Grigor'ev, E.V. *Zh.Obshch.Khim.* 1998, *68*, 514-516.

[66] Qian, C.; Huang, T. *J.Org.Chem.* 1998, *63*, 4125-4128.

[67] Manabe, K.; Kobayashi, S. *Chem.Commun.*, 2000; 669-670.

[68] Kaboudin, B.; Nazari, R. *Tetrahedron Lett.* 2001, *42*, 8211-8213.

[69] Takahashi, H.; Yoshioka, M.; Imai, N.; Onimura, K.; Kobayashi, S., *Synthesis*, 1994, 763-764.

[70] Lewkowski, J. *Conversions of Furfural, 5-Hydroxymethylfurfural and Some their Derivatives*, Ph.D. thesis, University of Lodz, Poland 1996

[71] Lewkowski, J. *Arkivoc*, 2001, *2*, 17-54.

[72] Cottier, L.; Descotes, G.; Lewkowski, J.; Skowroński, R. *Phosphorus Sulfur Silicon* 1996, *111*, 161.

[73] Cottier, L.; Descotes, G.; Lewkowski, J.; Skowroński, R. *Phosphorus Sulfur Silicon* 1996, *116*, 92-98.

[74] Cottier, L.; Descotes, G.; Gonera, G.; Grabowski, G.; Lewkowski, J.; Skowroński, R. *Phosphorus Sulfur Silicon* 1996, *118*, 181-190.

[75] Lewkowski, J.; Rzeźniczak, M.; Skowroński, R. *Org. Prep. Proced.Int.* 2000, *32*, 453-459.

[76] Yamamoto, Y.; Komatsu, T.; Maruyama, K. *J. Am. Chem. Soc.* 1984, *106*, 5031-5033.

[77] Latipov, S.K.; Seco, J.M.; Quinoa, E.; Riguera, R. *J. Org. Chem.* 1995, *60*, 1538-1547.

[78] Skowroński, R.; Grabowski, G.; Nazarski, R.; Lewkowski, J. *Phosphorus Sulfur Silicon* 1999, *144-146*, 449-452.

[79] Nazarski, R.B.; Lewkowski, J.A.; Skowroński, R. *Heteroatom Chem.* 2002, *13*, 120-125.

[80] Clayden, J.; Greeves, N.; Warren, S.; Wothers, P. *Organic Chemistry*, Oxford University Press, 2001, pp. 889-894.

[81] Baylis, E.K.; Campbell, C.D.; Dingwall, J.G., *J. Chem. Soc. Perkin Trans. I*, 1984, 2845-

[82] Bayer, E.; Gugel, K.H.; Haegele, M.; Hagenmaier, M.; Jessipov, S.; Konig, A.W.; Zahner, H., *Helv. Chim. Acta*, 1972, *55*, 224

[83] Rowley, G.L.; Greenleaf, A.L.; Kenyon, G.L.; *J. Am. Chem. Soc.*, 1971, *93*, 5542.

[84] Schmidt, H. *Chem. Ber.* 1948, *6*, 477-483.

[85] Grobelny, D., *Synthesis*, 1987, 942-943.

[86] Jiao, X.Y.; Verbruggen, C.; Barloo, M.; Bollaert, W.; DeGroot, A.; Dommisse, R.; Haemmers, A., *Synthesis*, 1994, 23-24.

[87] Hamilton, R; Walker, B.; Walker, B.J., *Tetrahedron Lett.*, 1995, *36*, 4451-4454.

[88] Skowroński, R.; Grabowski, G.; Lewkowski, J. *Phosphorus Sulfur Silicon* 1999, *147*, 207.

[89] Grabowski, G.; Lewkowski, J.; Nazarski, R.; Skowroński, R.; Karolak-Wojciechowska, J. *Polish J. Chem.* 1999, *73*, 307-312.

[90] Knerr, L.; Pannecouke, X.; Schmitt, G. and Luu, B., *Tetrahedron Lett.*, 1996, *37*, 5123-5126.

[91] Kashman, Y., *J. Org. Chem.*, 1972, *37*, 912-914.

[92] Benzera, C.; Bravet, J.-L. and Riess J., *Can. J. Chem.*, 1972, *50*, 2264-2269.

[93] Gibbs, D.E. and Larsen, C., *Synthesis*, 1984, 410-413.

[94] Puech, F.; Gosselin, G.; Balzarini, J.; De Clercq, E.; Imbach, J.-L., *J. Med. Chem.*, 1988, *31*, 1897-1907.

[95] Lesiak, K.; Stec, W.J. and Zieliński, W.S., *Polish J. Chem.*, 1978, *53*, 327-331.

[96] Lewkowski, J., *Synth. Commun.*, 2004, *34(4)*, 697-702.

INDEX

A

B

C

D

E

F

G

H

I

K

L

M

N

O

P

Q

R

S